Test-Taker's Checklist
ADVANTAGES TO THE TAKER OF THE COLLEGE BOARD ACHIEVEMENT TEST/CHEMISTRY

This book can help you prepare for the Chemistry Achievement Test by offering you the following:

✓ 1. *Detailed information* about the *test makeup—topics covered*, and to what extent.

✓ 2. A full-length *diagnostic test* to help you identify your areas of greatest need for review.

✓ 3. A *study plan* for the 6-week period before the test.

✓ 4. Advice about *how to prepare* for and take the test and how to register.

✓ 5. *Last-minute instructions* before taking the test.

✓ 6. A *concise but thorough review* of the most important concepts and principles of a good high school chemistry course.

✓ 7. *Chapter-end questions* to test your understanding of concepts and procedures.

✓ 8. *Four practice tests with* all *answers explained* that allow you to simulate the test experience. Each test has a diagnostic chart to help you pinpoint your areas of strength and weakness.

1991-92 Chemistry Achievement Test Dates

TEST DATES	REGISTRATION DEADLINES
November 2, 1991	September 27, 1991
December 7, 1991	November 1, 1991
January 25, 1992	December 20, 1991
May 2, 1992	March 27, 1992
June 6, 1992	May 1, 1992

The Chemistry Achievement Test is usually given as follows:

First Saturday in November
First Saturday in December
Last Saturday in January
First Saturday in May
First Saturday in June

The registration deadline is generally five weeks before the examination.

How to Prepare for College Board Achievement Tests

CHEMISTRY

FOURTH EDITION

JOSEPH A. MASCETTA
*Formerly Chemistry Teacher and Coordinator, Science Department
Mount Lebanon High School • Pittsburgh, Pennsylvania
Presently Principal
Mount Lebanon High School • Pittsburgh, Pennsylvania*

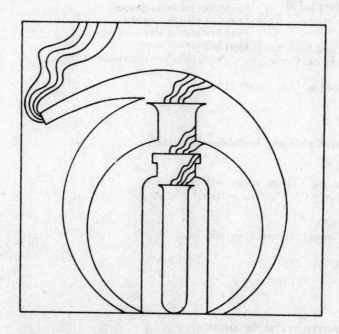

BARRON'S EDUCATIONAL SERIES, INC.
New York • London • Toronto • Sydney

All inquiries should be addressed to:
Barron's Educational Series, Inc.
250 Wireless Boulevard
Hauppauge, New York 11788

International Standard Book No. 0-8120-4082-1

Library of Congress Catalog Card No. 90-348

Library of Congress Cataloging-in-Publication Data

Mascetta, Joseph A.
  How to prepare for College Board achievement
tests. Chemistry / Joseph A. Mascetta.—4th ed.
      p.   cm.
  ISBN 0-8120-4082-1
  1. Chemistry—Examinations, questions, etc.   I.
Title.
QD42.M333   1990   90-348
540'.76—dc20   CIP

PRINTED IN THE UNITED STATES OF AMERICA

123   100   987654

Contents

Introduction

GENERAL INFORMATION ABOUT THE TEST

The College Board, which is a nonprofit membership organization, sponsors the Admissions Testing Program (ATP). Educational Testing Service develops and administers the tests for the College Board and also prepares a booklet about the testing program, of which the Achievement Tests are a part. Copies of this booklet, which describes each of the 14 Achievement Tests and gives sample questions, are automatically sent at the beginning of each academic year to secondary schools for free distribution to students who plan to register for the Achievement Tests. The name of this publication is *Taking the Achievement Tests*, and requests from schools or individuals about how to obtain free copies should be addressed to College Board ATP, CN 6200, Princeton, NJ 08541-6200. The phone number (Monday–Friday) is (609) 771-7600, 8:30 A.M. to 9:30 P.M.

All of the Achievement Tests are contained in the same test booklet. Each takes one hour of testing time, and you may choose any one, two, or three to take at one test sitting. They all consist of multiple-choice questions except the English Composition Test with Essay.

Some colleges require Achievement Tests for admission. The scores are used in conjunction with your high school record, results on the Scholastic Aptitude Test (SAT), teacher recommendations, and other background information to provide a reliable measure of your academic achievements.

In addition to obtaining a standardized assessment of your achievement from your scores, some colleges use the test results for placement into their particular programs in the freshman year. At others, advisers use the results to guide freshmen in the selection of courses.

Is the Chemistry Achievement Test Required?

The best information on whether Achievement Tests are required and, if so, which ones is found in the individual college catalogs or a directory of colleges. Some colleges specify which tests you must take, while others allow you to choose. Obviously, if you have a choice and you have done well in chemistry, you should pick it as one of your tests. The College Board publishes *The College Handbook*, which is a good source of information about colleges' requirements with regard to taking Achievement Tests.

When Should You Take the Test?

You will undoubtedly do best if you take the test after completing the high school chemistry course or courses that you plan to take. At this time, the material will be fresh in your mind. Forgetting begins very quickly after you are past a topic or have finished the course. You should plan a review program for at least the last 6 weeks before the test date. (A plan is provided later in this book for such a review.) Careful review definitely helps—cramming just will not do if you want to get the best score of which you are capable!

In general, colleges that use Achievement Test results as part of the admissions process usually require that you take the test no later than December or January of your senior year. For early decision programs, the test time is June of your junior year. Since chemistry is often a junior year course, June of that year is the optimum time to take the test.

When Is the Test Offered?

The chemistry test is available every time the Achievement Tests are given. These are offered five times during the school year: in November, December, January, May, and June. Be sure that the testing site for which you plan to register offers the Achievement Tests on each of these times. Remember that you may choose to take one or two additional tests besides chemistry on any one test day. You don't have to specify in advance which tests you plan to take or how many.

How Do You Register?

The *Registration Bulletin for the SAT and Achievement Tests* contains all the information you need to register and to have your scores sent to the college(s) of your choice. Copies of this publication should be available in every high school guidance office. If you have a problem getting one, write to College Board Admissions Testing Program (see the address in "General Information about the Test," above).

How Should You Prepare for the Test?

Barron's How to Prepare for College Board Achievement Tests/Chemistry will be especially helpful. The more you know about the test, the more likely you are to get the best score possible for you. This book provides you with a diagnostic test and four practice tests that allow you to become familiar with the question types and the wording of directions, and to gain a feel for the degree of emphasis on particular topics and the ways in which information may be tested. Each of these aspects should be consciously pursued as you use this book.

What Topics Appear on the Test, and to What Extent?

The following chart* shows the content of the test:

Topics Covered	Approximate Percentage of Test	Approximate Number of Questions
I. Atomic Theory and Structure, including periodic relationships	10	8
II. Nuclear Reactions	2	2
III. Chemical Bonding and Molecular Structure	11	9
IV. States of Matter and Kinetic Molecular Theory	9	8
V. Solutions, including concentration units, solubility, and colligative properties	6	5
VI. Acids and Bases	9	8
VII. Oxidation-reduction and Electrochemistry	7	6
VIII. Stoichiometry, including the mole concept, Avogadro's number, empirical and molecular formulas, percentage composition, stoichiometric calculations, and limiting reagents	11	9
IX. Reaction Rates, including rate equations and factors affecting rates	2	2
X. Equilibrium, including mass action expressions, ionic equilibria, and LeChatelier's prinicple	6	5
XI. Thermodynamics, including energy changes in chemical reactions, randomness, and criteria for spontaneity	4	3
XII. Descriptive Chemistry: physical and chemical properties of elements and their more familiar compounds, including simple examples from organic chemistry; periodic properties	16	14
XIII. Laboratory: equipment, procedures, observations, safety, calculations, and interpretation of results	7	6

*Reprinted with permission from **Taking the Achievement Tests,** copyright 1988, by College Entrance Examination Board, New York.

Note: Every edition contains approximately five questions on equation balancing and/or predicting products of chemical reactions. These are distributed among the various content categories.

Skills Specifications	Approximate Percentage of Test
Level I. *Essentially Recall:* remembering information and understanding facts	30
Level II. *Essentially Application:* applying knowledge to unfamiliar and/or practical situations; solving problems using mathematical relationships	55
Level III. *Essentially Interpretation:* inferring and deducing from qualitative and quantitative data and integrating information to form conclusions	15

The above charts give you a general overview of content of the test. These generalities are evaluated through 85 multiple-choice questions. This material is that generally covered in an introductory course in chemistry at a level suitable for college preparation. While the test covers the topics listed, different aspects of each topic are stressed from year to year. Add to this the differences that exist in high school courses with respect to the percentage of time devoted to each major topic and to the specific subtopics covered, you may find that there are questions on topics with which you have little or no familiarity.

Each of the sample tests in this book is constructed to match this distribution of topics closely so that you will gain a feel for the makeup of the actual test. After each test, a chart will show you which questions relate to each topic. This will be very helpful to you in planning your review because you can identify the areas on which you need to concentrate in your studies. Another chart is provided for you to see which chapters correspond to the various topic areas.

The level of mathematics required on this test is that found in math courses in which you are expected to solve simple algebraic relationships and to apply these to solving word problems. You will also need the ability to handle the concepts of ratio and direct and inverse proportions, scientific notation, and exponential functions. Since you are not permitted to use electronic calculators during the test, the numerical calculations are limited to simple arithmetic. The metric system of units is used in this test.

A periodic chart is provided as a resource and as the source of atomic numbers and atomic weights of elements that are needed in solving problems.

The test is composed of three types of questions. The directions and formats are identical to those used in all of the sample tests in this book. The questions are all of the multiple-choice type, in which you must choose from the five choices given.

How Can You Use This Book to Prepare for the Test?

The best way to use this book is a two-stage approach, and the next sections are arranged accordingly. First, you should take the diagnostic pretest. This will give you a preliminary exposure to the type of test you are planning to take, as well as a measure of how you achieve on each of the 12 areas of the test. You will also become aware of the types of questions that the test includes.

Having taken the diagnostic test, you should then follow a study program. A study plan covering the 6 weeks before the test has been developed for you and is given in detail on page xxxvii. It requires a minimum of 1 or 2 hours per night on weekdays but leaves your weekends free.

A DIAGNOSTIC TEST

The following test is a sample of the actual test you will take to measure your chemistry achievement. It has basically the same distribution of topics, directions, and number of questions. Before taking it, follow the advice given in the section entitled "Final Preparation—The Day before the Test" (p. xxxviii). Use the answer sheet provided below, and limit the test time to 1 hour.

A periodic chart has been included for your use on problems requiring this source of information. Use this chart with the practice tests, too.

Turn now to the test.

ANSWER SHEET FOR DIAGNOSTIC TEST

Determine the correct answer for each question. Then, using a No. 2 pencil, blacken completely the oval containing the letter of your choice.

1. Ⓐ Ⓑ Ⓒ Ⓓ Ⓔ	16. Ⓐ Ⓑ Ⓒ Ⓓ Ⓔ	31. Ⓐ Ⓑ Ⓒ Ⓓ Ⓔ
2. Ⓐ Ⓑ Ⓒ Ⓓ Ⓔ	17. Ⓐ Ⓑ Ⓒ Ⓓ Ⓔ	32. Ⓐ Ⓑ Ⓒ Ⓓ Ⓔ
3. Ⓐ Ⓑ Ⓒ Ⓓ Ⓔ	18. Ⓐ Ⓑ Ⓒ Ⓓ Ⓔ	33. Ⓐ Ⓑ Ⓒ Ⓓ Ⓔ
4. Ⓐ Ⓑ Ⓒ Ⓓ Ⓔ	19. Ⓐ Ⓑ Ⓒ Ⓓ Ⓔ	34. Ⓐ Ⓑ Ⓒ Ⓓ Ⓔ
5. Ⓐ Ⓑ Ⓒ Ⓓ Ⓔ	20. Ⓐ Ⓑ Ⓒ Ⓓ Ⓔ	35. Ⓐ Ⓑ Ⓒ Ⓓ Ⓔ
6. Ⓐ Ⓑ Ⓒ Ⓓ Ⓔ	21. Ⓐ Ⓑ Ⓒ Ⓓ Ⓔ	36. Ⓐ Ⓑ Ⓒ Ⓓ Ⓔ
7. Ⓐ Ⓑ Ⓒ Ⓓ Ⓔ	22. Ⓐ Ⓑ Ⓒ Ⓓ Ⓔ	37. Ⓐ Ⓑ Ⓒ Ⓓ Ⓔ
8. Ⓐ Ⓑ Ⓒ Ⓓ Ⓔ	23. Ⓐ Ⓑ Ⓒ Ⓓ Ⓔ	38. Ⓐ Ⓑ Ⓒ Ⓓ Ⓔ
9. Ⓐ Ⓑ Ⓒ Ⓓ Ⓔ	24. Ⓐ Ⓑ Ⓒ Ⓓ Ⓔ	39. Ⓐ Ⓑ Ⓒ Ⓓ Ⓔ
10. Ⓐ Ⓑ Ⓒ Ⓓ Ⓔ	25. Ⓐ Ⓑ Ⓒ Ⓓ Ⓔ	40. Ⓐ Ⓑ Ⓒ Ⓓ Ⓔ
11. Ⓐ Ⓑ Ⓒ Ⓓ Ⓔ	26. Ⓐ Ⓑ Ⓒ Ⓓ Ⓔ	41. Ⓐ Ⓑ Ⓒ Ⓓ Ⓔ
12. Ⓐ Ⓑ Ⓒ Ⓓ Ⓔ	27. Ⓐ Ⓑ Ⓒ Ⓓ Ⓔ	42. Ⓐ Ⓑ Ⓒ Ⓓ Ⓔ
13. Ⓐ Ⓑ Ⓒ Ⓓ Ⓔ	28. Ⓐ Ⓑ Ⓒ Ⓓ Ⓔ	43. Ⓐ Ⓑ Ⓒ Ⓓ Ⓔ
14. Ⓐ Ⓑ Ⓒ Ⓓ Ⓔ	29. Ⓐ Ⓑ Ⓒ Ⓓ Ⓔ	44. Ⓐ Ⓑ Ⓒ Ⓓ Ⓔ
15. Ⓐ Ⓑ Ⓒ Ⓓ Ⓔ	30. Ⓐ Ⓑ Ⓒ Ⓓ Ⓔ	45. Ⓐ Ⓑ Ⓒ Ⓓ Ⓔ

46. Ⓐ Ⓑ Ⓒ Ⓓ Ⓔ	61. Ⓐ Ⓑ Ⓒ Ⓓ Ⓔ	76. Ⓐ Ⓑ Ⓒ Ⓓ Ⓔ
47. Ⓐ Ⓑ Ⓒ Ⓓ Ⓔ	62. Ⓐ Ⓑ Ⓒ Ⓓ Ⓔ	77. Ⓐ Ⓑ Ⓒ Ⓓ Ⓔ
48. Ⓐ Ⓑ Ⓒ Ⓓ Ⓔ	63. Ⓐ Ⓑ Ⓒ Ⓓ Ⓔ	78. Ⓐ Ⓑ Ⓒ Ⓓ Ⓔ
49. Ⓐ Ⓑ Ⓒ Ⓓ Ⓔ	64. Ⓐ Ⓑ Ⓒ Ⓓ Ⓔ	79. Ⓐ Ⓑ Ⓒ Ⓓ Ⓔ
50. Ⓐ Ⓑ Ⓒ Ⓓ Ⓔ	65. Ⓐ Ⓑ Ⓒ Ⓓ Ⓔ	80. Ⓐ Ⓑ Ⓒ Ⓓ Ⓔ
51. Ⓐ Ⓑ Ⓒ Ⓓ Ⓔ	66. Ⓐ Ⓑ Ⓒ Ⓓ Ⓔ	81. Ⓐ Ⓑ Ⓒ Ⓓ Ⓔ
52. Ⓐ Ⓑ Ⓒ Ⓓ Ⓔ	67. Ⓐ Ⓑ Ⓒ Ⓓ Ⓔ	82. Ⓐ Ⓑ Ⓒ Ⓓ Ⓔ
53. Ⓐ Ⓑ Ⓒ Ⓓ Ⓔ	68. Ⓐ Ⓑ Ⓒ Ⓓ Ⓔ	83. Ⓐ Ⓑ Ⓒ Ⓓ Ⓔ
54. Ⓐ Ⓑ Ⓒ Ⓓ Ⓔ	69. Ⓐ Ⓑ Ⓒ Ⓓ Ⓔ	84. Ⓐ Ⓑ Ⓒ Ⓓ Ⓔ
55. Ⓐ Ⓑ Ⓒ Ⓓ Ⓔ	70. Ⓐ Ⓑ Ⓒ Ⓓ Ⓔ	85. Ⓐ Ⓑ Ⓒ Ⓓ Ⓔ
56. Ⓐ Ⓑ Ⓒ Ⓓ Ⓔ	71. Ⓐ Ⓑ Ⓒ Ⓓ Ⓔ	
57. Ⓐ Ⓑ Ⓒ Ⓓ Ⓔ	72. Ⓐ Ⓑ Ⓒ Ⓓ Ⓔ	
58. Ⓐ Ⓑ Ⓒ Ⓓ Ⓔ	73. Ⓐ Ⓑ Ⓒ Ⓓ Ⓔ	
59. Ⓐ Ⓑ Ⓒ Ⓓ Ⓔ	74. Ⓐ Ⓑ Ⓒ Ⓓ Ⓔ	
60. Ⓐ Ⓑ Ⓒ Ⓓ Ⓔ	75. Ⓐ Ⓑ Ⓒ Ⓓ Ⓔ	

Periodic Table of The Elements

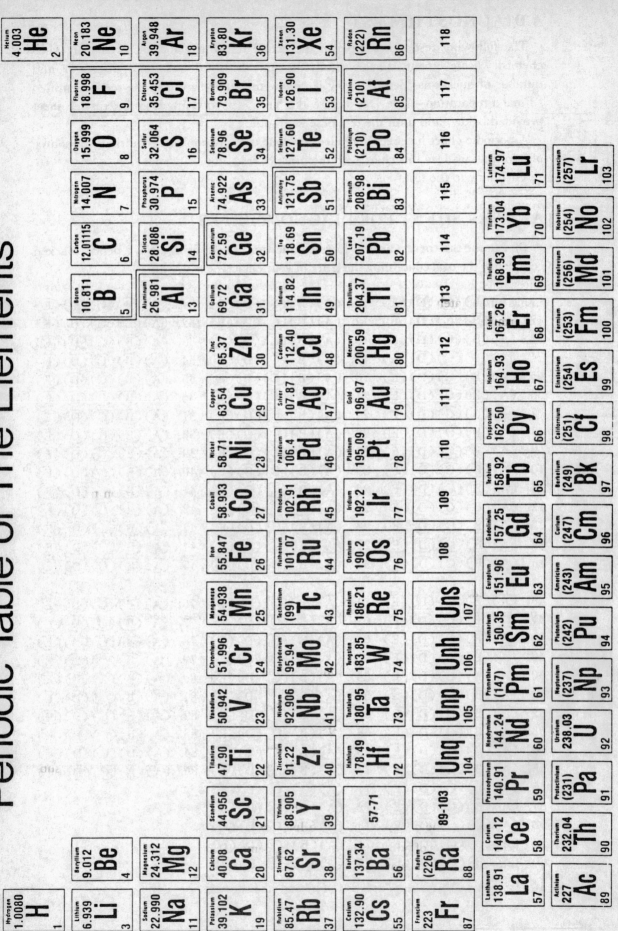

Diagnostic Test

Note: For all questions involving solutions and/or chemical equations, assume that the system is in water unless otherwise stated.

Part A

Directions: Each set of lettered choices below refers to the numbered statements or formulas immediately following it. Select the one lettered choice that best fits each statement or formula and then fill in the corresponding oval on the answer sheet. A choice may be used once, more than once, or not at all in each set.

Questions 1–5

          (A) Arrhenius Theory
          (B) Brönsted Theory
          (C) Lewis Theory
          (D) Hydrolysis 5
          (E) Neutralization 4

C 1. In this reaction: $NH_3 + H^+ \rightarrow NH_4^+$, the H^+ is acting as an electron pair acceptor.

A 2. In an acid the hydronium ion is always present in an aqueous solution.

B 3. In this reaction: $HCl(g) + H_2O(l) \rightarrow H_3O^+(aq) + Cl^-(aq)$, the water is acting as a base.

E 4. The reaction is that of an acid with a base.

D 5. Copper sulfate crystals give an acid solution when dissolved in water.

Questions 6–10

      7 (A) Ionic bonding
      8 (B) Covalent bonding
      9 (C) Polar covalent bonding
      6 (D) Hydrogen bonding
     10 (E) van der Waals forces

D 6. Explain(s) the high boiling point of water compared to other similar molecular substances

A 7. Is (are) the type of bond between a sodium and a chlorine atom

B 8. Is (are) the type of bond found in a hydrogen molecule

C 9. Is (are) the type of bond formed when two elements having a high difference in their respective electronegativities are joined

E 10. Is a general term used for weak intermolecular forces

Questions 11–15

12 (A) Alpha particle
13 (B) Neutron
14 (C) Electron
11 (D) Proton
15 (E) Gamma ray

D 11. Explains the difference between the atomic mass and the atomic number of an element

A 12. Has an atomic mass of 4

B 13. Is the "bullet" usually used to initiate the fission of ^{235}U

C 14. Has the smallest mass of the basic nuclear particles

E 15. Has the characteristics of an electromagnetic wave

Questions 16–19

19 (A) *s*
(B) *p*
16 (C) *d*
(D) *f*
18 (E) *sp*

C 16. The type of outermost orbital occupied by an electron in an atom with the atomic number 11

B 17. The type of orbital that is the last one filled in all inert gases except helium

E 18. The designation of a type of hybridized orbital

A 19. The type of orbital active in the oxidation of sodium

Questions 20–23

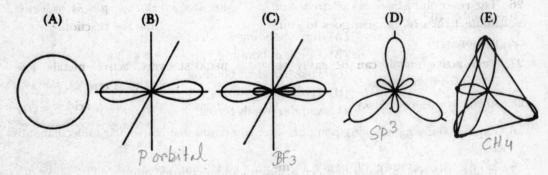

(A) (B) (C) (D) (E)

P orbital   BF₃   Sp³   CH4

D 20. Represents an sp^3 configuration D

C 21. Represents the boron configuration in BF_3 C

E 22. Represents the carbon configuration in CH_4 E

B 23. Represents a *p* orbital B

Part B

Directions: Each question below consists of an <u>assertion</u> (statement) in the left-hand column and a <u>reason</u> in the right-hand column. On the appropriate line of the answer sheet fill in oval

A if both assertion and reason are true statements and the reason is *a correct explanation* of the assertion;

B if both assertion and reason are true statements but the reason is *NOT a correct explanation* of the assertion;

C if the assertion is true but the reason is a false statement;

D if the assertion is false but the reason is a true statement;

E if both assertion and reason are false statements.

Directions Summarized			
	Assertion	*Reason*	
A	True	True	Reason is *a correct explanation*
B	True	True	Reason is *NOT a correct explanation*
C	True	False	
D	False	True	
E	False	False	

Assertion		*Reason*
24. Ionic solids have high melting points	BECAUSE	ionic bonds are strong bonds.
25. When $\triangle H$ is a negative quantity, the reaction is exothermic	BECAUSE	heat energy is released.
26. The reaction of calcium carbonate and dilute hydrochloric acid goes to completion	BECAUSE	chlorine gas is released from the reaction.
27. Very active metals can be easily reduced from their ores	BECAUSE	very active metals lose electrons readily.
28. Acetic acid is a weak acid	BECAUSE	glacial acetic acid is a very concentrated solution of acetic acid.
29. As the temperature of a gas is increased, the kinetic molecular velocity of the particles increases	BECAUSE	gases are compressible.
30. When reduction occurs to a metal ion, the charge on the ion increases	BECAUSE	reduction is a gain of electrons.

Directions Summarized

	Assertion	Reason	
A	True	True	Reason is *a correct explanation*
B	True	True	Reason is *NOT a correct explanation*
C	True	False	
D	False	True	
E	False	False	

Assertion		Reason
31. The condensation of steam is a chemical change	BECAUSE	it represents a new molecular structure.
32. Equilibrium can be expressed as the point at which the quantity of the reactants is equal to the quantity of the products	BECAUSE	the rate of the forward reaction is equal to the rate of the reverse reaction.
33. The formula CaO represents an acid anhydride	BECAUSE	CaO forms an acidic solution when it is dissolved in water.
34. Colloidal carbon adsorbs gases rapidly	BECAUSE	it has a large surface area.
35. The freezing point temperature of a substance is usually lower than its melting point	BECAUSE	the heat of fusion is usually greater than its heat of vaporization.
36. Astatine (At) has the highest ionization energy of the halogen family	BECAUSE	it has the largest atomic radius of the atoms in that family.
37. One mole of hydrogen gas at STP contains 2 moles of atoms	BECAUSE	a monoatomic gas at STP contains one atom in each molecule.
38. The pH of a solution that has a hydrogen ion concentration of 0.0001 mole/liter is 3	BECAUSE	pH is defined as $-\log[H^+]$.
39. As you proceed across the periodic chart, the electronegativity increases	BECAUSE	the ionic radius of a nonmetal ion is usually greater than the atomic radius of the atom.

Part C

Directions: Each of the questions or incomplete statements below is followed by five suggested answers or completions. Select the one that is best in each case and then fill in the correpsonding oval on the answer sheet.

40. The reaction between zinc and sulfuric acid goes to completion because of the condition called
    (A) volatility
    (B) solubility
    (C) ionization
    (D) nonionization
    (E) equilibrium

41. Two immiscible liquids when shaken together may form
    (A) a solution
    (B) a tincture
    (C) a colloidal dispersion
    (D) a hydrated solution
    (E) a dispersion of particles which do not separate on standing

Questions 42–45 refer to the following phase diagram:

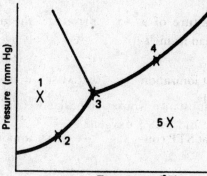

42. At which point can only the gas phase exist?
    (A) 1
    (B) 2
    (C) 3
    (D) 4
    (E) 5

43. Which point is called the triple point?
    (A) 1
    (B) 2
    (C) 3
    (D) 4
    (E) 5

44. At which point can only the liquid and vapor phases coexist?

    (A) 1
    (B) 2
    (C) 3
    (D) 4
    (E) 5

45. If point 4 represented the boiling point of a system and the system was at these conditions, what would happen to the boiling point temperature if the pressure were increased?

    (A) It would be increased.
    (B) It would be decreased.
    (C) It would remain the same.
    (D) It would remain the same at first and then decrease.
    (E) It would first decrease sharply and then increase slightly.

46. **Which of the following is a nonelectrolyte?**

    (A) a water solution of acetic acid
    (B) a water solution of ammonia
    (C) a water solution of ethyl alcohol
    (D) sodium chloride which is fused
    (E) an alloy of copper and silver

47. If you add 684 grams of sugar, $C_{12}H_{22}O_{11}$, to 2 liters of water, what will be the approximate boiling point of the resulting solution? (1 mole sugar = 342 g)

    (A) 1.04°C
    (B) 99.48°C
    (C) 100.51°C
    (D) 101.04°C
    (E) 212°F

48. What is the formula of a substance whose molecular weight is 230, and which is composed of 65.2% arsenic and 34.8% oxygen?

    (A) $As\,O_3$
    (B) As_2O_3
    (C) As_2O_4
    (D) As_2O_5
    (E) As_4O_{10}

49. Five liters of a gas (STP) weigh 6.25 grams. What is the gram-molecular weight of the gas?

    (A) 1.25 g
    (B) 14 g
    (C) 28 g
    (D) 56 g
    (E) 140 g

50. The gram-formula weight of $C_{12}H_{22}O_{11}$ (sugar) is 342 grams. A solution of 342 grams of $C_{12}H_{22}O_{11}$ in
    (A) 1 liter of solution is 1 molar
    (B) 1 liter of solution is 12 molar
    (C) 1 liter of solution is 1 molal
    (D) 500 mL of solution is 6 normal
    (E) 1000 grams of water is 0.5 molar

51. According to kinetic-molecular theory, molecules are in random motion. This motion is greatest when a substance is
    (A) being warmed as a solid
    (B) being warmed as a liquid
    (C) being warmed as a gas
    (D) changing from a solid to a liquid at its melting point
    (E) changing from a liquid to a gas at its boiling point

52. How many grams of $CaCO_3$ are needed to react with an excess of hydrochloric acid to produce 1 mole of CO_2 gas?
    (A) 44
    (B) 50
    (C) 88
    (D) 100
    (E) 200

Questions 53–56 refer to the following drawing:

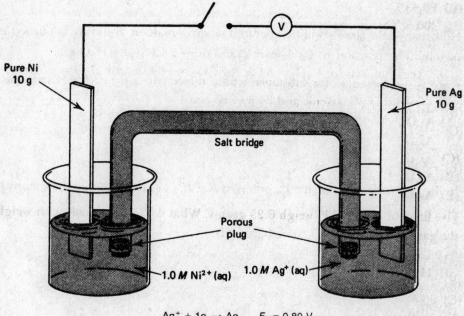

Pure Ni
10 g

Pure Ag
10 g

Salt bridge

Porous plug

1.0 M Ni^{2+} (aq)   1.0 M Ag^+ (aq)

$Ag^+ + 1e \rightarrow Ag$ $E_0 = 0.80$ V
$Ni^{2+} + 2e^- \rightarrow Ni$ $E_0 = -0.24$ V

53. What will be the voltmeter reading when the switch is closed?
    (A) 0
    (B) 0.24
    (C) 0.56
    (D) 0.80
    (E) 1.04

54. When equilibrium is reached, the voltmeter will read
    (A) 0
    (B) 0.24
    (C) 0.56
    (D) 0.80
    (E) 1.04

55. The anode in this cell is the
    (A) Ni
    (B) salt bridge
    (C) Ag
    (D) Ni^{2+}
    (E) Ag^+

56. After 0.1 mole of electrons passes through this voltaic cell, the nickel will weigh
    (A) 3.0 g
    (B) 4.1 g
    (C) 5.9 g
    (D) 7.0 g
    (E) 13 g

57. How many kilocalories will be liberated when 1 mole of acetylene is burned completely? The reaction is: $C_2H_2(g) + 2\frac{1}{2} O_2(g) \rightarrow 2\ CO_2(g) + H_2O(g)$.
    ΔH_f^0 (kcal/mole) is: for $C_2H_2 = 54.2$; for $CO_2 = -94.05$; for $H_2O = -57.8$.
    (A) +54.2
    (B) −97.6
    (C) −191.7
    (D) −245.2
    (E) −300.1

58. In the following reaction, if 85.2 grams of $NaClO_3$ are decomposed, how many liters of O_2 at STP are produced?

    $2\ NaClO_3 \rightarrow 2\ NaCl + 3\ O_2$ (not balanced)

    (A) 13.4
    (B) 25.6
    (C) 26.8
    (D) 38.4
    (E) 76.8

59. The balanced reaction in question 58 shows that the molar relationship between $NaClO_3$ and the O_2 produced is
    (A) 1:1
    (B) 2:1
    (C) 2:2
    (D) 2:3
    (E) 3:3

60. The Haber process for producing NH_3 is

    $$N_2 + 3 H_2 \rightarrow 2 NH_3$$

    What volume of ammonia, in liters, can be liberated by 3 liters of nitrogen?
    (A) 2
    (B) 3
    (C) 5
    (D) 6
    (E) 7

61. What is the formula weight of H_2SO_4? (H = 1, S = 32, O = 16)
    (A) 40
    (B) 50
    (C) 76
    (D) 84
    (E) 98

62. The structural formula for the second member of the alkene series is

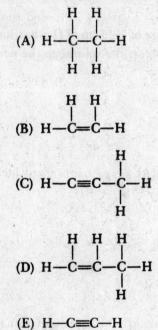

    (A)

    (B)

    (C)

    (D)

    (E) H—C≡C—H

63. What is the systematic (IUPAC) name of this structure?

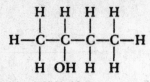

(A) butyl alcohol
(B) *n*-butanol
(C) 2-butanol
(D) butanol
(E) isobutyl alcohol

64. The oxidation state of carbon in oxalic acid, $H_2C_2O_4$, is
(A) +1
(B) +2
(C) +3
(D) +4
(E) +8

65. When dissolved in water, all of the following salts will result in a basic solution EXCEPT
(A) $CuSO_4$
(B) Na_2CO_3
(C) $K_2B_4O_7 \cdot 10\ H_2O$
(D) $Ca_3(PO_4)_2$
(E) $Mg(C_2H_3O_2)_2$

66. If a gas at STP is heated to 200°C and raised to a pressure of 78.0 cm of Hg, by what fractions must the original volume be multiplied to correct the volume to the new conditions?

(A) $\dfrac{273}{473} \times \dfrac{78}{76}$

(B) $\dfrac{473}{273} \times \dfrac{760}{780}$

(C) $\dfrac{780}{760} \times \dfrac{27.3}{47.3}$

(D) $\dfrac{100}{273} \times \dfrac{473}{273}$

(E) $\dfrac{200}{0} \times \dfrac{76}{78}$

67. Which of the following statements is (are) true of a substance which gives off a small amount of heat when it is formed from the elements that comprise it?
   I. It will be difficult to decompose by the addition of heat.
   II. It will be relatively easy to decompose by the addition of heat.
   III. It will be stable unless heated.

(A) I only
(B) III only
(C) I and II only
(D) II and III only
(E) I, II, and III

68. What is the molarity of a solution that contains 20 grams of NaOH in 500 ml of solution? ($Na = 23$, $O = 16$, $H = 1$)
(A) 0.25
(B) 0.5
(C) 1
(D) 20
(E) 40

69. In this reaction: $N_2 + 3H_2 \rightleftharpoons 2NH_3 + heat$, the equilibrium constant can be expressed as

(A) $K = \dfrac{[N_2][H_2]^3}{[NH_3]^2}$

(B) $K = \dfrac{[NH_3]^2}{[N_2][H_2]^3}$

(C) $K = \dfrac{[2NH3]}{[N_2][3H_2]}$

(D) $K = \dfrac{[NH_3]^2[H_2]^3}{[N_2]}$

(E) $K = \dfrac{[N_2H_6]}{[N_2][H_6]}$

70. The solubility of PbF_2 is 0.49 gram/liter at 18°C. What is the K_{sp} for PbF_2 at this temperature?

(A) 0.002
(B) 4×10^{-6}
(C) 8×10^{-6}
(D) 7.4×10^{-7}
(E) 3.2×10^{-8}

71. The reaction for burning hydrogen is

$$2 H_2 + O_2 \rightarrow 2 H_2O + 136.64 \text{ kcal}$$

How much heat is liberated if 10 grams of H_2 are burned?
(A) 54.65 kcal
(B) 273.28 kcal
(C) 312.6 kcal
(D) 341.6 kcal
(E) 683.2 kcal

72. Chemical properties include which of the following?
    I. Ability to conduct an electric current.
    II. Attraction to a magnet.
    III. Formation of an oxide coating in air.
        (A) I only
        (B) III only
        (C) I and II only
        (D) II and III only
        (E) I, II, and III

73. When a solid is melted, the energy absorbed is used for which of the following purposes?
    I. To decrease the kinetic energy of the system.
    II. To increase the volume of the solid.
    III. To overcome the attractive forces between molecules.
        (A) I only
        (B) III only
        (C) I and II only
        (D) II and III only
        (E) I, II, and III

74. What is the functional group of an ester? (R = a radical)

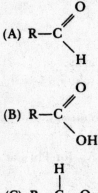

    (A) R—C with =O and —H

    (B) R—C with =O and —OH

    (C) R—C(H)(H)—O—C(H)(H)—R

    (D) R—C(=O)—O—R¹

    (E) R—C(H)(H)—OH

75. What is the name of this organic structure?

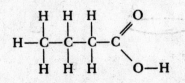

(A) ethylethanoate
(B) butanol
(C) butanal
(D) propyl acetate
(E) butanoic acid

76. The solubility of a solid is usually favored by a tendency of the system toward which of the following?
    I. Decreased energy.
    II. Increased entropy.
    III. Increased energy and increased entropy.
       (A) I only
       (B) III only
       (C) I and II only
       (D) II and III only
       (E) I, II, and III

77. A student took some sodium hydroxide crystals out of a bottle for an experiment. He then assembled his titration equipment. After making a solution of concentrated sulfuric acid, he weighed 40 grams of the sodium hydroxide and added enough water to it to make 1 liter of solution. What might be a source of error in his titration?
    (A) Some sulfuric acid evaporated.
    (B) The sulfuric acid became more concentrated.
    (C) The NaOH solution gained weight, thus increasing its molarity.
    (D) The NaOH crystals gained H_2O weight, thus making the solution less than 1 molar.
    (E) The NaOH crystals gained H_2O weight, thus making the solution more than 1 molar.

Questions 78–80 refer to the following drawings:

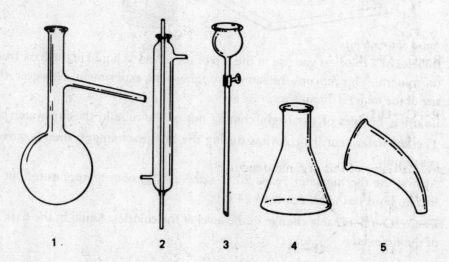

1.   2   3   4   5

78. Which piece of equipment condenses gases to liquids?
    (A) 1
    (B) 2
    (C) 3
    (D) 4
    (E) 5

79. Which piece of equipment allows the addition of more liquid to the reaction?
    (A) 1
    (B) 2
    (C) 3
    (D) 4
    (E) 5

80. When this equipment is assembled for an experiment, the best order to connect the pieces would be
    (A) 1-5-4-2-3
    (B) 1-2-3-4-5
    (C) 3-2-4-2-5
    (D) 2-4-3-5-1
    (E) 3-1-2-5-4

Questions 81–83 refer to the following diagram and observations:

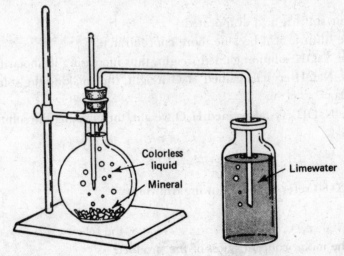

Recorded observations:

1. Bubbles of a colorless gas rise to the top of the colorless liquid in the flask from the mineral. This remains the same throughout the experiment. However, the size of the mineral diminishes.

2. Colorless bubbles of gas begin coming out of the tube in the limewater jar.

3. The limewater remains colorless during the first few minutes, then begins to get cloudly.

4. At first the thermometer reads 20°C, same as the room temperature, but in half an hour has risen slowly to 24°C.

5. There is no noticeable change in the level of the colorless liquid in the flask or of the limewater.

81. On the basis of these observations, which of the following hypotheses is most reasonable?
    (A) The mineral is undergoing a phase change in which energy is released.
    (B) The mineral is readily soluble in water.
    (C) The mineral is undergoing a chemical reaction with the colorless liquid.
    (D) The gas dissolves readily in the colorless liquid.
    (E) The mineral is not composed of any of the elements contained in limewater.

82. What is the most probable explanation of the fact that the limewater remained colorless for the first few minutes?
    (A) The composition of the gas given off by the mineral changed as the experiment progressed.
    (B) The bubbles of gas coming from the tube during the first few minutes were air bubbles.
    (C) The gas produced by the mineral does not react with the limewater.
    (D) The composition of the colorless liquid changed as the experiment progressed.
    (E) The temperature of the limewater was too low during the first few minutes.

83. Which of the following statements concerning the above experiment involves an *interpretation* of the data rather than an observed fact?
    (A) No liquid was transferred from the reaction bottle to the beaker.
    (B) The quantity of solid mineral decreased.
    (C) The cloudiness in the beaker was caused by the product of the reaction of the colorless gas and the limewater.
    (D) The bubbles of gas rising from the mineral remained colorless throughout the experiment.
    (E) There was a 4° rise in temperature at the end of 30 minutes.

84. Neutral atoms of neon have the same number of electrons as each of the following EXCEPT
    (A) O^{2-}
    (B) F^{1-}
    (C) Na^{1+}
    (D) Mg^{2+}
    (E) Ca^{2+}

85. In a chemical reaction at equilibrium, which of the following changes would always increase the molar concentration of the product?
    (A) Add a catalyst.
    (B) Increase the pressure.
    (C) Increase the temperature.
    (D) Decrease the temperature.
    (E) Increase the molar concentrations of the reactants.

STOP

IF YOU FINISH BEFORE ONE HOUR IS UP, YOU MAY GO BACK TO CHECK YOUR WORK OR COMPLETE UNANSWERED QUESTIONS.

Answers and
Explanations to Test

1. (C) Lewis Theory deals with electron pair acceptors.

2. (A) Arrhenius Theory deals with hydronium ions as acids.

3. (B) Brönsted Theory defines a base as a proton acceptor.

4. (E) When an acid reacts with a base, each loses its properties to form water and a salt. This is called neutralization.

5. (D) The interaction of a salt with water is hydrolysis.

6. (D) Hydrogen bonding between the polar water molecules must be broken down before boiling occurs. This explains the higher boiling point of water compared to similar but less polar molecules.

7. (A) Because of the electronegativity difference and noted difference of positions in the periodic table, these elements would form an ionic bond.

8. (B) The hydrogen molecule has equal sharing of electrons, typical of a covalent bond.

9. (A) An electronegativity difference of 7.5 or greater results in a bond that is essentially ionic.

10. (E) The general term used for all weak intermolecular forces is van der Waals forces.

11. (B) The atomic mass includes essentially the masses of the protons and neutrons. The atomic number of an element indicates the number of protons. So the atomic mass minus the atomic number equals the number of neutrons.

12. (A) When an alpha particle is emitted, the atomic mass decreases by 4 and the atomic number decreases by 2.

13. (B) The slow neutron is captured by the nucleus of ^{235}U and causes the disintegration.

14. (C) The electron has only 1/1837 the mass of a proton.

15. (E) The gamma ray has the characteristics of electromagnetic radiation identical to light and x-rays.

16. (A) An atom with atomic number 11 would have the orbital configuration

$$\begin{matrix} 2 & 2 & 6 & 1 \\ 1s & 2s & 2p & 3s. \end{matrix}$$

17. (B) The last-filled orbital for all inert gases except helium is the p orbital.

18. (E) The sp designation represents a hybridized orbital consisting of an s and a p orbital.

19. (A) The sodium atom oxidizes by losing an electron from its outer s orbital.

20. (E) The sp^3 configuration is a tetrahedron.

21. (C) The boron in BF_3 is trigonal planar.

22. (E) The carbon in CH_4 is a tetrahedron.

23. (B) The p orbital is a linear dumbbell shape composed of two equal lobes.

24. (A) Ionic solids have high melting points because of the strength of the ionic bonds which must be broken.

25. (A) Exothermic reactions release energy and have a negative $\triangle H$.

26. (C) This reaction does go to completion, but it does so because of the release of CO_2 gas, not Cl_2.

27. (D) Very active metals are difficult to reduce since reduction is the addition of electrons.

28. (B) Acetic acid is a weak acid and in its very concentrated form is called glacial acetic. However, it is a weak acid because it ionizes slightly in a water solution.

29. (B) Both statements are true, but they do not explain each other.

30. (D) Metal ions are positively charged, and as they gain electrons the charge decreases.

31. (E) The condensation of steam is a physical change with the same molecular structure.

32. (D) Whereas equilibrium is the point

where the forward and reverse reactions are equal, the quantity of reactants and the quantity of products are not.

33. (E) CaO is a basic anhydride and therefore forms a basic solution in water.

34. (A) Collodial carbon is used commercially because it has a large surface area to adsorb gases.

35. (E) Both statements are false. The freezing point is the same as the melting point.

36. (D) Astatine has the lowest ionization energy of the halogens because it has the largest atomic radius.

37. (B) Both statements are true, but the reason is not a correct explanation. Hydrogen is a diatomic molecule.

38. (D) A concentration of .0001 gives a pH of 4.

39. (C) Although electronegativity increases across the chart, it is not because of an increase in ionic radius.

40. (A) The hydrogen gas formed in the reaction is given off; thus volatility is the answer.

41. (C) The immiscible liquids form a colloidal dispersion which eventually separates. For (E) to be true, an emulsifying agent must be used.

42. (E) The vapor phase exists in the area of the phase diagram identified as 5.

43. (C) The triple point is at the intersection of the boundaries of the liquid, solid, and gas phases.

44. (D) Point 4 lies on the boundary between the liquid and gaseous phases, so they can coexist.

45. (A) When the pressure was increased, the system would have to be taken to a higher temperature for the liquid vapor pressure to equal the new pressure and the system again "boil."

46. (C) Ethyl alcohol solution does not contain sufficient ions to act as an electrolyte.

47. (C) Since the addition of 1 mole of nonionizing substance causes the boiling point of water solution to go up .51°C, the 2 moles of sugar in 2 liters (or 1 mole/liter) causes the boiling point to be approximately 100.51°C.

48. (D) 65.2% As ÷ 75 (at. mass of As)
$$= .867$$
34.8% O_2 ÷ 16 (at. mass of O)
$$= 2.18$$
Dividing by the smallest quotient:
.867 ÷ .867 = 1
2.18 ÷ .867 = 2.5
The ratio 1:2.5 can be made a whole-number ratio by multiplying both by 2; thus it becomes 2:5 or As_2O_5 (which agrees with the given molecular weight).

49. (C) $\dfrac{6.25 \text{ g}}{5 \text{ liters}} = \dfrac{x \text{ g}}{22.4 \text{ L}}$
$x = 28$ grams and the molecular weight since it is the weight of 22.4 L, the gram-molecular volume.

50. (A) 1 molar is defined as 1 mole of substance per liter of solution.

51. (C) The gaseous state has the highest kinetic energy imparted to the molecules.

52. (D) $\overset{x \text{ g}}{\underset{100 \text{ g}}{CaCO_3}}$ + 2 HCl
 1 mole
$\rightarrow H_2O + CO_2 + CaCl_2$
 1 mole
$x = 100$ g

53. (E) Cathode reaction:
$Ag^+ + 1e^- \rightarrow Ag \qquad E_0 = .80$
Anode reaction:
$\dfrac{Ni \rightarrow Ni^{2+} + 2e^- \qquad E_0 = .24}{ \text{Cell voltage} = 1.04 \text{ V}}$

54. (A) When equilibrium is reached, the voltage of a voltaic cell reaches 0 since the concentrations have stabilized for the reactants and products.

55. (A) The anode is the Ni plate since oxidation is occurring there.

56. (D) $Ni \rightarrow Ni^{2+} + 2e^-$
Since $2e^-$ are released per atom, .1 mole of e^- will result from forming .05 mole of Ni^{2+}.
.05 mole × 58.7 g/mole − 2.935 or 3 g lost; 10 g − 3 g = 7 g remaining on plate.

57. (E) $\Delta H_{\text{reaction}} = \Delta H_{\text{products}} - \Delta H_{\text{reactants}}$
$\Delta H_r = 2(-94.05) + (-57.8) - 54.2 + 0$
$\Delta H_r = -188.10 + (-57.8) - 54.2$
$\Delta H_r = -245.18 - 54.2 = -300.1 \text{ kcal}$

58. (C) 85.2 g   x L
    $2 NaClO_3 \rightarrow 2 NaCl + 3 O_2$
    213 g   67.2 L
    x L = 26.8 L O_2

59. (D) Balanced equation is $2 NaClO_3 \rightarrow 2$ $NaCl + 3O_2$. The coefficients give the molar relationship of 2:3.

60. (D) The coefficients indicate 1 mole of N_2 yields 2 moles of NH_3, so 3 L N_2 would yield 6 L NH_3.

61. (E) $(2 H = 2) + (1 S = 32) + (4 O = 64) = 98$.

62. (D) The alkene series starts with ethene:

    H   H
    | |
    H—C=C—H. The next member

    would be butene:
    H   H   H
    | | |
    H—C=C—C—H.
                |
                H

63. (C) By numbering the chain of carbons and beginning at the end closest to the added group (—OH), the identified carbon is the second one–hence the name of the chain (butane) with the prefix 2 and the -ol ending to indicate that the addition is an alcohol group.

64. (C) Total molecule charge is 0. H = +1, O = −2 are known. Hence 2(+1) + 4(−2) = −6 charge. So +6 is needed to counter this charge, and +6 divided by 2 carbons gives each a charge of +3.

65. (A) Salts of weak bases and strong acids hydrolyze to form an acid solution. $CuSO_4$ is such a salt.

66. (B) Since heating increases volume, the temperatures expressed in kelvins (K) must form the fraction $\frac{473}{273}$.
    Since increasing the pressure from 76 cm (standard) to 78 cm decreases volume, this fraction must be $\frac{76}{78}$ or $\frac{760}{780}$. You could use the formula $\frac{V_1 P_1}{T_1} = \frac{V_2 P_2}{T_2}$ and solve for $V_2 = V_1 \times \frac{T_2}{T_1} \times \frac{P_2}{P_2}$.

67. (D) II and III are correct because the reverse reaction will only need the small amount of energy that was released to make it happen. This can be seen graphically:

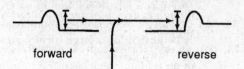

    forward   reverse

    This amount of energy becomes the activation energy for the reverse reaction.

68. (C) 20 grams of NaOH in 500 mL is equivalent to 40 grams in 1000 mL or 1 liter. 40 grams is 1 mole of NaOH. Therefore the molarity = 1.

69. (B) The equilibrium constant is
    $K =$
    $$\frac{[product]^{coefficient} \; [product]^{coefficient} \; etc.}{[reactant]^{coefficient} \; [reactant]^{coefficient} \; etc.}$$

70. (E) $PbF_2 \rightarrow Pb^2 + 2 F^-$ when it goes into solution.
    $K_{sp} = [Pb^{2+}][F^-]^2$
    Since $.49 \; g = \frac{.49 \; g}{245 \; g/mole \; PbF_2} = .002$
    mole of PbF_2, then, when the ions are in solution,
    $[Pb^{2+}] = .002$ mole/liter
    $[F^-] = 2 \times .002$ (because each molecule releases 2 F^-) = .004 mole /liter
    $K_{sp} = [.002][.004]^2 = 3.2 \times 10^{-3}$

71. (D) The heat shown in the equation is for the formation of 2 moles of H_2O from 2 moles of H_2, as shown by the coefficients of the equation. Two moles of H_2 = 4 grams of hydrogen, and this releases 136.64 kcal, or 68.32 kcal per mole of H_2 burned. 10 g of H_2 = 5 moles, so it would release 5 × 68.32 kcal = 341.60 kcal.

72. (B) Formation of an oxide is a chemical change. Only III is correct. Conducting an electric current and being attracted by a magnet does not change the chemical composition of a substance. This must occur in a chemical change. An example: aluminum metal forming aluminum oxide when exposed to air.

73. (B) Only III is correct since the energy used to melt a solid is used solely to break attractive forces between molecules. This is confirmed by the temperature of the material remaining constant, while melting occurs even though energy is being continuously supplied.

74. (D) An ester forms from the dehydration of an acid and an alcohol:

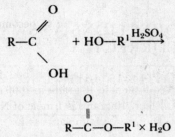

(R^1 indicates another radical–not necessarily the same as R.)

75. (E) This acid name is derived from the alkane, butane, by dropping the -e and adding -oic acid.

76. (C) When solids dissolve, the energy in the system usually decreases as particles dissociate. Because the system has a more random distribution, the entropy is increased. Both I and II are correct.

77. (D) Sodium hydroxide is hygroscopic and gains weight when exposed to air by attracting water to itself. This added weight of water caused the student to add an insufficient amount of sodium hydroxide, thus making his solution more dilute than 1 molar.

78. (B) This is the condenser.

79. (C) The separatory funnel has a stopcock which allows the addition of liquid.

80. (E) Correct order of assembly is shown below.

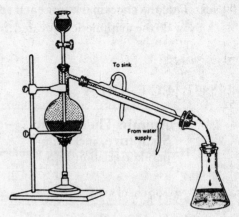

81. (C) The evolution of a gas and the temperature change are indicative of chemical change.

82. (B) Air is in the system and is expelled before CO_2 begins to be delivered.

83. (C) The attempt to explain the *cause* of the cloudiness makes this statement an interpretation of the data.

84. (E) Ca^{2+} has 18, not 10, electrons.

85. (E) (A) only speeds up the reaction; (B), (C), and (D) may either lower or raise the molar concentration of the product depending on the reaction.

DIAGNOSING YOUR NEEDS

This section will help you to diagnose your need to review the various categories tested by the College Board Achievement Test in Chemistry.

After taking the diagnostic test, check your answers against the correct ones. Then fill in the chart below. In the space under each question number, place a check (✔) if you answered that question correctly.

Next, total the checkmarks for each section and insert the number in the designated block. Now do the arithmetic indicated, and insert your percentage for each area.

SUBJECT AREA

(✔) QUESTIONS ANSWERED CORRECTLY

I. **Atomic Theory and Structure,** including periodic relationships	11	14	15	16	17	18	20	23	84

[ ] No. of checks ÷ 9 × 100 = _____%

II. **Nuclear Reactions**	12	13

[ ] No. of checks ÷ 2 × 100 = _____%

III. **Chemical Bonding and Molecular Structure**	6	7	8	9	10	21	22	24	37

[ ] No. of checks ÷ 9 × 100 = _____%

IV. **States of Matter and Kinetic Molecular Theory**	25	29	35	42	43	44	45	51	66

[ ] No. of checks ÷ 9 × 100 = _____%

V. **Solutions,** including concentration units, solubility, and colligative properties	41	46	47	50	68

[ ] No. of checks ÷ 5 × 100 = _____%

VI. **Acids and Bases**	1	2	3	4	5	28	38	65

[ ] No. of checks ÷ 8 × 100 = _____%

VII. **Oxidation-Reduction and Electrochemistry**	19	27	30	53	55	56

[ ] No. of checks ÷ 6 × 100 = _____%

VIII. Stoichiometry	48	49	52	58	59	60	61	71

☐ No. of checks ÷ 8 × 100 = _____%

IX. Reaction Rates							26	40

☐ No. of checks ÷ 2 × 100 = _____%

X. Equilibrium				32	54	69	70	85

☐ No. of checks ÷ 5 × 100 = _____%

XI. Thermodynamics: energy changes in chemical reactions, randomness, and criteria for spontaneity						57	67	76

☐ No. of checks ÷ 3 × 100 = _____%

XII. Descriptive Chemistry: physical and chemical properties of elements and their familiar compounds; organic chemistry; periodic properties	31	33	34	36	39	62	63	64	72
					73	74	75	77

☐ No. of checks ÷ 13 × 100 = _____%

XIII. Laboratory: equipment, procedures, observations, safety, calculations, and interpretation of results			78	79	80	81	82	83

☐ No. of checks ÷ 6 × 100 = _____%

PLANNING YOUR STUDY

The percentages give you an idea of how you have done on the various major areas of the test. Because of the limited number of questions on some parts, these percentages may not be as reliable as the percentages for parts with larger numbers of questions. However, you should now have at least a rough idea of the areas in which you have done well and those in which you need more study. (There are four more practice tests in the back of this book which may be used in a diagnostic manner as well.)

Start your study with the areas in which you are the weakest. The corresponding chapters are indicated below:

Subject Area	Chapters to Review
I. Atomic Theory and Structure, including periodic relationships	2, 15
II. Nuclear Reactions	15
III. Chemical Bonding and Molecular Structure	3, 4
IV. States of Matter and Kinetic Molecular Theory	1
V. Solutions, including concentration units, solubility, and colligative properties	7
VI. Acids and Bases	11
VII. Oxidation-reduction and Electrochemistry	12
VIII. Stoichiometry	5, 6
IX. Reaction Rates	9
X. Equilibrium	10
XI. Thermodynamics, including energy changes in chemical reactions, randomness, and criteria for spontaneity	8
XII. Descriptive Chemistry: physical and chemical properties of elements and their familiar compounds; organic chemistry; periodic properties	1, 2, 13, 14
XIII. Laboratory: equipment, procedures, observations, safety, calculations, and interpretation of results	All lab diagrams, 16

After you have spent some time in reviewing your weaker areas, plan a schedule of work that spans the 6 weeks before the test. Unless you set up a regular study pattern and goals, you probably will not prepare sufficiently.

The following schedule provides such a plan. Note that weekends are left free, and the time spans are held to 1- or 2-hour blocks. This will be time well spent!

	Monday	Tuesday	Wednesday	Thursday	Friday
First Week:	Ch. 1: 1 hr	Ch. 2: 2 hr	Ch 2: 1 hr	Ch. 3: 2 hr	Ch. 3: 1 hr
Second Week:	Ch. 4: 1 hr	Ch. 4: 1 hr	Ch. 5: 2 hr	Ch. 5: 1 hr	Ch. 6: 1 hr
Third Week:	Ch. 6: 2 hr	Ch. 7: 2 hr	Ch. 8: 1 hr	Ch. 9: 2 hr	Ch. 10: 1 hr
Fourth Week:	Ch. 10: 1 hr	Ch. 11: 2 hr	Ch. 12: 2 hr	Ch. 13: 1 hr	Ch. 14: 2 hr
Fifth Week:	Chs. 15, 16: 1 hr	Take Practice Test 1: 1 hr Study Ans.: 1 hr	Review weakest areas: 1 hr.	Take Practice Test 2: 1 hr Study Ans.: 1 hr	Review weakest areas: 1 hr
Sixth Week:	Take Practice Test 3: 1 hr Study Ans.: 1 hr	Review weakest areas: 1 hr.	Take Practice Test 4: 1 hr Study Ans.: 1 hr	Review weakest areas: 1 hr	Review a practice test already taken: 1 hr *Go to bed early.*

FINAL PREPARATION—THE DAY BEFORE THE TEST

The day before the test, review one of the practice tests you have already taken. Study again the directions for each type of question. Long hours of study at this point will probably only heighten your anxiety, so just look over the answer section of the practice test and refer to any chapter in the book if you need more information. This type of limited, relaxed review will probably make you feel more comfortable and better prepared.

Get together the materials that you will need. They are:

- Your admission ticket.
- Your identification. (You will not be admitted without some type of positive identification such as a student I.D. card with picture or a driver's license.)
- Two No. 2 pencils with erasers.

You should also go over this checklist:

A. Plan your time and activities so that you will have time for a good night's sleep.
B. Lay out comfortable clothes for the next day.
C. Review the following things to remember about the test:

1. Read the directions carefully.
2. In each group of questions, answer first those that you know. Temporarily skip difficult questions, but mark them in the margin so you can go back if you have time. Keep in mind that an easy question counts as much as a difficult one.
3. Avoid haphazard guessing since this will probably lower your score. However, if you can eliminate one or more of the choices to a question, it will generally be to your advantage to guess which of the remaining answers is correct. Your score will be based on the number right minus a fraction of the number answered incorrectly.
4. Remember that you neither gain nor lose credit for questions you do not answer.
5. Pace your time throughout the hour.
6. Mark the answer grid clearly and correctly. Be sure each answer is placed in the proper space and within the oval. *Erase all stray marks completely.*
7. You may write as much as you like in the test booklet. Use it as a scratch pad. Only the answers on the answer sheet are scored for credit.

D. Set your alarm clock so as to allow plenty of time to dress, eat your usual (or even a better) breakfast, and reach the test center without haste or anxiety.

Introduction to Chemistry

MATTER

Definition of Matter

Matter is defined as anything that occupies space and has mass. *Mass* is the quantity of matter which a substance possesses and, depending on the gravitational force acting on it, has a unit of weight assigned to it. Although the *weight* then can vary, the mass of the body is a constant and can be measured by its resistance to a change of position or motion. This property of mass to resist a change of position or motion is called *inertia*. Since matter does occupy space, we can compare the masses of various substances that occupy a particular unit volume. This relationship of mass to a unit volume is called the *density* of the substance. It can be shown in a mathematical formula as $D = \dfrac{m}{V}$. The basic unit of mass (m) in chemistry is the gram (g), and of volume (V) is the cubic centimeter (cm³) or milliliter (mL).

States of Matter

Matter occurs in three states: solid, liquid, and gas. A *solid* has both a definite size and shape. A *liquid* has a definite volume but takes the shape of the container and a *gas* has neither definite shape nor definite volume. These states of matter can often be changed by the addition of heat energy. An example of this is ice changing to liquid water and finally steam.

Composition of Matter

Matter can be subdivided into two general categories: distinct substances and mixtures. The distinct substances are either elements or compounds. If a substance is made up of only one kind of atom, it is called an *element*. If, however, it is composed of two or more kinds of atoms joined together in a definite grouping, it is classified as a *compound*. That a compound then always occurs in a definite composition is called the *Law of Definite Composition* or *Proportion*. An example of this is water: it always occurs in a two hydrogen atoms to one oxygen atom relationship to form the compound water. *Mixtures*, however, can vary in their composition.

In general, then:

Mixtures	*Distinct Substances*
	ELEMENTS
1. Indefinite composition *(generally heterogeneous) (Example: marble)	1. Composition is made up of one kind of atom (Examples: nitrogen, gold, neon)
2. Properties of the constituents are retained	2. All parts are the same throughout (homogeneous)
3. Parts of the mixture react differently to changed conditions	**COMPOUNDS**
*Solutions are mixtures (like sugar in water) but since the substance, like sugar, is distributed evenly throughout the water, it can be said to be an homogeneous mixture.	1. Definite composition (homogeneous) (Examples: water, carbon dioxide)
	2. All parts react the same.
	3. Properties of the compound are distinct and different from the properties of the individual type of elements that are combined in its make-up

Chemical and Physical Properties

Physical properties of matter are those properties of matter that can be usually observed with our senses. They include everything about a substance that can be noted when no change is occurring in the type of structure that makes up its smallest component. Some common examples are physical state, color, odor, solubility in water, density, melting point, taste, boiling point, and hardness.

Chemical properties are those properties which can be observed in regard to whether or not a substance reacts with other substances. Some common examples are: iron rusts in moist air, nitrogen does not burn, gold does not rust, sodium reacts with water, silver does not react with water, and water can be decomposed by an electric current.

Chemical and Physical Changes

The changes matter undergoes are classified as either physical or chemical. In general, a *physical change* alters physical properties but the composition remains constant. The most often altered properties are form and state. Some examples are: breaking glass, cutting wood, melting ice, and magnetizing a piece of metal. In some cases, the process that caused the change can be easily reversed and we again have the substance in its original form.

Chemical changes are always changes in the composition and structure of a substance. They are always accompanied by energy changes. If the energy released from the formation of a new structure exceeds the chemical energy in the original substances, energy will be given off usually in the form of heat or light or both. This is called an *exothermic reaction*. If the new structure, however, needs to absorb more energy than is available from the reactants, we have an *endothermic reaction*. This can be shown graphically.

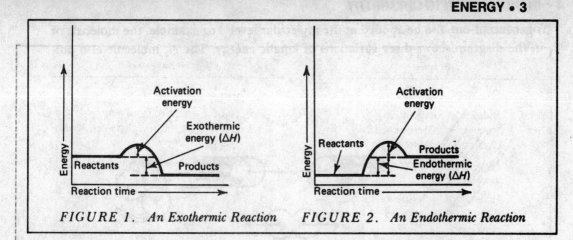

FIGURE 1. *An Exothermic Reaction*   FIGURE 2. *An Endothermic Reaction*

Notice that in Figures 1 and 2 the term *activation energy* is used. The activation energy is the energy necessary to get the reaction going by increasing the energy of the reactants so they can combine. You know you have to heat paper before it burns. This heat raises the energy of the reactants so that the burning can begin; then enough energy is given off from the burning so that an external source of energy is no longer necessary.

Conservation of Mass

When ordinary chemical changes occur, the mass of the reactants equals the mass of the products. This can be stated another way; that is, in a chemical change, matter can neither be created nor destroyed, but only changed from one form to another. This is referred to as the *Law of Conservation of Matter* (Lavoisier—1785). This law is contradicted by the Einstein mass-energy relationship, which states that matter and energy are interchangeable.

ENERGY

Definition of Energy

Energy is usually defined as the ability to do work.

Forms of Energy

Energy may appear in a variety of forms. Most commonly, energy in reactions is evolved as *heat*. Some other forms of energy are *light, sound, mechanical energy, electrical energy*, and *chemical energy*. Energy can be converted from one form to another. An example of this is the heat from burning fuel being used to vaporize water to steam. The energy of the steam is used to turn the turbine wheels to produce mechanical energy. The turbine turns the generator armature to produce electricity. The electricity is then available in homes for use as light, heat, or the operation of many modern appliances.

Two general classifications of energy are *potential energy* and *kinetic energy*. Potential energy is due to position; kinetic energy is energy of motion. The difference can be illustrated by a boulder sitting on the side of a mountain. It has a high potential energy due to its position above the valley floor. If it falls, however, its potential energy is converted to kinetic energy. This illustration is very similar to the situation of electrons cascading to lower energy levels in the atomic model.

This concept can also be applied at the molecular level. For example, the molecule of N_2 in the diagram shows three variations of kinetic energy. The N_2 molecule also pos-

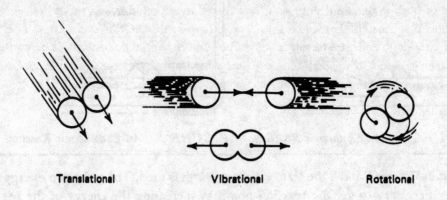

Translational   Vibrational   Rotational

sesses potential energy because of the chemical bond within the molecule. To break this bond would require energy. The potential energy in each isolated atom of nitrogen would then be greater. This would be similar to raising the boulder to a higher position on the mountainside. Both systems have a higher potential energy and have a tendency to fall back to the lower state, which is more stable.

Types of Reactions (Exothermic versus Endothermic)

When physical or chemical changes occur, energy is involved. If the heat content of the product(s) is higher than that of the reactants, the reaction is *endothermic*. If, on the other hand, the heat content of the product(s) is less than that of the reactants, the reaction is called *exothermic*. This change of heat content can be designated as ΔH. The heat content (H) is sometimes referred to as the *enthalpy*. Every system has a certain amount of heat. This changes during the course of a physical or chemical change. The change in heat content, ΔH, is the difference between the heat content of the products and that of the reactants. The equation is:

$$\Delta H = H_{\text{products}} - H_{\text{reactants}}$$

If the heat content of the products is greater than the heat content of the reactants, ΔH is a positive quantity ($\Delta H > 0$) and the reaction is endothermic. If, however, the heat content of the products is less than the heat content of the reactants, ΔH is a negative quantity ($\Delta H < 0$) and the reaction is exothermic. This relationship is shown graphically in Figures 1 and 2.

Conservation of Energy

Experiments have shown that energy is neither gained nor lost in physical or chemical changes. This principle is known as the *Law of Conservation of Energy* and is often stated as follows: Energy is neither created nor destroyed in ordinary physical and chemical changes.

CONSERVATION OF MATTER AND ENERGY

With the introduction of atomic theory and a more complete understanding of the nature of both matter and energy, it was found that a relationship exists between these two concepts. Einstein formulated the *Law of Conservation of Mass and Energy*. This states that matter and energy are interchangeable under special conditions. The conditions have been created in nuclear reactors and accelerators, and the law has been verified. This relationship can be expressed by Einstein's famous equation:

$$E = mc^2$$
$$\text{Energy} = \text{Mass} \times (\text{Velocity of light})^2$$

MEASUREMENTS AND CALCULATIONS

The student of chemistry must be able to use the correct measurement terms accurately and to solve problems that require mathematical skill, as well as proper terminology, for their correct solution. The following sections review these topics.

Metric System

It is important that scientists around the world use the same units when communicating information. For this reason, scientists use the modernized metric system, designated in 1960 by The General Conference on Weights and Measures as the International System of Units. This is commonly known as The SI system, an abbreviation for the French name Le Systeme International d'Unites. It is now the most common system of measurement in the world.

The reason it is so widely accepted is twofold. First, SI uses the decimal system as its base. Secondly, in many cases units for various quantities are defined in terms of units for simpler quantities.

There are seven basic units that can be used to express the fundamental properties of measurement. These are called the SI base units and are shown in the table below.

SI BASE UNITS		
Property	*Unit*	*Abbreviation*
mass	kilogram	kg
length	meter	m
time	second	s
electric current	ampere	A
temperature	kelvin	K
amount of substance	mole	mol
luminous intensity	candela*	cd

*The candela is rarely used in chemistry.

Other SI units are derived by combining prefixes with a root unit. The prefixes represent multiples or fractions of 10. The following table gives some basic prefixes used in the metric system.

Prefix	Multiple	Scientific Notation	Abbreviation
mega-	1,000,000	10^6	M
kilo-	1,000	10^3	k
hecto-	100	10^2	h
deka-	10	10^1	da
deci-	.1	10^{-1}	d
centi-	.01	10^{-2}	c
milli-	.001	10^{-3}	m
nano-	.000,000,001	10^{-9}	n

For an example of how the prefix works in conjunction with the root word, consider the term *kilometer*. The prefix *kilo-* means "multiply the root word by 1000," so a kilometer is 1000 meters. By the same reasoning, a millimeter is 1/1000 of a meter.

Because of the prefix system, all units and quantities can be easily related by some factor of 10. Here is a brief table of some metric unit equivalents.

Length
10 millimeters (mm) = 1 centimeter (cm)
100 cm = 1 meter (m)
1000 m = 1 kilometer (km)
Volume
1000 milliliters (mL) = 1 liter (L)
1000 cubic centimeters (cm^3) = 1 liter
1 mL = 1 cm^3
Mass
1000 milligrams (mg) = 1 gram (g)
1000 g = 1 kilogram (kg)

A unit of length, used especially in expressing the length of light waves, is the angstrom, abbreviated as Å and equal to 10^{-8} cm.

Because measurements are often reported in units of the English system, it is important to be able to convert them to metric units. Some common conversion factors are shown in the following table.

2.54 cm = 1 inch
1 meter = 39.37 inches (10% longer than a yard)
1 ounce = 28.35 grams
1 pound = 454 grams
2.2 pounds = 1 kilogram
1 quart = .946 liter
1.06 quarts = 1 liter (5% larger than a quart)

The metric system standards were chosen as natural standards. The meter was first described as the distance marked off on a platinum-iridium bar but now can be reproduced as 1,650,763.73 times the wavelength of an isotope of krypton when it is excited to give off an orange-red spectral line.

There are some interesting relationships between volume and mass units in the metric system. Since water is most dense at 4°C, the gram was intended to be 1 cubic centimeter of water at this temperature. This means, then, that:

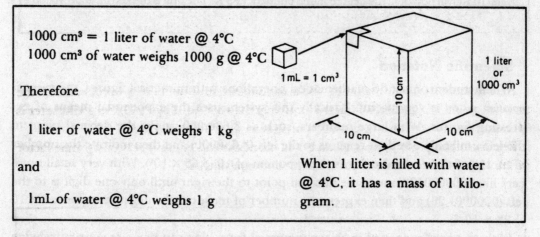

1000 cm^3 = 1 liter of water @ 4°C
1000 cm^3 of water weighs 1000 g @ 4°C

Therefore

1 liter of water @ 4°C weighs 1 kg

and

1 mL of water @ 4°C weighs 1 g

1 mL = 1 cm³

10 cm
10 cm
10 cm

1 liter
or
1000 cm³

When 1 liter is filled with water @ 4°C, it has a mass of 1 kilogram.

Temperature Measurements

The most commonly used temperature scale in scientific work is the Celsius scale. It gets its name from the Swedish astronomer Anders Celsius and dates back to 1742. For a long time it was called the centigrade scale because it is based on the concept of dividing the distance on a thermometer between the freezing point of water and its boiling point into 100 equal markings or degrees.

There is another scale that is based on the lowest theoretical temperature (called absolute zero). This temperature has never actually been reached, but scientists in laboratories have reached temperatures within about a thousandth of a degree above absolute zero. Lord William Kelvin proposed this scale, which would have the same size degrees as the Celsius degree and is now referred to as the Kelvin scale. Through experiments and calculations, it has been determined that absolute zero is 273.15 degrees below zero on the Celsius scale. This figure is usually rounded off to 273.

The diagram and formulas that follow give the graphic and algebraic relationship between the temperature scales encountered in chemistry.

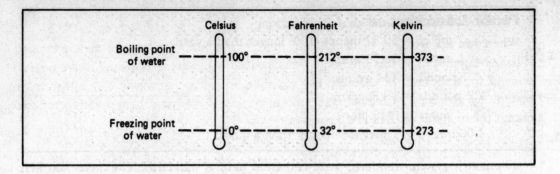

Conversion Formulas:

$$°F = \tfrac{9}{5}°C + 32°$$
$$°C = \tfrac{5}{9}(°F - 32°)$$
$$K = °C + 273°$$

• **EXAMPLES:**

$30°C = ?°F$ $°F = \tfrac{9}{5}(30°) + 32° = 86°$

$68°F = ?°C$ $°C = \tfrac{5}{9}(68° - 32°) = 20°$

$10°C = ? K$ $K = 10 + 273 = 283$

Note: In Kelvin notation, the degree sign is omitted: 283 K. The unit is the kelvin, abbreviated as K.

Scientific Notation

When students must do mathematical operations with numerical figures, the *scientific notation system* is very useful. Basically this system uses the exponential means of expressing figures. With large numbers, such as 3,630,000, move the decimal point to the left until only one digit remains to the left (3.630000) and then indicate the number of moves of the decimal point as the exponent of 10 (3.63×10^6). With very small numbers like .000000123, move the decimal point to the right until only one digit is to the left (0000001.23) and then express the number of moves as the negative exponent of 10 (1.23×10^{-7}).

With numbers expressed in this exponential form, you can now use your knowledge of exponents in mathematical operations. Remember: in multiplication you add exponents and in divisions you subtract the exponents of 10.

• **EXAMPLES:**

Multiplications:

$(2.3 \times 10^5)(5.0 \times 10^{-12})$. Multiplying the first number in each, you get 11.5 and the addition of the exponents gives 10^{-7}. Now changing to a number with only one digit to the left of the decimal point gives you 1.15×10^{-6} for the answer.

Try these:

$(5.1 \times 10^{-6})(2 \times 10^{-3}) = 10.2 \times 10^{-9} = 1.02 \times 10^{-8}$

$(3 \times 10^5)(6 \times 10^3) = 18 \times 10^8 = 1.8 \times 10^9$

Divisions:

$(1.5 \times 10^3) \div (5.0 \times 10^{-2}) = .3 \times 10^5 = 3 \times 10^4$

$(2.1 \times 10^{-2}) \div (7.0 \times 10^{-3}) = .3 \times 10^1 = 3$

(Notice that in division the exponents of 10 are subtracted.)

Factor-Label Method of Conversion

When you are working problems that involve numbers with units of measurement, it is convenient to use this method so that you do not become confused in the operations of multiplication or division. For example, if you are changing .001 kilogram to milligrams, you set up each conversion as a fraction so that all the units will factor out except the one you want in the answer.

$$1 \times 10^{-3} \, \text{kilogram} \times \frac{1 \times 10^3 \, \text{grams}}{1 \, \text{kilogram}} \times \frac{1 \times 10^3 \, \text{milligrams}}{1 \, \text{gram}} = 1 \times 10^3 \, \text{mg}$$

Notice that the kilogram is made the denominator in the first fraction to be factored with the original kilogram unit. The numerator is equal to the denominator except that the numerator is expressed in smaller units. The second fraction has the gram unit in the denominator to be factored with the gram unit in the preceding fraction. The answer is in milligrams since it is the only unit remaining and it assures you that the correct operations have been performed in the conversion.

• ANOTHER EXAMPLE:

1 ft. = ? cm

$$1 \, \text{ft.} \times \frac{12 \, \text{in.}}{1 \, \text{ft.}} \times \frac{2.54 \, \text{cm}}{1 \, \text{in.}} = 30.48 \, \text{cm}$$

This method is used in examples throughout this book.

Precision, Accuracy, and Uncertainty

Two other factors to consider in measurement are *precision* and *accuracy*. Precision indicates the reliability or reproducibility of a measurement. Accuracy indicates how close a measurement is to its known or accepted value.

For example, suppose you were taking a reading of the boiling point of pure water at sea level. Using the same thermometer in three trials, you record 96.8, 96.9, and 97.0 degrees Celsius. Since these figures show a high reproducibility, you can say that they are precise. However, the values are considerably off from the accepted value of 100 degrees Celsius, so we say they are not accurate. In this example we probably would suspect that the inaccuracy was the fault of the thermometer.

Regardless of precision and accuracy, all measurements have a degree of *uncertainty*. This is usually dependent on two factors—the limitation of the measuring instrument and the skill of the person making the measurement. This can best be shown by example.

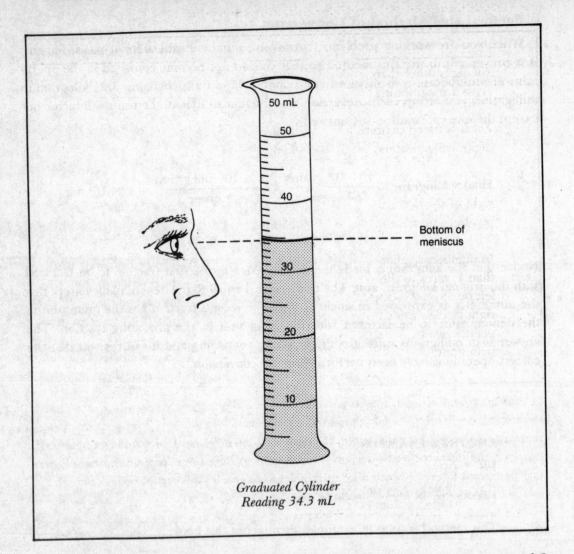

Graduated Cylinder
Reading 34.3 mL

The graduated cylinder in the illustration contains a quantity of water to be measured. It is obvious that the quantity is between 30 and 40 milliliters because the meniscus lies between these two marked quantities. Now, checking to see where the bottom of the meniscus lies with reference to the ten intervening subdivisions, we see that it is between the fourth and fifth. This means that the volume lies between 34 and 35 milliliters. The next step introduces the uncertainty. We have to guess how far the reading is between these two markings. We can make an approximate guess, or estimate, that the level is more than .2 but less than .4 of the distance. We therefore report the volume as 34.3 mL (milliliters). The last digit in any measurement is an estimate of this kind and is uncertain.

Significant Figures

Any time a measurement is recorded, it includes all the digits that are certain plus one uncertain digit. These certain digits plus the one uncertain digit are referred to as *significant figures*. The more digits you are able to record in a measurement, the less relative uncertainty there is in the measurement. The following table summarizes the rules of significant figures.

Rule	Example	Number of Significant Figures	
All digits other than zeros are significant.	25 g 5.471 g	2 4	
Zeros between nonzero digits are significant.	309 g 40.06 g	3 4	
Final zeros to the right of the decimal point are significant.	6.00 mL 2.350 mL	3 4	
In numbers smaller than 1, zeros to the left or directly to the right of the decimal point are not significant.	0.05 cm	1	The zeros merely mark the position of the decimal point.
	0.060 cm	2	The first two zeros mark the position of the decimal point. The final zero is significant.

One last rule deals with final zeros in a whole number. These zeros may or may not be significant, depending on the measuring instrument. For instance, if an instrument that measures to the nearest mile is used, the number 3000 miles has four significant figures. If, however, the instrument in question records miles to the nearest thousands, there is only one significant figure. The number of significant figures in 3000 could be one, two, three, or four, depending on the limitation of the measuring device.

This problem can be avoided by using the system of scientific notation. For this example, the following notations would indicate the numbers of significant figures:

$$3 \times 10^3 \quad \text{one significant figure}$$
$$3.0 \times 10^3 \quad \text{two significant figures}$$
$$3.00 \times 10^3 \quad \text{three significant figures}$$
$$3.000 \times 10^3 \quad \text{four significant figures}$$

• QUESTIONS FOR CHAPTER 1

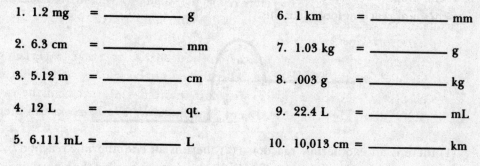

1. 1.2 mg = _____ g

2. 6.3 cm = _____ mm

3. 5.12 m = _____ cm

4. 12 L = _____ qt.

5. 6.111 mL = _____ L

6. 1 km = _____ mm

7. 1.03 kg = _____ g

8. .003 g = _____ kg

9. 22.4 L = _____ mL

10. 10,013 cm = _____ km

11. The density of CCl_4 (carbon tetrachloride) is 1.58 g/mL. What will 100 mL of CCl_4 weigh?

12. A piece of sulfur weighs 8 oz. When it was submerged in a graduated cylinder containing 50 mL of H_2O, the level rose to 150 mL. What is the density (g/mL) of the sulfur? (16 oz. = 454 g)

13. (a) A box 20 cm × 20 cm × 5.08 inches has what volume in cubic centimeters?
    (b) What weight of H_2O @ 4°C will the box hold?

14. Set up the following using the *factor-label method:*

$$\frac{5 \text{ cm}}{\text{sec}} = \frac{\text{mi}}{\text{hr}}$$

15. How many significant figures are in each of the following? (a) 1.01   (b) 200.0 (c) .0021   (d) .0230

• **ANSWERS:**

1. .0012
2. 63
3. 512
4. 12.7
5. .006111
11. 158 grams
*12. 2.27 grams/mL
*13. (a) 5160 cm³ (b) 5160 grams

6. 1,000,000 or 1×10^6
7. 1030 or 1.03×10^3
8. .000003 or 3×10^{-6}
9. 22400 or 2.24×10^4
10. .10013

14. $5\dfrac{\text{cm}}{\text{sec}} = \dfrac{60 \text{ sec.}}{1 \text{ min.}} \times \dfrac{60 \text{ min.}}{1 \text{ hr.}} \times \dfrac{1 \text{ in.}}{2.54 \text{ cm}} \times \dfrac{1 \text{ ft.}}{12 \text{ in.}} \times \dfrac{1 \text{ mi}}{5280 \text{ ft.}}$

15. (a) 3   (b) 4   (c) 2   (d) 3

Chapter

1

Review

1. If the graphic representation of the energy levels of the reactants and products in a chemical reaction looks like this:

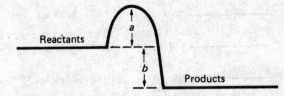

(1) there is an exothermic reaction   (2) there is an endothermic reaction   (3) the (a) portion is the energy given off   (4) the (b) portion is called the activation energy.

*Answer explained in section at the end of the book.

2. The amount of mass per unit volume refers to the (1) density   (2) specific weight   (3) volume   (4) weight.

3. A baking powder can carries the statement, "Ingredients: corn starch, sodium bicarbonate, calcium acid phosphate, and sodium aluminum sulfate." Therefore, this baking powder is (1) a compound   (2) a mixture   (3) a molecule   (4) a mixture of elements.

4. Which of the following is a physical property of sugar? (1) It decomposes readily.   (2) Its composition is carbon, hydrogen, and oxygen.   (3) It turns black with concentrated H_2SO_4.   (4) It can be decomposed with heat.   (5) It is a white crystalline solid.

5. A substance that can be further simplified may be either (1) an element or a compound   (2) an element or a mixture   (3) a mixture or a compound   (4) a mixture or an atom.

6. A substance composed of two or more elements chemically united is called (1) an isotope   (2) a compound   (3) an element   (4) a mixture.

7. An example of a chemical change is the (1) breaking of a glass bottle   (2) sawing of a piece of wood   (3) rusting of iron   (4) melting of an ice cube.

8. A substance which cannot be further decomposed by ordinary chemical means is (1) water   (2) air   (3) sugar   (4) silver.

9. An example of a physical change is the (1) fermenting of sugar to alcohol   (2) rusting of iron   (3) burning of paper   (4) solution of sugar in water.

10. Chemical action may involve all of the following except (1) combining of atoms of elements to form a molecule   (2) separation of the molecules in a mixture   (3) breaking down compounds into elements   (4) reacting a compound and an element to form a new compound and a new element.

11. The energy of a system can be (1) easily changed to mass   (2) transformed into a different form   (3) measured only as potential energy   (4) measured only as kinetic energy.

12. If the ΔH of a reaction is a negative quantity, the reaction is definitely (1) endothermic   (2) unstable   (3) exothermic   (4) reversible.

• **ANSWERS:**

1. (1)		
2. (1)	5. (3)	9. (4)
3. (2)	6. (2)	10. (2)
4. (5)	7. (3)	11. (2)
	8. (4)	12. (3)

In the following list, write E for the elements, C for the compounds, and M for the mixtures.

Water	Aluminum oxide	Hydrochloric acid
Wine	Hydrogen	Nitrogen
Soil	Carbon dioxide	Tin
Silver	Air	Potassium chloride

• ANSWERS:

C		C		C
M		E		E
M		C		E
E		M		C

Terms You Should Know

accuracy
activation energy
angstrom
Celsius
chemical change
chemical property
compound
density
element
endothermic
enthalpy

exothermic
Fahrenheit
gas
heterogeneous
homogeneous
inertia
Kelvin
kinetic energy
liquid
mass
matter

mixture
physical change
physical property
potential energy
precision
scientific notation
significant figures
SI units
solid
uncertainty
weight

Law of Conservation of Energy
Law of Conservation of Matter
Law of Conservation of Matter and Energy
Law of Definite Composition or Proportion

2

Atomic
Structure

HISTORY

Around 1805, John Dalton proposed some basic assumptions about atoms. Some of these ideas dated back to the ancient Greeks. They were:

1. All matter is made up of very small, discrete particles called atoms.
2. All atoms of an element are alike in weight and different from the weight of any other kind of atom.
3. Atoms unite in chemical changes to form compounds. The atom loses its characteristic properties when this combination occurs.
4. The atom is not entirely altered in the formation of a compound, and its weight remains constant.

ELECTRIC NATURE OF ATOMS

For the past 90 years, scientists have been gathering evidence about the structure of atoms and fitting the information into a model of the atomic structure.

Basic Electric Charges

The discovery of the electron as the first subatomic particle is credited to J. J. Thomson (England, 1897). He used an evacuated tube connected to a spark coil as shown in Figure 3. As the voltage across the tube was increased, a beam of light became visible. This was referred to as a cathode ray. Thomson found that the beam was deflected by both electrical and magnetic fields. From this, he concluded that the cathode rays were made up of very small negatively charged particles, which he named *electrons*.

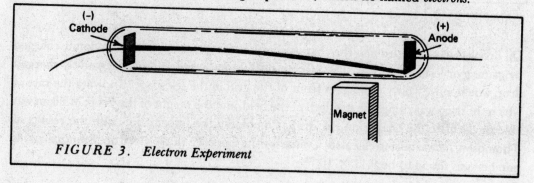

FIGURE 3. Electron Experiment

Further experimentation led Thompson to find the ratio of the electrical charge of the electron to its mass. This was a major step toward understanding the nature of the particle. He was awarded a Nobel Prize in 1906 for his accomplishment.

It was an American scientist, Robert Millikan, who in 1909 was able to measure the charge on an electron using the apparatus pictured below.

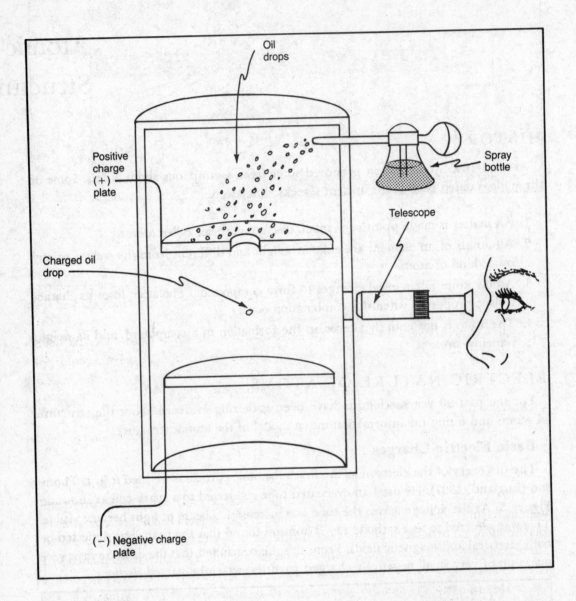

Oil droplets were sprayed into the chamber and, in the process, became randomly charged by gaining or losing electrons. The electric field was adjusted so that a negatively charged drop would move slowly upward in front of the grid in the telescope. Knowing the rate at which the drop was rising, the strength of the field, and the weight of the drop, Millikan was able to calculate the charge on the drop. Combining the information with the results of Thompson, he was able to calculate a value for the mass of a single electron. Eventually, this number was found to be 9.11×10^{-28} g.

Ernest Rutherford (England, 1911) performed a gold foil experiment (Figure 4) which had tremendous implications for atomic structure.

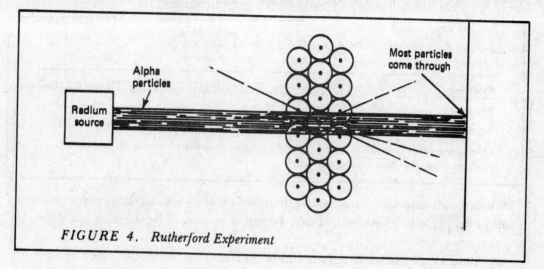

FIGURE 4. Rutherford Experiment

Alpha particles (helium nuclei) passed through the foil with few deflections. However, some deflections (1 per 8000) were almost directly back toward the source. This was unexpected and suggested an atomic model with mostly empty space between a *nucleus,* in which most of its mass was located and which was positively charged, and the electrons that defined the volume of the atom.

Further experiments showed that the nucleus was made up of still smaller particles called *protons.* Rutherford realized, however, that protons, by themselves, could not account for the entire mass of the nucleus. He predicted the existence of a new nuclear particle that would be neutral and would account for the missing mass. In 1932, James Chadwick (England) discovered this particle, the *neutron.*

Today the number of subatomic particles identified as discrete units has risen to well over 30 named units.

Bohr Model

In 1913, Neils Bohr (Denmark) proposed his model of the atom. This pictured the atom as having a dense, positively charged nucleus and negatively charged electrons in specific spherical orbits around this nucleus which are also called energy levels or shells. These shells are arranged concentrically around the nucleus. Each shell is designated by a letter K, L, M, N, O... The closer to the nucleus, the less energy an electron needed in one of these shells, but it had to gain energy to go from one shell to another that was further away from the nucleus.

Because of its simplicity and general ability to explain chemical change the Bohr model still has some usefulness today.

Shell	Letter Designation	Maximum Number of Electrons
1	K	2
2	L	8
3	M	18
4	N	32
5	O	32

Components of Atomic Structure

The chart below lists the basic particles of the atom.

Particle	Charge	Actual Mass	Relative Mass Compared to Proton	Discovery
Electron	$- (e^-)$	9.109×10^{-28} g	1/1837	J. J. Thomson–1897
Proton	$+ (p^+)$	1.673×10^{-24} g	1	——— early 1900's
Neutron	$0(n^0)$	1.675×10^{-24} g	1	J. C. Chadwick–1932

(There are now some 30 or more named particles or units of atomic structure, but the above are the most commonly used.)

When these components are used in the Bohr model, we show the protons and neutrons in the nucleus. These particles are known as *nucleons*. The electrons are shown in the shells.

The number of protons in the nucleus of an atom determines the *atomic number*. All atoms of the same element have the same number of protons and therefore the same atomic number; atoms of different elements have different atomic numbers. Thus, the atomic number identifies the element. An English scientist, Henry Moselely, first determined the atomic numbers of the elements through the use of X-rays.

Since the actual masses of subatomic particles and atoms themselves are very small numbers when expressed in grams, scientists use atomic mass units (amu) instead. An *atomic mass unit* is defined as 1/12 the mass of a carbon-12 atom. Thus, the mass of any atom is expressed relative to the mass of one atom of carbon-12.

The sum of the number of protons and neutrons in the nucleus is called the *mass number*.

Table 1 summarizes the relationships just discussed. Notice that the outermost shell can contain no more than eight electrons. The explanation of this is given in the next section.

Element	Atomic No.	Mass No.	Number of Protons	Number of Neutrons	Number of Electrons	K	L	M	N
Hydrogen	1	1	1	0	1	1			
Helium	2	4	2	2	2	2			
Lithium	3	7	3	4	3	2	1		
Beryllium	4	9	4	5	4	2	2		
Boron	5	11	5	6	5	2	3		
Carbon	6	12	6	6	6	2	4		
Nitrogen	7	14	7	7	7	2	5		
Oxygen	8	16	8	8	8	2	6		
Fluorine	9	19	9	10	9	2	7		
Neon	10	20	10	10	10	2	8		
Sodium	11	23	11	12	11	2	8	1	
Magnesium	12	24	12	12	12	2	8	2	
Aluminum	13	27	13	14	13	2	8	3	
Silicon	14	28	14	14	14	2	8	4	
Phosphorus	15	31	15	16	15	2	8	5	
Sulfur	16	32	16	16	16	2	8	6	
Chlorine	17	35	17	18	17	2	8	7	
Argon	18	40	18	22	18	2	8	8	
Potassium	19	39	19	20	19	2	8	8	1
Calcium	20	40	20	20	20	2	8	8	2
Scandium	21	45	21	24	21	2	8	9	2

TABLE 1. Table of the First 21 Elements

If atoms of the same element have different masses, they are called *isotopes*. The variance in mass and mass number is due to a different number of neutrons in the nucleus. Isotopes of an element occur in the same relative percentages wherever a natural sample of the element is found. The atomic mass (formerly atomic weight) of an element is defined as the weighted average of the masses of all its isotopes as they occur in a natural sample of the element. It is due to the occurrence of isotopes that the atomic mass of an element is not a whole number, but rather a number containing a decimal. When the atomic mass is rounded off to the nearest whole number, the resulting number represents the atomic mass of the most common isotope. This number of amu will be the same as the mass number of that isotope.

Drawings like those in Figure 5 give valuable aid in determining oxidation numbers (see below); metallic, nonmetallic, and inert elements; chemical reactivity; and the periodic arrangement of the elements.

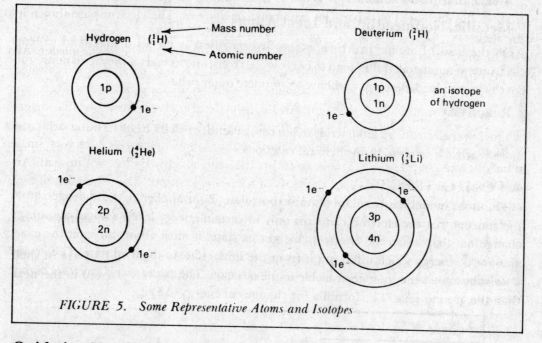

FIGURE 5. *Some Representative Atoms and Isotopes*

Oxidation Number and Valence

Each atom attempts to have its outer orbit complete and accomplishes this by borrowing, lending, or sharing its electrons. The electrons found in the outermost orbit are called *valence electrons*. The absolute number of electrons gained, lost, or borrowed is referred to as the valence of the atom. When valence electrons are lost or partially lost by sharing, the valence number is assigned a + sign for that element and is called its *oxidation number*. If valence electrons are gained or partially gained by an atom, its valence is assigned a − sign for its oxidation number.

• **EXAMPLE:**

$_{17}Cl =$ ●)2)8)7 ← valence electrons

This picture can be simplified to ·C̈l:, showing only the valence electrons as dots. This is called the Lewis dot structure of the atom. To complete its outer orbit to eight electrons, Cl must borrow one from another atom. Its valence number then is 1. When electrons are gained, we assign a − sign to this number so its oxidation number is − 1.

• ANOTHER EXAMPLE:

$$_{11}Na = \overset{\text{nucleus}}{\bullet} \;)2 \;)8 \;)1 \leftarrow \text{valence electron}$$

Na· (Lewis dot structure)

Since Na tends to lose this electron, its oxidation number is +1.

Some other general rules for oxidation numbers are:
1. Atoms of free elements have an oxidation number of zero.
2. Hydrogen has an oxidation number of +1 except in metallic hydrides, where it is −1.
3. Oxygen has an oxidation number of −2 except in peroxides, where it is −1. In combination with fluorine, it is +2.

A table of oxidation numbers is given on page 480

Metallic, Nonmetallic, and Inert Atoms

On the basis of atomic structure, atoms are classified as metals if they tend to lend electrons; as nonmetals if they tend to borrow electrons; and as inert if they tend neither to borrow nor to lend electrons and have a complete outer orbit.

Reactivity

The fewer electrons an atom tends to borrow, or lend, or share to fill its outer orbit, the more reactive it tends to be in chemical reactions.

ATOMIC SPECTRA

The Bohr model was based on a simple postulate. Bohr applied to the hydrogen atom the concept that the electron can exist only in certain energy levels without an energy change but that, when the electron changes its state, it must absorb or emit the exact amount of energy which will bring it from the initial state to the final state. The *ground state* is the lowest energy state available to the electron. The *excited state* is any level higher than the ground state. The formula for changes in energy (ΔE) is:

$$\Delta E_{\text{electron}} = E_{\text{final}} - E_{\text{initial}}$$

When the electron moves from the ground state to an excited state, it must absorb energy. When it moves from an excited state to the ground state, it emits energy. This release of energy is the basis for *atomic spectra*. (See Figure 6.)

The energy values were calculated from Bohr's equation:

$$E_{\text{mole of electrons}} = \frac{-313.6 \text{ kcal}}{n^2}$$

n (indicating each orbit) = 1, 2, 3, 4, ...

These values are also expressed in electron volt (eV)* units by using this relationship:

$$\frac{\text{kcal}}{\text{mole of atoms}} = 23.1 \times \frac{\text{eV}}{\text{atom}}$$

Both values are given in Figure 6.

*In this book, the names of units and the abbreviations for them are written in accordance with the preferred style of the American Chemical Society.

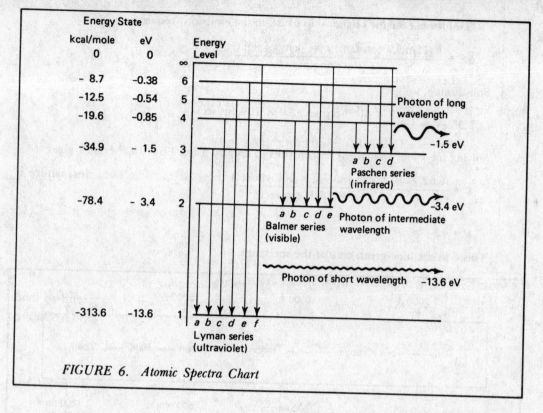

FIGURE 6. *Atomic Spectra Chart*

When energy is released in the "allowed" values, it is released in the form of discrete radiant energy called *photons*. Each of the first three levels has a particular name associated with the emissions that occur when an electron reaches its ground state on that level. The Lyman series, consisting of ultraviolet radiation, is the name for emissions that occur when an electron cascades from a level higher than the first level down to $n = 1$. Note in Figure 6 that the next two higher levels have the names Balmer (for $n = 2$) and Paschen ($n = 3$) series, respectively.

When the light emitted by energized atoms is examined with an instrument called a *spectroscope*, the prism or diffraction grating in the spectroscope disperses the light to allow an examination of the *spectra* or distinct colored lines. Since only particular energy jumps are available in each type of atom, each element has its own unique emission spectra made up of only the lines of specific wavelength that correspond to its atomic structure.

• **EXAMPLE:**

Hydrogen can have an electron drop from the $n = 4$ to $n = 2$ level. What visible spectral line in the Balmer series will result from this emission of energy?

$$\Delta E_{evolved} = E_{n=4} - E_{n=2}$$

From Figure 6, $E_4 = -.85$ eV, $E_2 = -3.4$ eV.

$$\Delta E_{evolved} = -.85 \text{ eV} - (-3.4 \text{ eV}) = 2.55 \text{ eV}$$

(Since ΔE is positive, this indicates energy released.)

The formula for the relationship of ΔE to the emission frequency is

$$\Delta E = \frac{h \text{ (Planck's constant) } c \text{ (velocity of light)}}{\lambda \text{ (wavelength)}}$$

Substituting, we get

$$2.55 \text{ eV} = \frac{(6.62 \times 10^{-27} \text{ erg sec})(3.0 \times 10^{10} \text{ cm/sec})}{\lambda}$$

Solving for λ and changing electron volts to ergs by using 1.6×10^{-12} erg/eV gives

$$\lambda = \frac{(6.62 \times 10^{-27} \text{ erg sec})(3.0 \times 10^{10} \text{ cm/sec})}{(2.55 \text{ eV})(1.6 \times 10^{-12} \text{ erg/sec})}$$

$$\lambda = 4.87 \times 10^{-5} \text{ cm}$$

This is in the blue-green area of the spectrum.

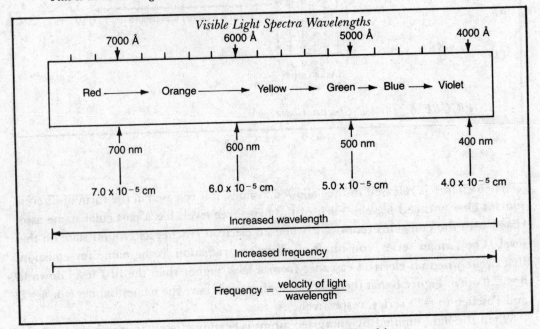

A partial atomic spectrum for hydrogen would look like this:

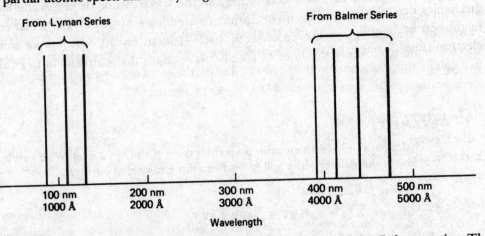

The right-hand group is in the visible range and is part of the Balmer series. The left-hand group is in the ultraviolet region and belongs to the Lyman series.

Investigating spectral lines like these can be used in the identification of unknown specimens.

Mass Spectroscopy

Another tool used to identify specific atomic structures is the process of mass spectroscopy. This is based on the concept that differences in mass cause differences in the degree of bending that occurs in a beam of ions passing through a magnetic field. This is shown in Figure 7.

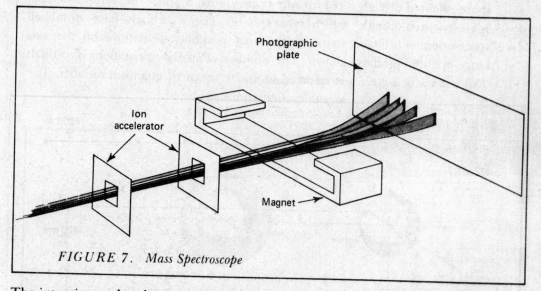

FIGURE 7. *Mass Spectroscope*

The intensity on the photographic plate indicates the amount of each particular *isotope*. Other collectors may be used in place of the photographic plate to collect and interpret these data.

THE WAVE-MECHANICAL MODEL

In the 1920's, a more mathematical model of the atom was developed through the work of such men as deBroglie, Heisenberg, and Schroedinger. Their work in quantum mechanics described a probability picture of where the electrons at each energy level might best be found. This region of greatest probability is sometimes referred to as an electron cloud. Unlike the previous model, the wave-mechanical model does not attempt to explain the path an electron travels. DeBroglie attributed wave properties to the electron, and these were confirmed later by experimentation.

Quantum Numbers

Each electron orbital of the atom may be described by a set of four quantum numbers in this model. They give the position with respect to the nucleus, the shape of the orbital, its spatial orientation, and the spin of the electron in the orbital.

Principal quantum number
   1, 2, 3, 4, 5, etc.

Refers to average distance of the orbital from the nucleus. 1 is closest to the nucleus and has the least energy. These numbers correspond to the orbits in the previous model. They are called energy levels.

Secondary or azimuthal
quantum number
s, p, d, f
(in order of
increasing energy)

Shapes (see Figure 8)

Magnetic quantum number

$s = 1$ space-oriented orbital
$p = 3$ space-oriented orbitals
$d = 5$ space-oriented orbitals
$f = 7$ space-oriented orbitals

Refers to the shape of the orbital. The number of possible shapes is limited by the principal quantum number. The first energy level has only one possible shape, the *s* orbital. The second has two possible, *s* and *p*.

The drawings in Figure 8 show the *s*-orbital shape, which is a sphere. The *p* orbitals have dumbbell shapes with three possible orientations on the axis shown. The number of spatial orientations of orbitals is referred to as the magnetic quantum number. The possible orientations are listed.

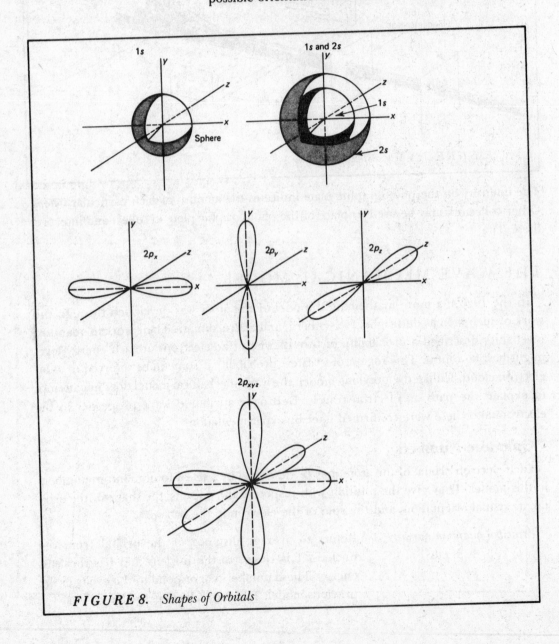

FIGURE 8. Shapes of Orbitals

Spin quantum number

+ spin − spin

Electrons are assigned one more quantum number, called the spin quantum number. This describes the spin in either of two possible directions. Each orbital can be filled by only two electrons, each with an opposite spin. This is referred to as the *Pauli Exclusion Principle*. Therefore each orbital in Figure 8 can hold only two electrons.

It is important to remember that, when there is more than one orbital at a particular energy level, such as three *p* orbitals or five *d* orbitals, only one electron will fill each orbital until each has one electron. After this, pairing will occur with the addition of one more electron to each orbital. This is called *Hund's Rule of Maximum Multiplicity* and is shown in Table 2. Each slant line indicates an electron (∅).

If each orbital is shown in an energy diagram as a ○, we can show the relative energies in a chart like Figure 9. If this drawing represented a ravine with the energy levels as ledges onto which stones could come to rest only in numbers equal to the circles for orbitals, then pushing stones into the ravine would cause the stones to lose their kinetic energy as they dropped to the lowest level available to them. Much the same is true for electrons.

SUBSHELLS AND ELECTRON CONFIGURATION

Order of Filling and Notation

The subshells do not fill up in numerical order, and the pattern of filling is shown on the right side of the Approximate Relative Energy Levels chart (Figure 9). The first instance of this is the 4*s* filling before the 3*d*. (Study Figure 9 carefully before going on.)

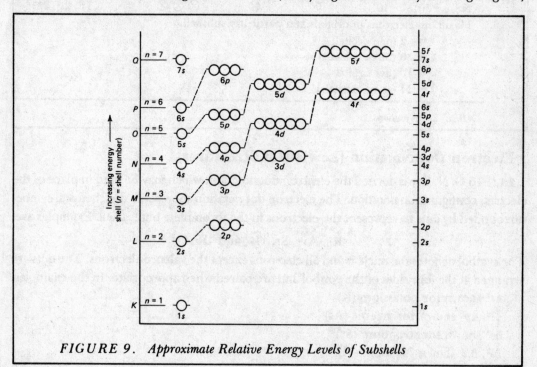

FIGURE 9. Approximate Relative Energy Levels of Subshells

$_{19}$K $1s^2, 2s^2, 2p^6, 3s^2, 3p^6, 4s^1$

$_{20}$Ca $1s^2, 2s^2, 2p^6, 3s^2, 3p^6, 4s^2$

$_{21}$Sc $1s^2, 2s^2, 2p^6, 3s^2, 3p^6, 4s^2, 3d^1$ (note $4s$ filled before $3d$)

There is a more stable configuration to a half-filled or filled subshell so at atomic number 24 the $3d$ subshell becomes half-filled by taking a $4s$ electron;

$_{24}$Cr $1s^2, 2s^2, 2p^6, 3s^2, 3p^6, 3d^5, 4s^1$

and at atomic number 29 the $3d$ becomes filled by taking a $4s$ electron.

$_{29}$Cu $1s^2, 2s^2, 2p^6, 3s^2, 3p^6, 3d^{10}, 4s^1$

Table 3 shows the electron configurations of the elements. When a triangular mark ▶ appears, it indicates the phenomenon of an outer-shell electron dropping back to a lower unfilled orbital.

Chemical Symbol	Atomic No.	Orbital Notation			Electron Configuration Notation
		1s	2s	2p	
H	1	∅	○	○ ○ ○	$1s^1$
He	2	⊗	○	○ ○ ○	$1s^2$
Li	3	⊗	∅	○ ○ ○	$1s^2, 2s^1$
Be	4	⊗	⊗	○ ○ ○	$1s^2, 2s^2$
B	5	⊗	⊗	∅ ○ ○	$1s^2, 2s^2, 2p^1$
C	6	⊗	⊗	∅ ∅ ○	$1s^2, 2s^2, 2p^2$
N	7	⊗	⊗	∅ ∅ ∅	$1s^2, 2s^2, 2p^3$
O	8	⊗	⊗	⊗ ∅ ∅	$1s^2, 2s^2, 2p^4$
F	9	⊗	⊗	⊗ ⊗ ∅	$1s^2, 2s^2, 2p^5$
Ne	10	⊗	⊗	⊗ ⊗ ⊗	$1s^2, 2s^2, 2p^6$

Maximum electrons in orbitals at a particular subshell:

$s =$ 2 (one orbital)
$p =$ 6 (three orbitals)
$d =$ 10 (five orbitals)
$f =$ 14 (seven orbitals)

TABLE 2. Notations

Electron Dot Notation (Lewis Dot Structures)

In 1916 G. N. Lewis devised the electron dot notation, which may be used in place of the electron configuration notation. The electron dot notation shows only the chemical symbol surrounded by dots to represent the electrons in the incomplete outer shell. Examples are:

$$K̇, \cdot \ddot{A}s\cdot, \ddot{S}r, :\ddot{I}:, \text{ and } :\ddot{R}n:$$

The symbol denotes the nucleus and all electrons except the valence electrons. The dots are arranged at the four sides of the symbol and are paired when appropriate. In the examples:

   $4s^1$ shown for potassium (K)
   $4s^2, 4p^3$ shown for arsenic (As)
   $5s^2$ shown for strontium (Sr)
   $5s^2, 5p^5$ shown for iodine (I)
   $6s^2, 6p^6$ shown for radon (Rn)

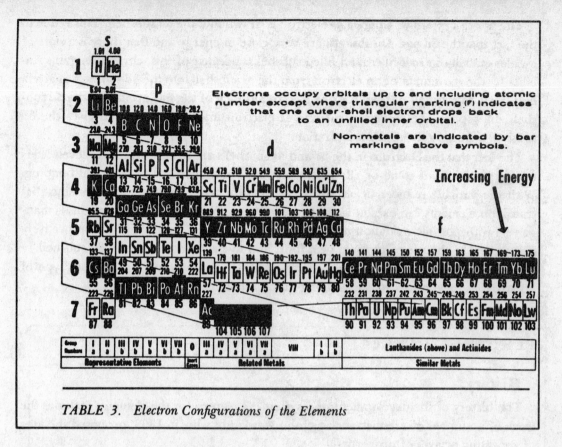

TABLE 3. *Electron Configurations of the Elements*

TRANSITION ELEMENTS AND VARIABLE OXIDATION NUMBERS

The elements involved with the filling of a *d* subshell with electrons after two electrons are in the *s* subshell of the next major shell are often referred to as the *transition elements*. The first example of these is the elements between calcium, atomic number 20, and gallium, atomic number 31. Their electron configurations are the same in the 1*s*, 2*s*, 2*p*, 3*s*, and 3*p* subshells. It is the filling of the 3*d* and changes in the 4*s* that are of interest. This is shown in the table below.

		Electron Configuration	
Element	Atomic No.	$1s^2, 2s^2, 2p^6, 3s^2, 3p^6, 3d$	$4s$
Scandium	21	1	2
Titanium	22	All	2	2
Vanadium	23	3	2
Chromium	24	the	5	1*
Manganese	25	same	5	2
Iron	26	6	2
Cobalt	27	7	2
Nickel	28	8	2
Copper	29	10	1*
Zinc	30	10	2

The asterisk (*) shows where a 4s electron is drawn into the 3d subshell. This is due to the fact that the 3d and 4s subshells are very close in energy and that there is a state of greater stability in half-filled and filled subshells. Because of this, chromium gains stability by the movement of an electron from the 4s subshell into the 3d subshell to give a half-filled 3d subshell. It then has one electron in each of the five orbitals of the 3d subshell. In copper, the movement of one 4s electron into the 3d subshell gives the 3d subshell a completely filled configuration.

The fact that the electrons in the 3d and 4s subshells are so close in energy levels leads to the possibility of some or all the 3d electrons being involved in chemical bonding. With the variable number of electrons available for bonding, it is not surprising that transition elements can exhibit variable oxidation numbers. An example of this is manganese with possible oxidation numbers of +2, +3, +4, +6, and +7, which respectively correspond to the use of no, one, two, four, and five electrons from the 3d subshell.

The transition elements in the other periods of the table show this same type of anomaly, as they have d subshells filling in the same manner.

PERIODIC TABLE

History

The history of the development of a systematic pattern for the elements includes the work of a number of scientists such as John Newlands, who, in 1863, proposed the idea of repeating octaves of properties.

Dmitri I. Mendeleev (or Mendeleyev) in 1869 proposed a table containing seventeen columns and is usually given credit for the first periodic table since he arranged elements in groups according to their atomic weights and properties. In a revision of his table in 1871 he rearranged some elements and proposed a table of eight columns, obtained by splitting each of the long periods into a period of seven elements, an eighth group containing the three central elements (such as Fe, Co, Ni), and a second period of seven elements. The first and second periods of seven were later distinguished by use of the letters a and b attached to the group symbols, which are Roman numerals. This nomenclature of periods (Ia, IIa, etc.) appears slightly revised in the present periodic tables, even in the extended form. It is interesting to note that Lothar Meyer proposed a similar arrangement about the same time.

Mendeleev's table had the elements arranged by atomic weights with recurring properties in a periodic manner. Where atomic weight placement disagreed with the properties that should occur in a particular spot in the table, he gave preference to the element with the correct properties. He even predicted elements for places that were not yet occupied in the table. These predictions proved to be amazingly accurate.

Periodic Law

Moseley stated, after his work with X-ray spectra in the early 1900's, that the properties of elements are a periodic function of their atomic numbers, thus changing the basis of the periodic law from atomic weight to atomic number. This is the present statement of the periodic law.

The Table

The horizontal rows of the periodic chart are called *periods* or *rows*. There are seven periods each of which begins with an atom having only one valence electron and ends with a complete outer shell structure of an inert gas. The first three periods are short periods consisting of 2, 8, and 8 elements respectively. Periods 4 and 5 are long periods of 18 each, while period 6 has 32 elements, and period 7 is incomplete with only 17 elements most of which are radioactive and do not occur in nature.

The vertical columns of the periodic chart are called *groups* or *families*. The elements in a group exhibit similar or related properties. The Roman numeral group number gives an indication of the number of electrons probably found in the outer shell of the atom and thus an indication of one of its possible valence numbers.

PROPERTIES RELATED TO THE PERIODIC CHART

Metals are found on the left of the chart with the most active metal in the lower left corner. Nonmetals are found on the right side with the most active nonmetal in the upper right-hand corner of the chart. The noble or inert gases are on the far right. Since the most active metals react with water to form bases, the Group I metals are called alkali metals. As you proceed to the right, the base-forming property decreases and the acid-forming properties increase. The metals in the first two groups are the light metals, and those toward the center are called heavy metals.

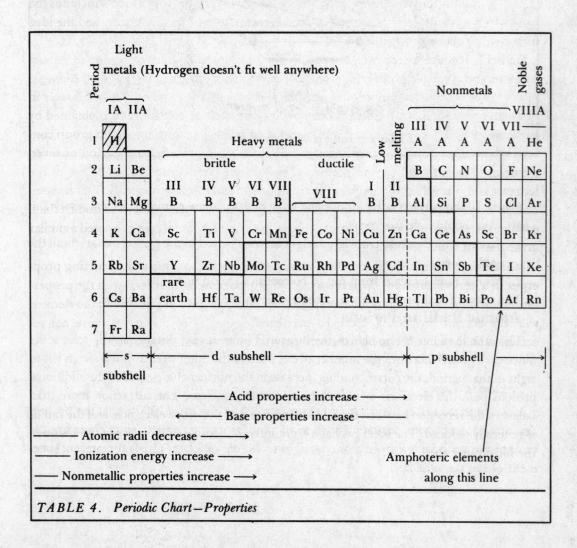

TABLE 4. *Periodic Chart — Properties*

The elements found along the dark line in the periodic chart (Table 4) are called metalloids. They are elements which have certain characteristics of metals and other characteristics of nonmetals. Some examples of metalloids are boron, silicon, arsenic, and tellurium. Study Table 4 carefully because it summarizes many of these properties.

Radii of Atoms

The size of an atom is difficult to describe because atoms have no definite outer boundary. Unlike a volleyball, the atom does not have a definite circumference.

To overcome this problem, the size of an atom is estimated by describing its radius. In metals, this is done by measuring the distance between two nuclei in the solid state and dividing this distance by 2. Such measurements can be made with X-ray diffraction. For nonmetallic elements that exist in pure form as molecules, such as chlorine, measurements can be made of the distance between nuclei for two atoms covalently bonded together. Half of this distance is referred to as the *covalent radius*. The method for finding the covalent radius of the chlorine atom is illustrated below.

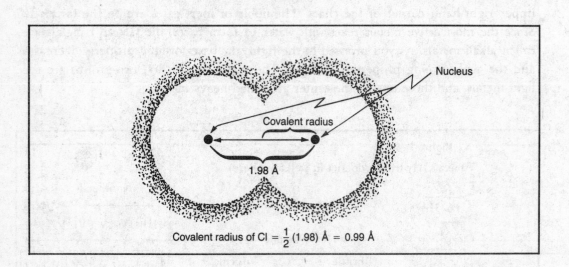

Covalent radius of Cl $= \frac{1}{2}$ (1.98) Å = 0.99 Å

Figure 10 shows the relative atomic and ionic radii for some elements. As you review this chart, you should note two trends:

1. Atomic radii decrease from left to right across a period in the periodic chart (until the noble gases).
2. Atomic radii increase from top to bottom in a group or family.

The reason for these trends will become clear in the following discussions.

Atomic Radii in Periods

The atomic radius is one-half of the measured internuclear distance in the solid state. Since each period increases the number of electrons in the outer shell as you go from left to right in that period, the corresponding increase in the nuclear charge due to the additional protons pulls the electrons more tightly around the nucleus. This attraction more than balances the repulsion between the added electrons and the other electrons, and the radius is generally reduced. The inert gas has a slight increase because of the electron repulsion in the filled outer shell.

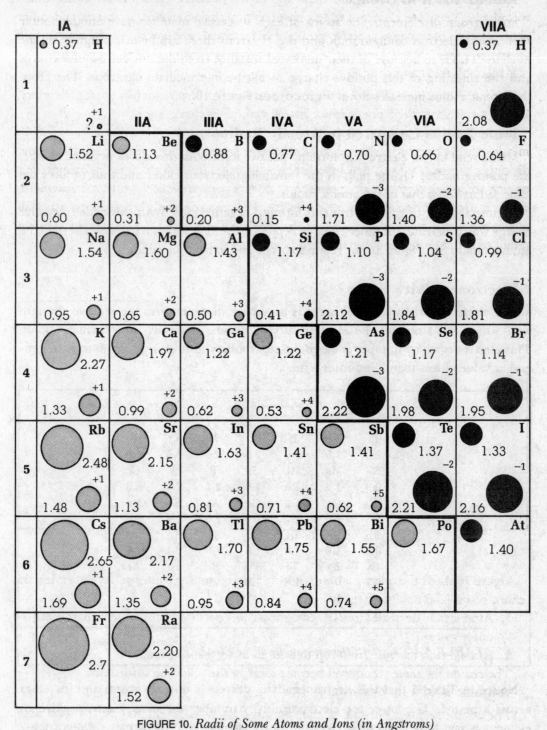

FIGURE 10. *Radii of Some Atoms and Ions (in Angstroms)*

Note: The atomic radius is usually given for metal atoms, which are shown in gray, and the covalent radius is usually given for atoms of nonmetals, which are shown in black.

Atomic Radii in Groups

For a group of elements, the atoms of each successive member have another outer shell in the electron configuration and the electrons there are held less tightly by the nucleus. This is so because of their increased distance from the nuclear positive charge and the shielding of this positive charge by all the intermediate electrons. Therefore the atomic radius increases down a group. See Figure 10.

Ionic Radius Compared to Atomic Radius

Metals tend to lose electrons in forming positive ions. With this loss of negative charge, the positive nuclear charge pulls in the remaining electrons closer and thus reduces the ionic radius below that of the atomic radius.

Nonmetals tend to gain electrons in forming negative ions. With this added negative charge which increases the inner electron repulsion the ionic radius is increased beyond the atomic radius. See Figure 10 for relative atomic and ionic radii values.

Electronegativity

The electronegativity of an element is a number that measures the relative strength with which the atoms of the element attract valence electrons in a chemical bond. This electronegativity number is based on an arbitrary scale going from 0 to 4. In general, a value of less than 2 indicates a metal.

H 2.1						
Li 1.0	Be 1.5	B 2.0	C 2.5	N 3.0	O 3.5	F 4.0
Na 0.9	Mg 1.2	Al 1.5	Si 1.8	P 2.1	S 2.5	Cl 3.0
K 0.8	Ca 1.0	Ga 1.7	Ge 1.8	As 2.0	Se 2.4	Br 2.8
Rb 0.8	Sr 1.0	In 1.7	Sn 1.8	Sb 1.9	Te 2.1	I 2.5
Cs 0.7	Ba 0.9	Tl 1.8	Pb 1.8	Bi 1.9	Po 2.0	At 2.2
Fr 0.7	Ra 0.9					

TABLE 5. Chart of Electronegativity of the Elements

Notice in Table 5 that the electronegativity decreases down a group and increases across a period. The lower the electronegativity number the more electropositive an element is said to be. The most electronegative element is in the upper right corner — F. The most electropositive is in the lower left corner of the chart — Fr.

Ionization Energy

The atoms do hold their valence electrons then with different amounts of energy. If enough energy is supplied to one outer electron to remove it from its atom, this amount of energy is called the first ionization energy. With the first electron gone, the removal of succeeding electrons becomes more difficult because of the imbalance between the positive nuclear charge and the remaining electrons. The lowest ionization energy would be found with the least electronegative atom. Study the following charts carefully and notice the patterns established.

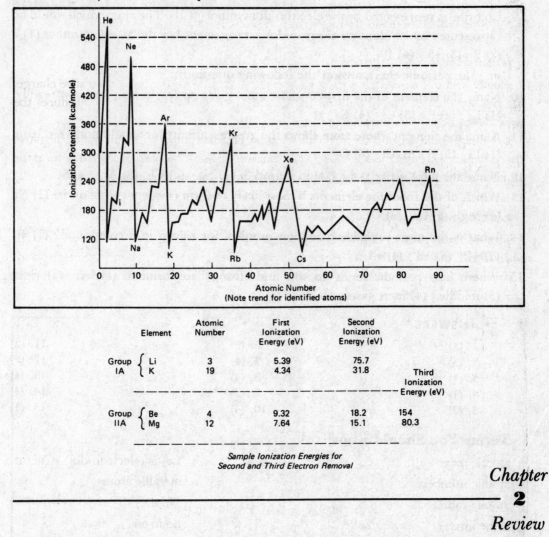

Element		Atomic Number	First Ionization Energy (eV)	Second Ionization Energy (eV)	
Group IA	Li	3	5.39	75.7	
	K	19	4.34	31.8	Third Ionization Energy (eV)
Group IIA	Be	4	9.32	18.2	154
	Mg	12	7.64	15.1	80.3

Sample Ionization Energies for
Second and Third Electron Removal

Chapter
2
Review

1. The two main parts of an atom are (1) the shells and energy levels   (2) nucleus and kernel   (3) nucleus and energy levels   (4) planetary electrons and energy levels.

2. The lowest energy level which an electron can have is the (1) O shell   (2) L shell   (3) K shell   (4) E shell.

3. The sublevel which has only one orbital is identified by the letter (1) s (2) p (3) d (4) f.

4. The sublevel which can be occupied by a maximum of 10 electrons is identified by the letter (1) d (2) f (3) p (4) s.

5. An orbital may never be occupied by (1) 1 electron   (2) 2 electrons   (3) 3 electrons   (4) 0 electron.

6. An atom of beryllium consists of 4 protons, 5 neutrons, 2 K electrons, and 2 L electrons. The mass number of this atom is (1) 13   (2) 9   (3) 8   (4) 5.

7. The number of orbitals in the L shell of an atom is (1) 1   (2) 9   (3) 16   (4) 4.

8. Electron-dot symbols consist of the symbol representing the element and an arrangement of dots which usually shows (1) the atomic number   (2) the atomic mass   (3) the number of neutrons   (4) the electrons in the outermost energy level.

9. Chlorine is represented by the electron-dot symbol ($:\overset{\cdot}{\underset{\cdot\cdot}{Cl}}:$). The atom which would be represented by an identical electron-dot arrangement has the atomic number (1) 7   (2) 9   (3) 15   (4) 19.

Using the periodic chart, answer the following questions:

10. Name the element of the first 20 whose atom gives up an electron the most readily. (1) Li   (2) F   (3) Cl   (4) K

11. Name the element whose atom shows the greatest affinity for an additional electron. (1) Li   (2) F   (3) Cl   (4) O

12. Name the most active nonmetal in Period 3. (1) Na   (2) Cl   (3) Ar   (4) S

13. Which of the following elements is most likely to form covalent compounds? (1) Na   (2) Mg   (3) Cs   (4) C

14. What is the most probable oxidation number for silicon in a compound? (1) +1   (2) +2   (3) +3   (4) +4

15. Where in a periodic series do you find strong base-formers: (1) left   (2) right   (3) middle   (4) inert gases

• ANSWERS:

1. (3)	6. (2)	11. (2)
2. (3)	7. (4)	12. (2)
3. (1)	8. (4)	13. (4)
4. (1)	9. (2)	14. (4)
5. (3)	10. (4)	15. (1)

Terms You Should Know

atomic mass	Lewis (electron dot) structure
atomic number	metallic atoms
atomic radii	Moseley
Bohr model	neutron
covalent radius	nonmetallic atoms
Dalton's atomic theory	nucleus
electron	oxidation number
electronegativity	Pauli Exclusion Principle
group or family	periodic law
Hund's Rule	period or row
inert atoms	proton
ionic radii	quantum numbers
ionization energy	s, p, d, f, orbitals
isotopes	transition elements
K, L, M, N, O	valence electrons

Bonding

Some elements show no tendency to combine with either like atoms or other kinds of elements. They are said to be monoatomic molecules, like He, Ne, and Ar. A molecule is defined as the smallest particle of an element or a compound that retains the characteristics of the original substance. Water is a triatomic molecule since two hydrogen atoms and one oxygen atom must be combined to have the substance water with its characteristic properties. When atoms do combine to form molecules, there is a shifting of valence electrons, that is, the electrons in the outer shell of each atom. This results in a usual completion of the outer shell of each atom. This more stable form may be achieved by the gain or loss of electrons or the sharing of pairs of electrons. The resulting attraction of the atoms involved is called a chemical bond. When a chemical bond forms, energy is released; when this bond is broken, energy is absorbed.

TYPES OF BONDS

Ionic Bonds

When the electronegativity values of two kinds of atoms differ by 1.7 or more (espe-

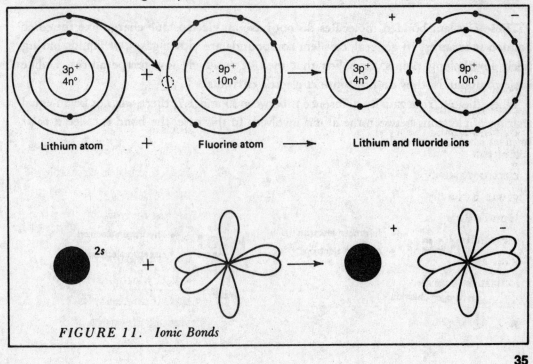

FIGURE 11. *Ionic Bonds*

cially greater differences than 1.7), the more electronegative atom will borrow the electrons it needs to fill its outer shell and the other kind of atom will lend electrons until it, too, has a complete outer shell. Because of this exchange, the borrower becomes negatively charged and the lender becomes positively charged. They are now referred to as *ions*, and the bond or attraction between them is called an *ionic* or *electrovalence* bond. These ions do not retain the properties of the original atoms. An example can be seen in Figure 11.

These ions do not form an individual molecule in the liquid or solid phase but are arranged into a crystal lattice or giant ion-molecule containing many of these ions. Ionic solids like this tend to have high melting points and will not conduct a current of electricity until they are in the molten state.

Covalent Bonds

When the electronegativity difference is zero or very small (not greater than about .5) between two or more atoms, they tend to share the valence electrons in the respective outer shells. This is called a *nonpolar covalent bond.* An example using electron-dot notation:

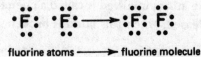

fluorine atoms ⟶ fluorine molecule

These covalent bonded molecules do not have an electrostatic charge like the ionic bonded substances. In general, covalent compounds are either gases, or liquids having fairly low boiling points, or solids that melt at relatively low temperatures. Unlike ionic compounds, they do not conduct electric currents.

When the electronegativity difference is between .5 and 1.7, there will not be an equal sharing of electrons between the atoms involved. In this case, the bond is called a *polar covalent bond.* An example:

H :Cl: • hydrogen electron
         ○ chlorine electrons

Hydrogen chloride

H :O: • hydrogen electron
   H   ○ oxygen electrons

Water

Notice the electron pair in the bond is shown closer to the more electronegative atom. Because of this unequal sharing, the molecules shown are said to be polar molecules, or a *dipole*. There are cases of polar covalent bonds existing in nonpolar molecules. Some examples are CO_2, CH_4, and CCl_4. (See Figure 12.)

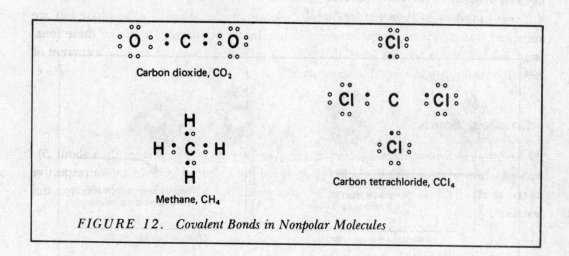

Carbon dioxide, CO_2

Methane, CH_4

Carbon tetrachloride, CCl_4

FIGURE 12. Covalent Bonds in Nonpolar Molecules

In all these examples the bonds are polar covalent bonds, but the important thing is that they are symmetrically arranged in the molecule. This results in a nonpolar molecule.

In the *covalent* bonds described so far, the shared electrons in the pair were contributed one each from the atoms bonded. There are cases where both electrons for the shared pair are supplied by only one of the atoms. These are called *coordinate covalent bonds*. Two examples are NH_4^+ and H_2SO_4. (See Figure 13.)

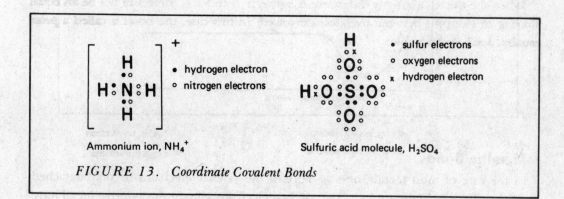

• hydrogen electron
○ nitrogen electrons

• sulfur electrons
○ oxygen electrons
× hydrogen electron

Ammonium ion, NH_4^+

Sulfuric acid molecule, H_2SO_4

FIGURE 13. Coordinate Covalent Bonds

The formation of a covalent bond can be described in a graphic form and related to the potential energy of the atoms involved. Using the formation of the hydrogen molecule as an example, we can show how the potential energy changes as the two atoms approach and form a covalent bond. In the illustration that follows, frames (1), (2), and (3) show the effect on potential energy as the atoms move closer to each other. In frame (3), the atoms have reached the condition of lowest potential energy, but the inertia of the atoms pulls them even closer, as shown in frame (4). The repulsion between them then forces the two nucleii to a stable position, as shown in frame (5).

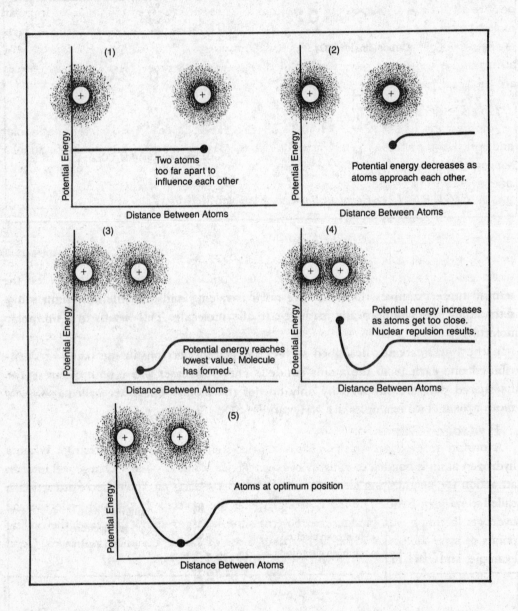

Metallic Bonds

In the case of most metals, one or more of the valence electrons become detached from the atom and migrate in a "sea" of free electrons among the positive metal ions. The strong attraction between these different charged particles forms the metallic bond. Because of this firm bonding, metals usually have high melting points, show great strength, and are good conductors of electricity.

INTERMOLECULAR FORCES OF ATTRACTION

Intermolecular forces refer to attractions *between* molecules. Although it is proper to refer to all intermolecular forces as *van der Waals forces,* it is clearer to expand on this concept.

Dipole-Dipole Attraction

One component of van der Waals forces is dipole-dipole attraction. It was shown under polar covalent bonding that the unsymmetrical distribution of electronic charges leads to positive and negative charges in the molecules. These are referred to as *dipoles.* In polar molecular substances, the dipoles line up so that the positive pole of one molecule attracts the negative pole of another. This is much like the lineup of small bar magnets. The force of attraction between polar molecules is called *dipole-dipole attraction.* These attractive forces are less than the full charges carried by ions in ionic crystals.

London Forces

Another component of van der Waals forces is called *London force.* Found in both polar and nonpolar molecules, it can be attributed to the fact that an atom that usually is nonpolar sometimes becomes polar. This is because the constant motion of its electrons may cause uneven charge distribution at any one instant. When this occurs, the atom has a temporary dipole. This dipole can then cause a second, adjacent atom to be distorted and to have its nucleus attracted to the negative end of the first atom. London forces are about one tenth the force of most dipole interactions and are the weakest of all the electrical forces that act between atoms or molecules. These forces help to explain why nonpolar substances such as noble gases and the halogens condense into liquids and freeze into solids when the temperature is lowered sufficiently. In general, they also explain why liquids composed of discrete molecules with no permanent dipole attraction have low boiling points relative to their molecular weights. It is also true that compounds of the solid state that are bound mainly by this type of attraction have rather soft crystals, are easily deformed, and vaporize easily. Because of the low intermolecular forces, the melting points are low and evaporation takes place so easily that it may occur at room temperature. Examples are iodine crystals and moth balls (paradichlorobenzene and naphthalene).

Hydrogen Bonds

A proton or hydrogen nucleus has a high concentration of positive charge. When a hydrogen atom is bonded to a highly electronegative atom, its positive charge will have an attraction for neighboring electron pairs. This special kind of dipole-dipole attraction is called a *hydrogen* bond. The more strongly polar the molecule is, the more effective the hydrogen bonding is in binding the molecules into a larger unit. This causes the boiling points of such molecules to be higher than those of similar nonpolar molecules. Good examples are water and hydrogen fluoride (Figure 14).

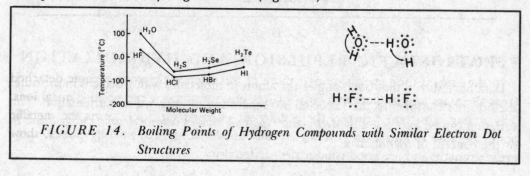

FIGURE 14. Boiling Points of Hydrogen Compounds with Similar Electron Dot Structures

DOUBLE AND TRIPLE BONDS

To achieve the *octet* structure, which is an outer shell resembling the noble gas configuration of eight electrons, it is necessary for some atoms to share two or even three pairs of electrons. Sharing two pairs of electrons produces a *double bond*. An example:

$$\ddot{O}\!:\!C\!:\!\ddot{O}$$, and by a line formula O=C=O

carbon dioxide

In the line formula, only the shared pair of electrons is indicated by a bond (—). The sharing of three electron pairs results in a *triple* bond. An example:

$$H\!:\!C\!:\!:\!C\!:\!H$$, and by a line formula H—C≡C—H

acetylene

It can be assumed from these structures that there is a greater electron density between the nuclei involved and hence a greater attractive force between the nuclei and the shared electrons. Experimental data verify that greater energy is required to break double bonds than single bonds, and triple bonds than double bonds. Also, since these stronger bonds tend to pull atoms closer together, the atoms joined by double and triple bonds have smaller interatomic distances and greater bond strengths, respectively.

RESONANCE STRUCTURES

It is not always possible to represent the bonding structure by either the Lewis dot structure or the line drawing because data about the bonding distance and bond strength are between possible drawing configurations and really indicate a hybrid condition. To represent this, the possible alternatives are drawn with arrows in between. Classic examples are sulfur trioxide and benzene. These are shown in Chapters 13 and 14, respectively, but are repeated here as examples.

Sulfur trioxide resonance structures:

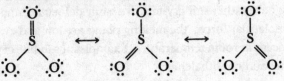

Benzene resonance structures:

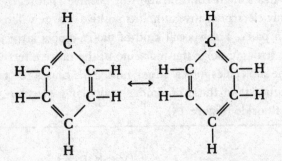

ELECTROSTATIC REPULSION AND HYBRIDIZATION

Data have shown that bond angles for atoms in molecules with *p* orbitals in the outer shell do not conform to the expected 90° separation of an *x, y, z* axis orientation. This variation can be expressed by *electrostatic repulsion between valence electron charge clouds* or by the concept of *hybridization*.

Electrostatic repulsion uses as its basis the fact that like charges will orient themselves in such a way as to diminish the repulsion between them.

1. Mutual repulsion of two electron clouds forces them to the opposite sides of a sphere. This is called a *linear* arrangement.

• EXAMPLE: BeF$_2$

2. Minimum repulsion between three electron pairs occurs when each is at the vertices of an equilateral triangle inscribed in a sphere. This is called *trigonal planar*.

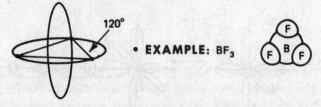

• EXAMPLE: BF$_3$

3. Four electron pairs are farthest apart at the vertices of a tetrahedron inscribed in a sphere. This is called a *tetrahedral* shape.

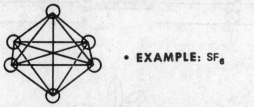

• EXAMPLE: CH$_4$

4. Mutual repulsion of six identical electron clouds directs them to the corners of an inscribed regular octahedron. This is said to have *octahedral* geometry.

• EXAMPLE: SF$_6$

These same configurations can also be arrived at through the concept of *hybridization*. Briefly stated, this means that two or more pure atomic orbitals (usually *s*, *p*, and *d*) can be mixed to form two or more hybrid atomic orbitals that are identical. This can be illustrated as follows:

1. *sp* Hybrid Orbitals
   Beryllium fluoride spectroscopic measurements reveal a bond angle of 180° and equal bond lengths.

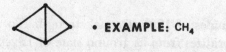

The ground state of beryllium is:

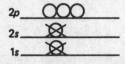

To accommodate the experimental data we theorize that a 2*s* electron is excited to a 2*p* orbital; then the two orbitals hybridize to yield two identical orbitals called *sp* orbitals. Each contains one electron but is capable of holding two electrons.

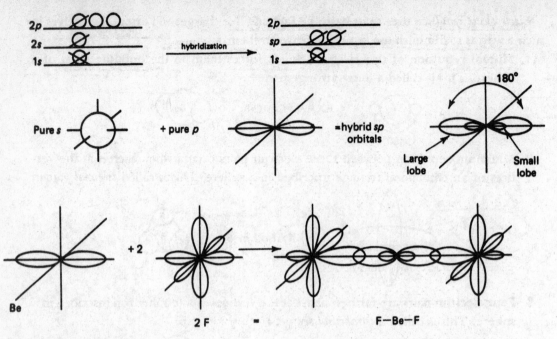

2. sp^2 Hybrid Orbitals

Boron trifluoride has bond angles of 120° of equal strength. To accommodate these data the boron atom hybridizes from its ground state of $1s^2 2s^2 2p^1$ to:

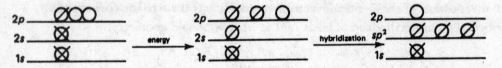

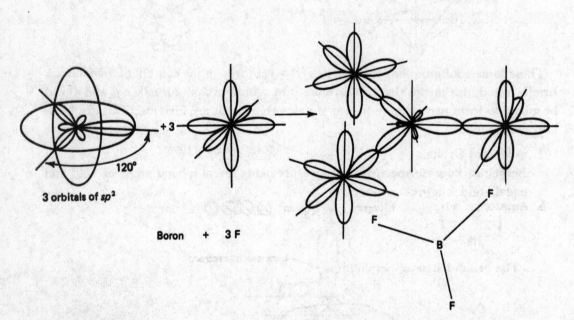

3. sp^3 Hybrid Orbitals

Methane, CH_4, can be used to illustrate this hybridization. Carbon has a ground state of $1s^2 2s^2 2p^2$. One $2s$ electron is excited to a $2p$ orbital, and the four involved orbitals then form four new identical sp^3 orbitals.

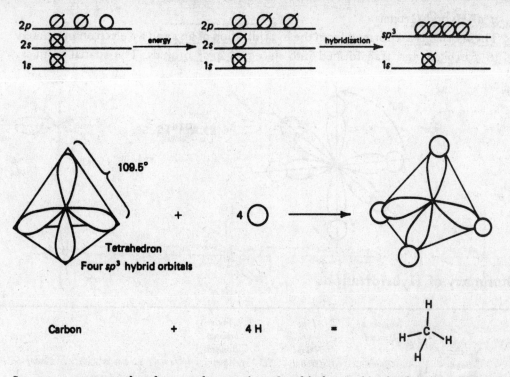

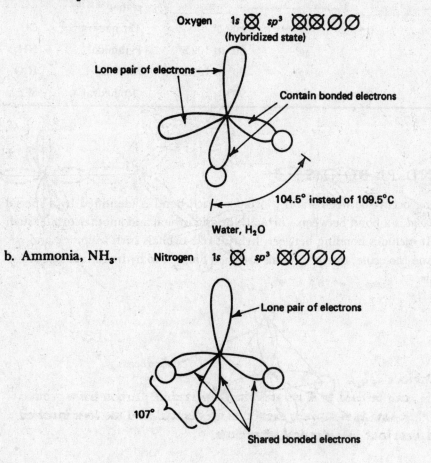

Carbon   +   4 H   =

In some compounds where only certain sp^3 orbitals are involved in bonding, distortion in the bond angle occurs because of unbonded electron repulsion. Examples:

a. Water, H_2O.

Oxygen 1s sp^3 (hybridized state)

Lone pair of electrons →

Contain bonded electrons

104.5° instead of 109.5°C

Water, H_2O

b. Ammonia, NH_3.   Nitrogen 1s sp^3

Lone pair of electrons

107°

Shared bonded electrons

4. sp^3d^2 Hybrid Orbitals

These orbitals are formed from the hybridization of an s and a p electron promoted to d orbitals and transformed into six equal sp^3d^2 orbitals. The spatial form is:

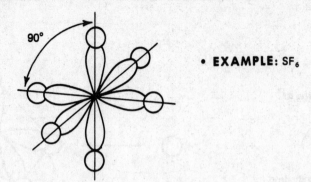

• **EXAMPLE:** SF_6

Summary of Hybridization

Number of Bonds	Number of Unused e Pairs	Type of Hybrid Orbital	Angle between Bonded Atoms	Geometry	Example
2	0	sp	180°	Linear	BeF_2
3	0	sp^2	120°	Trigonal planar	BF_3
4	0	sp^3	109.5°	Tetrahedral	CH_4
3	1	sp^3	90° to 109.5°	Pyramidal	NH_3
2	2	sp^3	90° to 109.5°	Angular	H_2O
6	0	sp^3d^2	90°	Octahedral	SF_6

SIGMA AND PI BONDS

When bonding occurs between s and p orbitals, each bond is identified by a special term. A *sigma bond* is a bond between s orbitals, or an s orbital and another orbital such as a p orbital. It includes bonding between hybrids of s orbitals such as sp, sp^2, and sp^3.

In the methane molecule, the sp^3 orbitals are each bonded to hydrogen atoms. These are sigma bonds.

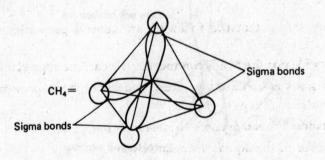

$CH_4 =$

Sigma bonds

Sigma bonds

When two *p* orbitals share electrons in a covalent bond, this is called a *pi bond*. An example:

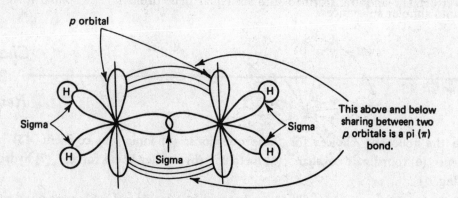

Chapter 14 gives more examples of sigma and pi bonding.

PROPERTIES OF IONIC SUBSTANCES

Laboratory experiments reveal that, in general, ionic substances are characterized by the following properties:

1. In the solid phase at room temperature they do not conduct appreciable electric current.

2. In the liquid phase they are relatively good conductors of electric current. The conductivity of ionic substances is much smaller than that of metallic substances.

3. They have relatively high melting and boiling points. There is a wide variation in the properties of different ionic compounds. For example, potassium iodide (KI) melts at 686°C and boils at 1330°C, while magnesium oxide (MgO) melts at 2800°C and boils at 3600°C. Both KI and MgO are ionic compounds.

4. They have relatively low volatilities and low vapor pressures. In other words, they do not vaporize readily at room temperatures.

5. They are brittle and easily broken when a stress is exerted on them.

6. Those that are soluble in water form electrolytic solutions which are good conductors of electricity. There is, however, a wide range in the solubilities of ionic compounds. For example, at 25°C, 92 g of sodium nitrate ($NaNO_3$) dissolves in 100 g of water, while only 0.0002 g of $BaSO_4$ dissolves in the same mass of water.

PROPERTIES OF MOLECULAR CRYSTALS AND LIQUIDS

Experiments have shown that the following are general properties of these substances:

1. Neither the liquids nor the solids conduct electric current appreciably.

2. Many exist as gases at room temperature and atmospheric pressure, and many solids and liquids are relatively volatile.

3. The melting points of solid crystals are relatively low.

4. The boiling points of the liquids are relatively low.

5. The solids are generally soft and have a waxy consistency.

6. A large amount of energy is often required to decompose the substance chemically into simpler substances.

Chapter

3

Review

Use the following choices for these questions: (1) ionic (2) covalent (3) polar covalent (4) coordinate covalent (5) metallic (6) van der Waals forces (7) hydrogen bonding.

1. When the electronegativity difference between two atoms is 2, what type bond can be predicted?

2. If two atoms are bonded such that one member of the pair is supplying both electrons that are shared, what is this type of bond called?

3. If all the above replies were ordered from the strongest bond to the weakest, which would be the last one?

4. Which of the above bonds explains water's abnormally high boiling point?

5. When electron pairs are shared equally between two atoms, what is the bond called?

6. If a sharing of an electron pair is unequal, what is it called?

7. The force of attraction of the electrons of one atom for the protons of another atom in close proximity is which choice above?

- **ANSWERS:**

1. (1)	4. (7)	6. (3)
2. (4)	5. (2)	7. (6)
3. (6)		

Terms You Should Know

covalent bond	ionic bond	resonance structure
dipole-dipole attraction	London forces	sigma bond
hybridization	metallic bond	van der Waals forces
hydrogen bond	pi bond	

Four different kinds of intermolecular forces

a) Van der Waal forces proportional to the square of the polarizability.

b) dipole-dipole forces, proportional to the fourth power of the dipole moments.

c) induced dipolar forces. proportional to the polarizability and the square of dipole moment

d) hydrogen-bonding forces, usually presence of O-H, or N-H groups in the molecule.

Chemical Formulas

WRITING FORMULAS

From the knowledge of oxidation numbers and valence and the understanding of atomic structure, it is now possible to write chemical formulas. Table 6 is a list of oxidation numbers often encountered in a first-year chemistry course.

• GENERAL OBSERVATIONS:

1. The symbols of the metals have + signs while those of the nonmetals and all the radicals *except* the ammonium radical have − signs.
2. When an element exhibits two possible oxidation states, the lower state can be indicated with the suffix -ous and the higher one with -ic. Another method of indicating this difference is to use the Roman numeral of the oxidation state in parentheses after the name of the element. For example: Ferrous iron or iron (II) for the +2 oxidation state of iron.
3. A radical is a group of elements which act like a single atom in the formation of a compound. The bonds within these radicals are predominantly covalent, but the groups of atoms as a whole have an excess of electrons when combined, and thus are negative ions.

When you attempt to write a formula, it is important to know whether it actually exists. For example, one can easily write the formula of carbon nitrate but no chemist has ever prepared this compound. Here are the basic rules for writing formulas with three examples carried through each step:

1. Represent the symbols of the components using the positive part first, and then the negative part.

Sodium chloride	Calcium oxide	Ammonium sulfate
$NaCl$	CaO	NH_4SO_4

		MONOVALENT I	BIVALENT II	TRIVALENT III	TETRAVALENT IV	V
METALS	Cations (+)	Hydrogen H Potassium K Sodium Na Silver Ag Mercury (ous) Hg Copper (ous) Cu Gold (aurous) Au *Ammonium (NH$_4$)	Barium Ba Calcium Ca Cobalt Co Magnesium Mg Lead Pb Zinc Zn Mercury (ic) Hg Copper (cupric) Cu Iron (ferrous) Fe Manganese (ous) Mn Tin (stannous) Sn	Aluminum Al Gold (auric) Au Arsenic (ous) As Chromium Cr Iron (ferric) Fe Phosphorus (ous) P Antimony (ous) Sb Bismuth (ous) Bi	Carbon C Silicon Si Manganese (ic) Mn Tin (stannic) Sn Platinum Pt Sulfur S	Arsenic (ic) As Phosphorus (ic) P Antimony (ic) Sb Bismuth (ic) Bi
NONMETALS†	Anions (−)	Fluorine F Chlorine Cl Bromine Br Iodine I	Oxygen O Sulfur S	Nitrogen N Phosphorus P	Carbon C	
RADICALS	(−)	Hydroxide (OH) Bicarbonate (HCO$_3$) Nitrite (NO$_2$) Nitrate (NO$_3$) Hypochlorite (ClO) Chlorate (ClO$_3$) Chlorite (ClO$_2$) Perchlorate (ClO$_4$) Acetate (C$_2$H$_3$O$_2$) Permanganate (MnO$_4$) Bisulfate (HSO$_4$)	Carbonate (CO$_3$) Sulfite (SO$_3$) Sulfate (SO$_4$) Tetraborate (B$_4$O$_7$) Silicate (SiO$_3$) Chromate (CrO$_4$) Oxalate (C$_2$O$_4$)	Borate (BO$_3$) Phosphate (PO$_4$) Phosphite (PO$_3$) Ferricyanide [Fe(CN)$_6$]	Ferrocyanide [Fe(CN)$_6$]	

*Radical

†Last syllable of nonmetal is changed to -ide in binary compound.

TABLE 6. Table of Oxidation Numbers

2. Indicate the respective oxidation numbers above and to the right of each symbol. (Enclose radicals in parentheses for the time being.)

$$Na^{1+}Cl^{1-} \qquad Ca^{2+}O^{2-} \qquad (NH_4)^{1+}(SO_4)^{2-}$$

3. Write a subscript number equal to the oxidation number of the other element or radical. This is the same as the mechanical criss-cross method.

$$Na_1{}^{1+}Cl_1{}^{1-} \qquad Ca_2{}^{2+}O_2{}^{2-} \qquad (NH_4)_2{}^{1+}(SO_4)_1{}^{2-}$$

Since the positive oxidation number shows the number of electrons that may be lost or shared and the negative oxidation number shows the number of electrons that may be gained or shared, you must have just as many electrons lost (or partially lost in sharing) as are gained (or partially gained in sharing).

4. Now rewrite the formulas, omitting the subscript 1, the parentheses of the radicals which have the subscript 1, and the plus and minus numbers.

$$NaCl \qquad Ca_2O_2 \qquad (NH_4)_2SO_4$$

5. As a general rule, the subscript numbers in the final formula are reduced to their lowest terms. There are, however, certain exceptions, such as hydrogen peroxide (H_2O_2), and acetylene (C_2H_2). For these exceptions, you must know more specific information about the compound.

The only way to become proficient at writing formulas is to memorize the oxidation numbers of common elements (or learn to use the period chart group numbers) and practice writing formulas.

NAMING COMPOUNDS

Binary compounds consist of two elements. The name of the compound consists of the two elements, the second name having its ending changed to -ide, such as NaCl = sodium chloride; AgCl = silver chloride. If the metal has two different oxidation numbers, this can be indicated by the use of the suffix -ous for the lower one and -ic for the higher one. The more modern way is to use a Roman numeral after the name to indicate the oxidation state.

- **EXAMPLES:**

  $FeCl_2$ = ferrous chloride or iron (II) chloride
  $FeCl_3$ = ferric chloride or iron (III) chloride

If elements combine in varying proportions, thus forming two or more compounds of varying composition, the name of the second element may be preceded by a prefix, such as: mono- (one), di- (two), tri- (three), pent- (five). Some examples are: carbon dioxide, CO_2; carbon monoxide, CO; phosphorous trioxide, P_2O_3; phosphorous pentoxide, P_2O_5. Notice that, when these prefixes are used, it is not necessary to indicate the oxidation state of the first element in the name since it is given indirectly by the prefixed second element.

Ternary compounds, consisting of three elements, are usually made up of an element and a radical. To name them, you merely name each in the order of positive first and negative second.

Binary acids use the prefix hydro- in front of the stem or full name of the nonmetallic element, and add the ending -ic. Examples are *hydro*chlor*ic* acid (HCl) and *hydro*sulfur*ic* acid (H_2S).

Ternary acids usually contain hydrogen, a nonmetal, and oxygen. Since the amount of oxygen often varies, the most common form of the acid in the series consists of merely the stem of the nonmetal with the ending -ic. The acid containing one less atom of oxygen than the most common acid has the ending -ous. The acid containing one more atom of oxygen than the most common acid has the prefix per- and the ending -ic. The acid containing one less atom of oxygen than the -ous acid has the prefix hypo- and the ending -ous.

You can remember the names of the common acids and their salts by learning the following simple rules:

Rule	Example
-ic acids form -ate salts.	Sulfuric acid forms sulfate salts.
-ous acids form -ite salts.	Sulfurous acid forms sulfite salts.
hydro-(stem)-ic acids form -ide salts.	Hydrochloric acid forms chloride salts.

Where the name of the ternary acid has the prefix hypo- or per-, that prefix is retained in the name of the salt (hypochlorous acid = sodium hypochlorite).

ACIDS, BINARY		ACIDS, TERNARY	
Name	*Formula*	*Name*	*Formula*
Hydrofluoric	HF	Nitric	HNO_3
Hydrochloric	HCl	Nitrous	HNO_2
Hydrobromic	HBr	Hypochlorous	HClO
Hydriodic	HI	Chlorous	$HClO_2$
Hydrosulfuric	H_2S	Chloric	$HClO_3$
		Perchloric	$HClO_4$
BASES		Sulfuric	H_2SO_4
		Sulfurous	H_2SO_3
Sodium hydroxide	NaOH	Phosphoric	H_3PO_4
Potassium "	KOH	Phosphorous	H_3PO_3
Ammonium "	NH_4OH	Carbonic	H_2CO_3
Calcium "	$Ca(OH)_2$	Acetic	$HC_2H_3O_2$
Magnesium "	$Mg(OH)_2$	Oxalic	$H_2C_2O_4$
Barium "	$Ba(OH)_2$	Boric	H_3BO_3
Aluminum "	$Al(OH)_3$	Silicic	H_2SiO_3
Ferrous "	$Fe(OH)_2$		
Ferric "	$Fe(OH)_3$		
Zinc "	$Zn(OH)_2$		
Lithium "	LiOH		

TABLE 7. *Formulas of Common Acids and Bases*

CHEMICAL FORMULAS

The chemical formula is an indication of the make-up of a compound in terms of the kinds of atoms and their relative numbers. It also has some quantitative applications. By using the atomic weights assigned to the elements, we can find the *formula weight* of a compound. If we are sure that the formula represents the actual make-up of one molecule of the substance, the term *molecular weight* may be used as well. In some cases the formula represents an ionic lattice and no discrete molecule exists, such as table salt, NaCl, or the formula merely represents the simplest ratio of the combined substances and not a molecule of the substance, such as CH_2. (This is the simplest ratio of carbon and hydrogen united to form the actual compound ethylene, C_2H_4.) This simplest ratio formula is called the *empirical* formula, and the actual formula is the *true* formula. The formula weight is determined by multiplying the atomic weight (in whole numbers) by the subscript for that element in the formula. For example:

$Ca(OH)_2$ means (one calcium atomic weight + two hydrogen and two oxygen atomic weights = formula weight).

$$
\begin{array}{l}
1 \ Ca \ (at. \ wt. \ = \ 40) = 40 \\
2 \ O \ \ (at. \ wt. \ = \ 16) = 32 \\
\underline{2 \ H \ \ (at. \ wt. \ = \ \ \ 1) = \ \ 2} \\
Formula \ wt. \ Ca \ (OH)_2 = 74
\end{array}
$$

Fe_2O_3

$$
\begin{array}{l}
2 \ Fe \ (at. \ wt. \ = \ 56) = 112 \\
\underline{3 \ O \ (at. \ wt. \ = \ 16) = \ \ 48} \\
Formula \ wt. \ Fe_2O_3 = 160
\end{array}
$$

It is sometimes useful to know what percent of the total weight of a compound is made up of a particular element. This is called finding the *percentage composition*. The simple formula for this is:

$$
\frac{\text{Total wt. of the element in the compound}}{\text{Total formula wt.}} \times 100\% = \text{Percentage composition of that element}
$$

To find the percent composition of calcium in calcium hydroxide in the example above, we set the formula up as follows:

$$
\frac{Ca = 40}{\text{Formula wt.} = 74} \times 100\% = 54\% \ Calcium
$$

To find the percent composition of oxygen in calcium hydroxide:

$$
\frac{O = 32}{\text{Formula wt.} = 74} = 100\% = 43\% \ Oxygen
$$

To find the percent composition of hydrogen in calcium hydroxide:

$$
\frac{H = 2}{\text{Formula wt.} = 74} \times 100\% = 2.7\% \ Hydrogen
$$

• **ANOTHER EXAMPLE:**

Find the percent composition of Cu and H_2O in the compound $CuSO_4 \cdot 5\ H_2O$ (the dot is read "with").

First, we calculate the formula weight:

$$
\begin{array}{rl}
1 \quad Cu = & 64 \\
1 \quad S = & 32 \\
4 \quad O = & 64\ (4 \times 16) \\
5\ H_2O = & 90\ (5 \times 18) \\
\hline
& 250
\end{array}
$$

and then find the percentages:

Percentage Cu:

$$\frac{Cu = 64}{\text{Formula wt.} = 250} \times 100\% = 26\%$$

Percentage H_2O:

$$\frac{5\ H_2O = 90}{\text{Formula wt.} = 250} \times 100\% = 36\%$$

When you are given the percentage of each element in a compound, you can find the empirical formula as shown with the following example:

Given that a compound is composed of 60.0% Mg and 40.0% O, find the simple formula of the compound.

1. It is easiest to think of 100 weight units of this compound. In this case, the 100 weight units are composed of 60 wt. units of Mg and 40 wt. units of O. Since you know that 1 unit of Mg is 24 wt. units (from its atomic weight) and, likewise, 1 unit of O is 16, you can divide 60 by 24 to find the number of units of Mg in the compound and divide 40 by 16 to find the number of units of O in the compound.

$$
\begin{array}{cc}
Mg & O \\
24)\overline{60} & 16)\overline{40} \\
\text{2.5 units Mg} & \text{2.5 units O}
\end{array}
$$

2. Now, since we know formulas are made up of whole number units of the elements which are expressed as subscripts, we must manipulate these numbers to get whole numbers. This is usually accomplished by dividing these numbers by the smallest quotient. In this case they are equal, so we divide by 2.5.

$$
\begin{array}{cc}
Mg & O \\
2.5)\overline{2.5} & 2.5)\overline{2.5} \\
1 & 1
\end{array}
$$

Our simple formula, then, is one Mg and one O. Therefore, MgO is the formula.

• **ANOTHER EXAMPLE:**

Given: Ba = 58.81%, S = 13.73%, and O = 27.46%.
Find the simple formula.

1. Divide each percent by the atomic weight of the el

Ba	S	O
137)58.8	32)13.7	16)27.5
.43	.43	1.72

2. Manipulate numbers to get small whole num
   smallest first.

Ba	S	O
.43).43	.43).43	.43)1.72
1	1	4

3. The formula is $BaSO_4$.

In some cases you may be given the true formula weight of the compc if your empirical formula is correct, add up the formula weight of the empirica and compare it to the given formula weight. If it is *not* the same, multiply the emp formula by the small, whole number which gives you the correct formula weight. For example, if your simple formula was CH_2 (which has a formula weight of 14) and the true formula weight is given as 28, you can see that you must double the simple formula by doubling all the subscripts. The true formula is C_2H_4.

LAWS OF DEFINITE COMPOSITION AND MULTIPLE PROPORTIONS

In the problems involving percent composition, we have depended on two things: each unit of an element has the same atomic weight, and every time the particular compound forms, it forms in the same percent composition. That this latter statement is true no matter the source of the compound is the *Law of Definite Composition*. There are some compounds formed by the same two elements in which the weight of one element is constant, and the weight of the other varies. In every case, the weights of the other element are in the ratio of small whole numbers. This is called the *Law of Multiple Proportions*. An example is H_2O and H_2O_2.

> In H_2O the proportion of H:O = 2:16 or 1:8
> In H_2O_2 the proportion of H:O = 2:32 or 1:16

The ratio of the weight of oxygen in each is 8:16 or 1:2 (a small whole number ratio).

WRITING AND BALANCING SIMPLE EQUATIONS

An equation is a simplified way of recording a chemical change. Instead of words, chemical symbols and formulas are used to represent the reactants and the products. Here is an example of how this can be done. The following is the word equation of the reaction of burning hydrogen with oxygen:

Hydrogen + oxygen yields water.

Replacing the words with the chemical formulas, we have

$$H_2 + O_2 \rightarrow H_2O$$

RMULAS

drogen and oxygen with the formulas for their diatomic molecular states appropriate formula for water based on the respective oxidation (valence) hydrogen and oxygen. Note that the word *yields* was replaced with the arrow. the chemical statement tells what happened, it is not an equation because the are not equal. While the left side has two atoms of oxygen, the right side has only nowing that the Law of Conservation of Matter dictates that matter cannot easily be ed or destroyed, we must get the number of atoms of each element represented on the side to equal the number on the right. To do this, we can only use numbers, called *efficients*, in front of the formulas. It is important to note that in attempting to balance equations THE SUBSCRIPTS IN THE FORMULAS MAY NOT BE CHANGED.

Looking again at the skeleton equation, we notice that by placing a 2 in front of H_2O the number of oxygen atoms represented on both sides of the equation would be equal. However, we have now introduced four hydrogens on the right side with only two on the left. This can be corrected by using a coefficient of 2 in front of H_2. Now we have a balanced equation:

$$2 H_2 + O_2 = 2 H_2O$$

This equation tells us more than merely that hydrogen reacts with oxygen to form water. It has quantitative meaning as well. It tells us that two molecular weights of hydrogen react with one molecular weight of oxygen to form two molecular weights of water. Because molecular weights are indirectly related to grams, we may also relate the weights of reactants and products in grams.

$$2 H_2 + O_2 \rightarrow 2 H_2O$$
$$2(2) \quad 32 \quad 2(18)$$
$$4 \text{ units} + 32 \text{ units} = 36 \text{ units}$$
$$4 \text{ grams of } H_2 + 32 \text{ grams of } O_2 = 36 \text{ grams of water}$$

This aspect will be important in solving problems related to the weight of substances in a chemical equation.

Here is another, more difficult example: Write the balanced equation for the burning of butane (C_4H_{10}) in oxygen. First, we write the skeleton equation:

$$C_4H_{10} + O_2 \text{ yields } CO_2 + H_2O.$$

Looking at the oxygens, we see that there are an even number on the left but an odd number on the right. This is a good place to start. If we use a coefficient of 2 for H_2O, that will even out the oxygens but introduce four hydrogens on the right while there are ten on the left. A coefficient of 5 will give us the right number of hydrogens but introduces an odd number of oxygens. Therefore we have to go to the next even multiple of 5, which is 10. Ten gives us twenty hydrogen atoms on the right. By placing another coefficient of 2 in front of C_4H_{10}, we also have twenty hydrogen atoms on the left. Now the carbons need to be balanced. By placing an 8 in front of CO_2, we have eight carbons on both sides. The remaining step is to balance the oxygens. We have twenty-six on the right side, so we need a coefficient of 13 in front of the O_2 on the left to give us twenty-six oxygens on both sides. Our balanced equation is:

$$2 C_4H_{10} + 13 O_2 \rightarrow 8 CO_2 + 10 H_2O$$

For more practice in balancing equations, see the section on page 347 in the back of the book. (The balancing of equations using the electron exchange method is introduced at the end of Chapter 12.)

Chapter
4
Review

Complete the formula or name:

1. AgCl _Silver chloride_

2. CaSO$_4$ _Calcium Sulfate_

3. Al$_2$(SO$_4$)$_3$ _aluminium Sulfate_

4. NH$_4$NO$_3$ _ammonium nitrate_

5. FeSO$_4$ _Ferrous Sulfate_

6. Potassium chromate _K$_2$CrO$_4$_

7. Sodium fluoride _NaF_

8. Magnesium sulfite _MgSO$_3$_

9. Copper (II) sulfate _CuSO$_4$_

10. Ferric chloride _FeCl$_3$_

- **ANSWERS:**

  1. Silver chloride
  2. Calcium sulfate
  3. Aluminum sulfate
  4. Ammonium nitrate

  5. Ferrous sulfate or iron(II) sulfate
  6. K$_2$CrO$_4$
  7. NaF

  8. MgSO$_3$
  9. CuSO$_4$
  10. FeCl$_3$

11. Find the percentage of sulfur in H$_2$SO$_4$.
12. What are the empirical formula and the true formula of a compound composed of 85.7% C and 14.3% H with a true formula weight of 42 amu?

- **ANSWERS:**

  *11. 32.65%
  *12. CH$_2$ and C$_3$H$_6$

Terms You Should Know

binary compound
coefficient
empirical formula
Law of Definite Composition
Law of Multiple Proportions

formula weight
molecular weight
percentage composition

radical
ternary compound

*Answer explained in section at the end of the book.

5

Gases

SOME REPRESENTATIVE GASES

Oxygen

Of the gases that occur in the atmosphere, the most important one to us is oxygen. Although it makes up only approximately 21% of the atmosphere, by volume, the oxygen found on the earth is equal in weight to all the other elements combined. About 50% of the earth's crust (including the waters on the earth, and the air surrounding it) is oxygen. (Note Figure 15.)

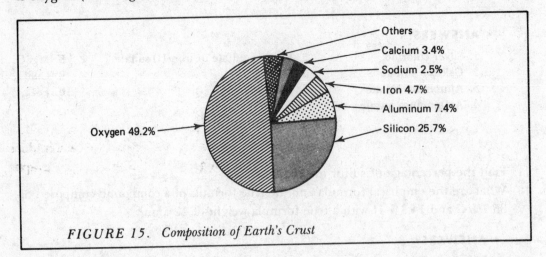

FIGURE 15. *Composition of Earth's Crust*

The composition of air varies slightly from place to place because it is a mixture of gases. The composition by volume is approximately as follows: nitrogen, 78%; oxygen, 21%; argon 1%. There are also small amounts of carbon dioxide, water vapor, and trace gases.

Preparation of oxygen. In 1774, an English scientist named Joseph Priestley discovered oxygen by heating mercuric oxide in an enclosed container with a magnifying glass. That mercuric oxide decomposes into oxygen and mercury can be expressed in an equation: $2HgO = 2Hg + O_2$. After his discovery, Priestley visited one of the greatest of all scientists, Antoine Lavoisier, in Paris. As early as 1773 Lavoisier had carried on experiments concerning burning, and they had caused him to doubt the phlogiston theory (that a substance called phlogiston was released when a substance burned; the theory went through several modifications before it was finally abandoned). By 1775, Lavoisier had demonstrated the true nature of burning and called this gas "oxygen."

Today oxygen is usually prepared in the lab by heating an easily decomposed oxygen compound such as potassium chlorate ($KClO_3$). The equation for this reaction is:

$$2KClO_3 + MnO_2 \rightarrow 2KCl + 3O_2 \uparrow + MnO_2$$

The usual laboratory setup is shown in Figure 16.

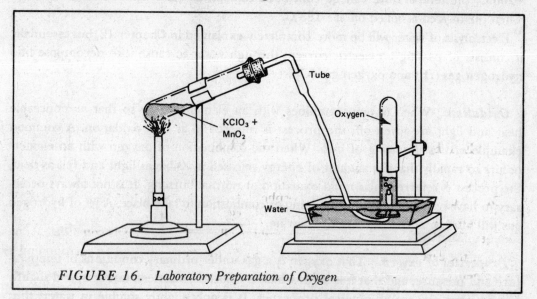

FIGURE 16. Laboratory Preparation of Oxygen

In this preparation manganese dioxide (MnO_2) is often used. This compound is not used up in the reaction and can be shown to have the same composition as it had before the reaction occurred. The only effect it seems to have is that it lowers the temperature needed to decompose the $KClO_3$, and thus speeds up the reaction. Substances that behave in this manner are referred to as catalysts. Catalysts can also be used to slow down a reaction as well. The mechanism by which a catalyst does this is not completely understood in all cases, but it is known that in some reactions the catalyst does change its structure, temporarily. Its effect is shown graphically in the reaction graphs in Figure 17.

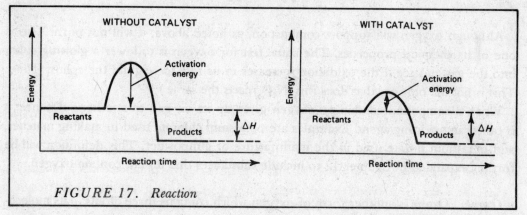

FIGURE 17. Reaction

In this reaction you notice that the activation energy necessary to make the reaction take place is lower so that the reaction can take place at a more rapid rate.

Commercially, oxygen is prepared by either fractional distillation of liquid air or the electrolysis of water. The fractional distillation of liquid air is typical of all fractional distillation processes in that it attempts to separate different components of a mixture in solution by taking advantage of the fact that each component has a distinct boiling point. By raising the temperature slowly and holding it constant when any boiling is taking place, the various components can be separated. With liquid air at

−200°C, the temperature can be controlled so that nitrogen will boil off at −196°C. Then the oxygen is boiled off at −183°C.

Electrolysis of water will be more completely explained in Chapter 12, but essentially it consists of passing an electric current through water to cause it to decompose into hydrogen gas (H_2) and oxygen gas (O_2).

Oxidation. When oxygen combines with an element slowly so that *no* noticeable heat and light are given off, the process is referred to as slow oxidation. A common example of this is rusting of iron. When the combination of oxygen with an element occurs so rapidly that the amount of energy released is visible as light and felt as heat, the process is referred to as rapid oxidation or normal burning. It is not always necessary to have oxygen present for burning or combustion to take place. A jet of hydrogen gas will burn in an atmosphere of chlorine.

Properties of oxygen. That oxygen is a gas under ordinary conditions of temperature and pressure, and that it is a gas which is colorless, odorless, tasteless, and slightly heavier than air, are all physical properties. It is only slightly soluble in water, thus making it possible to collect the gas over water, as shown in Figure 16. Oxygen at −183°C changes to a pale blue liquid with magnetic properties. This attraction by a magnet (paramagnetism) is believed to be due to an unpaired group of electrons shown like this in a dot structure:

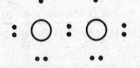

Although oxygen will support combustion, as noted above, it will not burn. This is one of its chemical properties. The usual test for oxygen is to lower a glowing splint into the gas and see if the oxidation increases in its rate to reignite the splint. (Note: This is not the only gas that does this. N_2O reacts the same.)

When a compound containing oxygen gives up its oxygen to another substance, it is called an oxidizing agent. Examples are potassium chlorate used in making matches, and potassium nitrate used in the manufacture of gunpowder. This definition will be further expanded in Chapter 12 to include substances that do not contain oxygen.

Ozone. Ozone is another form of oxygen which contains three atoms in its molecular structure (O_3). Since ordinary oxygen and ozone differ in energy content and form, they have slightly different properties. They are called allotropic forms of oxygen. Ozone occurs in small quantities in the upper layers of the earth's atmosphere, and can be formed in the lower atmosphere where high-voltage electricity in lightning passes through the air. This formation of ozone also occurs around machinery using high voltage. The reaction can be shown by this equation:

$$3O_2 + elec. = 2O_3$$

Because of its higher energy content, ozone is more reactive chemically than oxygen.

SOME REPRESENTATIVE GASES • 59

Hydrogen

Of all the 103 elements in the periodic chart, hydrogen is number one by reason of the fact that it is the lightest element known. Its lightness makes it useful in balloons, but care must be taken because of its extreme combustibility. Its ability to burn well makes it an important part of most fuels. Its powerful attraction to oxygen makes it a good reducing agent in some important industrial reactions. Hydrogen can be added to the structure of liquid fats, under the proper conditions, to make solid shortenings, or it can be added to certain fuels to improve their burning properties. Its isotopes are used in making the hydrogen bomb.

Preparation of hydrogen. Although there is evidence of the preparation of hydrogen before 1766, Henry Cavandish was the first person to recognize this gas as a separate substance. He observed that, whenever it burned, it produced water. Lavoisier named it "hydrogen," which means "water former."

Electrolysis of water, which is the process of passing an electric current through water to cause it to decompose, is one method of obtaining hydrogen. This is a widely used commercial method, as well as a laboratory method.

Another method of producing hydrogen is to displace it from the water molecule by using a metal. To choose the metal you must be familiar with the activity of the metal with respect to hydrogen. This is shown in the form of an activity chart of the common metals (Table 8).

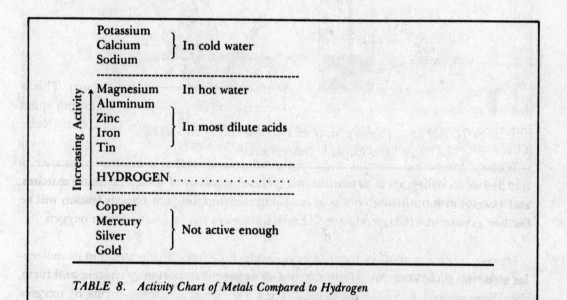

TABLE 8. *Activity Chart of Metals Compared to Hydrogen*

As noted in Table 8, the first three metals will react with water in the following manner:

Very active metal + Water = Hydrogen + Metal hydroxide

Using sodium as an example:

$$2\,Na + 2\,HOH \rightarrow H_2 \uparrow + 2\,NaOH$$

With the metals that react slower, a dilute acid reaction is needed to produce hydrogen in sufficient quantities to collect in the laboratory. This general equation is:

Active metal + Dilute acid → Hydrogen + Salt of the acid

An example:

$Zn + dil. H_2SO_4 \rightarrow H_2\uparrow + ZnSO_4$

This equation shows the usual laboratory method of preparing hydrogen. Mossy zinc is used in a setup as shown in Figure 18. The acid is introduced down the thistle tube after the zinc is placed in the reacting bottle. In this sort of setup, you would not begin collecting the gas that bubbles out of the delivery tube for a few minutes so that the air in the system has a chance to be expelled and you can collect a rather pure volume of the gas generated.

$Zn + H_2SO_4 = H_2 + ZnSO_4$

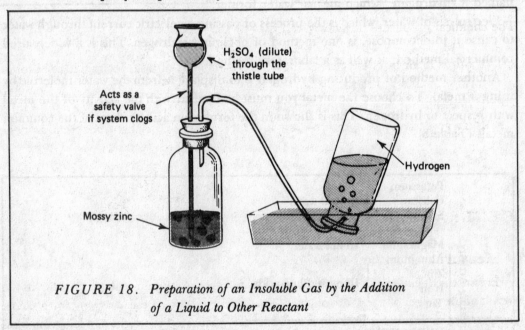

H$_2$SO$_4$ (dilute) through the thistle tube

Acts as a safety valve if system clogs

Hydrogen

Mossy zinc

FIGURE 18. Preparation of an Insoluble Gas by the Addition of a Liquid to Other Reactant

In industry, hydrogen is produced by (1) the electrolysis of water, (2) passing steam over red-hot iron or through hot coke, or (3) by decomposing natural gas (mostly methane, CH_4) with heat ($CH_4 + H_2O = CO + 3H_2$).

Properties of hydrogen. Hydrogen has the following important physical properties:

1. It is ordinarily a gas; colorless, odorless, tasteless when pure.
2. It weighs 0.9 gram per liter at 0°C and 1 atmosphere pressure. This is 1/14 as heavy as air.
3. It is slightly soluble in water.
4. It becomes a liquid at a temperature of −240°C and a pressure of 13 atmospheres.
5. It diffuses (moves from place to place in gases) more rapidly than any other gas. This property can be demonstrated as shown in Figure 19.

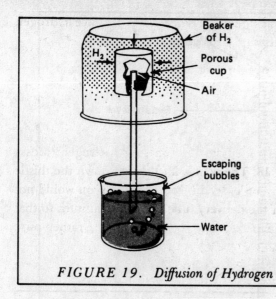

Here the H_2 in the beaker that is placed over the porous cup diffuses faster through the cup than the air can diffuse out. Consequently, there is a pressure build-up in the cup, which pushes the gas out through the water in the lower beaker.

FIGURE 19. Diffusion of Hydrogen

The chemical properties of hydrogen are:

1. It burns in air or in oxygen, giving off large amounts of heat. Its high heat of combustion makes it a good fuel.
2. It does not support ordinary combustion.
3. It is a good reducing agent in that it withdraws oxygen from many hot metal oxides.

GENERAL CHARACTERISTICS OF GASES

Measuring Pressure of Gases

Pressure is defined as force per unit area. With respect to the atmosphere, pressure is the result of the weight of a mixture of gases. This pressure, which is called *atmospheric pressure, air pressure,* or *barometric pressure,* is approximately equal to the weight of a kilogram mass on every square centimeter of surface exposed to it. This weight is about 10 newtons.

The pressure of the atmosphere varies with altitude. At higher altitudes, the weight of the overlying atmosphere is less, so the pressure is less. Air pressure also varies somewhat with weather conditions as low- and high-pressure areas move with weather fronts. On the average, however, the air pressure at sea level can support a column of mercury 760 mm in height. This average sea-level air pressure is known as *normal atmospheric pressure,* also called *standard pressure.*

The instrument most commonly used for measuring air pressure is the *mercury barometer.* The diagram that follows shows how it operates. Atmospheric pressure is exerted on the mercury in the dish, and this in turn holds the column of mercury up in the tube. This column at standard pressure will measure 760 mm above the level of the mercury in the dish below. This method of measuring air pressure is still commonly used. The usual unit of pressure used in the solution of gas law problems is millimeter of mercury (also known as a torr.)

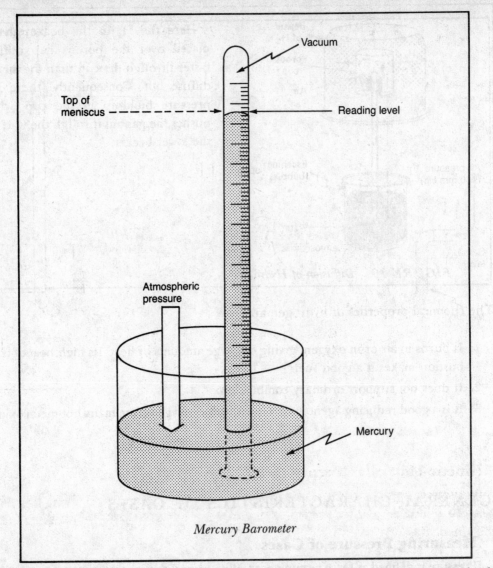

Mercury Barometer

There are several other means of expressing gas pressure. The unit of pressure in the metric system is the pascal (Pa), defined as the pressure of 1 newton per square meter. Since this is a very small unit, the kilopascal, which is abbreviated as kPa, is more convenient to use. Standard pressure is about 100 kPa.

A device similar to the barometer can be used to measure the pressure of gases in a confined container. This apparatus, called a *manometer*, is illustrated on the next page. A manometer is basically a U-tube containing mercury or some other liquid. When both ends are open in the air, as in (1) in the diagram, the level of the liquid will be the same on both sides since the same pressure is being exerted on both ends of the tube. In (2) and (3), a vessel is connected to one end of the U-tube. Now the height of the mercury column serves as a means of reading the pressure inside the vessel if the atmospheric pressure is known. When the pressure inside the vessel is the same as the atmospheric pressure outside, the levels are the same. When the pressure inside is greater than outside, the column of liquid will be higher on the side that is exposed to the air, as in (2). Conversely, when the pressure inside the vessel is less than the outside atmospheric pressure, the additional pressure will force the liquid to a higher level on the side near the vessel, as in (3).

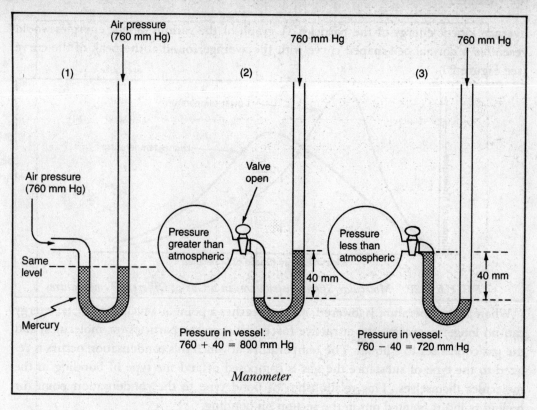

Manometer

Kinetic-Molecular Theory

By indirect observations, the kinetic theory has been arrived at to explain the forces between molecules and the energy they possess. There are three basic assumptions to the kinetic theory:

1. Matter in all its forms (solid, liquid and gases) is composed of extremely small particles. In many cases these are called molecules. The space occupied by the gas particles themselves is ignored in comparison with the volume of the space they occupy.

2. The particles of matter are in constant motion. In solids, this motion is restricted to a small space. In liquids, the particles have a more random pattern but still are restricted to a kind of rolling over one another. In a gas, the particles are in continuous, random, straight-line motion.

3. When these particles collide with each other or with the walls of the container, there is no loss of energy.

Some Particular Properties of Gases

As the temperature of a gas is increased, its kinetic energy is increased and this increases the random motion. It should be pointed out that at a particular temperature not all the particles have the same kinetic energy, but the temperature is a measure of the average kinetic energy of the particles. A graph of the various kinetic energies would resemble a normal bell-shaped curve with the average found at the peak of the curve (see Figure 20).

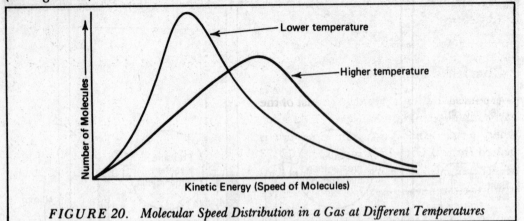

FIGURE 20. *Molecular Speed Distribution in a Gas at Different Temperatures*

When the temperature is lowered, the gas reaches a point at which the kinetic energy can no longer overcome the attractive forces between the particles (or molecules) and the gas condenses to a liquid. The temperature at which this condensation occurs is related to the type of substance the gas is composed of and the type of bonding in the molecules themselves. This relationship of bond type to the condensation point (or boiling point) is pointed out in the section on bonding.

That gases are moving in a random motion so that they may move from one position to another is referred to as diffusion. You know that, if a bottle of perfume is opened in one corner of a room, it—that is, its molecules—will move or diffuse to all parts of the room in time.

GAS LAWS AND RELATED PROBLEMS

Graham's Law

This law relates the rate at which a gas diffuses to the type of molecule in the gas. It can be expressed as follows: The rate of diffusion of a gas is inversely proportional to the square root of its molecular weight. Hydrogen, with the lowest molecular weight, can diffuse more rapidly than other gases under similar conditions. A sample problem:

Compare the rate of diffusion of hydrogen to that of oxygen under similar conditions.

The formula is

$$\frac{\text{rate A}}{\text{rate B}} = \frac{\sqrt{\text{Molecular wt. of B}}}{\sqrt{\text{Molecular wt. of A}}}$$

Let A be H_2 and B be O_2.

$$\frac{\text{rate } H_2}{\text{rate } O_2} = \frac{\sqrt{32}}{\sqrt{2}} = \frac{\sqrt{16}}{\sqrt{1}} = \frac{4}{1}$$

Therefore hydrogen diffuses four times as fast as oxygen.

In dealing with the gas laws, a student must know what is meant by standard conditions of temperature and pressure (abbreviated as STP). The standard pressure is defined as the height of mercury that can be held in an evacuated tube by one atmosphere of pressure (14.7 pounds per square inch). This is usually expressed as 760 mm of Hg. Standard temperature is defined as 273 Kelvin or absolute (which corresponds to 0° Celsius).

Charles' Law

Jacques Charles, a French chemist of the early nineteenth century, discovered that, when a gas under constant pressure is heated from 0°C to 1°C, it expands 1/273 of its volume. It contracts this amount when the temperature is dropped 1 degree to −1°C. Charles reasoned that, if a gas at 0°C was cooled to −273°C, its volume would be zero. Actually, all gases are converted into liquids before this temperature is reached. By using the Kelvin scale to rid the problem of negative numbers, Charles' Law can be stated as follows:

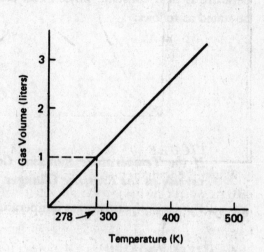

If the Pressure Remains Constant, the Volume of a Gas Varies Directly as the Absolute Temperature. Then

Initial $\dfrac{V_1}{T_1}$ = Final $\dfrac{V_2}{T_2}$ at constant pressure

• TYPE PROBLEM:

The volume of a gas at 20°C is 500 mL. Find its volume at standard temperature if pressure is held constant.

Convert temperatures:

$$20°C = 20° + 273° = 293K$$
$$0°C = 0° + 273° = 273K$$

If you know that cooling a gas decreases its volume, then you know that 500 mL will have to be multiplied by a fraction (made up of the Kelvin temperatures) which has a smaller numerator than the denominator. So

$$500 \text{ mL} \times \frac{273}{293} = 465 \text{ mL}$$

Or you can use the formula:

$$\frac{V_1}{T_1} = \frac{V_2}{T_2} \quad \text{SO} \quad \frac{500 \text{ mL}}{293} = \frac{X \text{ mL}}{273}$$

$$X \text{ mL} = 465 \text{ mL}$$

Boyle's Law

Robert Boyle, a seventeenth century English scientist, found that the volume of a gas decreases when the pressure on it is increased, and vice versa, when the temperature is held constant. Boyle's Law can be stated as follows:

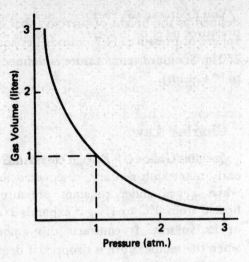

If the Temperature Remains Constant, the Volume of a Gas Varies Inversely as the Pressure Changes. Then

$P_1V_1 = P_2V_2$ at constant temperature

• TYPE PROBLEM:

Given the volume of a gas as 200 mL at 800 mm pressure. Calculate the volume of the same gas at 765 mm. Temperature is held constant.

If you know that this *decrease* in pressure will cause an increase in the volume, then you know 200 mL must be multiplied by a fraction (made up of the two pressures) which has a larger numerator than the denominator. So

$$200 \text{ mL } \frac{800}{765} = 209 \text{ mL}$$

Or you can use the formula:

$P_1V_1 = P_2V_2$

So, $V_2 = V_1 \times \dfrac{P_1}{P_2}$

$$V_2 = 200 \text{ mL} \times \frac{800}{765} = 209 \text{ mL}$$

Combined Gas Law

This is a combination of the two preceding gas laws and can be expressed as follows:

$$\frac{P_1V_1}{T_1} = \frac{P_2V_2}{T_2}$$ *Charles & Boyles Law*

• TYPE PROBLEM:

The volume of a gas at 780 mm pressure and 30°C is 500 mL. What volume would the gas occupy, at STP?

You again can use reasoning to determine the kind of fractions the temperatures and pressures must be to arrive at your answer. Since the pressure is going from 780 mm to 760 mm, this should cause the volume to increase. The fraction must then be $\frac{780}{760}$. Since the temperature is going from 30°C (303 K) to 0°C (273 K), the volume should decrease; this fraction must be $\frac{273}{303}$. So

$$500 \text{ mL} \times \frac{780}{760} \times \frac{273}{303} = 462 \text{ mL}$$

Or you can use the formula:

$$\frac{P_1 V_1}{T_1} = \frac{P_2 V_2}{T_2}$$

Solve for $V_2 = V_1 \times \dfrac{P_1}{P_2} \times \dfrac{T_2}{T_1}$

$$V_2 = 500 \text{ mL} \times \frac{780}{760} \times \frac{273}{303} = 462 \text{ mL}$$

Pressure versus Temperature

At Constant Volume, the Pressure of a Given Mass of Gas Varies Directly with the Absolute Temperature. Then

$$\frac{P_1}{T_1} = \frac{P_2}{T_2} \text{ at constant volume}$$

• TYPE PROBLEM:

A steel tank contains a gas at 27°C and a pressure of 12 atmospheres. Determine the gas pressure when the tank is heated to 100°C.

Reasoning that an increase of temperature will cause an increase in pressure at constant volume, you know the pressure must be multiplied by a fraction which has a larger numerator than denominator. The fraction must be $\frac{373}{300}$ (temperatures expressed in Kelvins). So

$$12 \text{ atm.} \times \frac{373}{300} = 14.9 \text{ atm.}$$

Or you can use the formula:

$$\frac{P_1}{T_1} = \frac{P_2}{T_2}$$

$$P_2 = P_1 \times \frac{T_2}{T_1}$$

So $P_2 = 12$ atm. $\times \dfrac{373}{300} = 14.9$ atm.

Dalton's Law of Partial Pressures

When a gas is made up of a mixture of different gases, the total pressure of the mixture is equal to the sum of the partial pressures of the components, that is, the partial pressure of the gas would be the pressure of the individual gas if it alone occupied the volume.

$$P_{total} = P_{gas\ 1} + P_{gas\ 2} + P_{gas\ 3} + \cdots$$

• TYPE PROBLEM:

A mixture of gases at 760 mm pressure contains 65% nitrogen, 15% oxygen, and 20% carbon dioxide by volume. What is the partial pressure of each gas?

$.65 \times 760 = 494$ mm pressure (N_2)
$.15 \times 760 = 114$ mm pressure (O_2)
$.20 \times 760 = 152$ mm pressure (CO_2)

Corrections of Pressure

Correction of pressure when a gas is collected over water. When a gas is collected over a volatile liquid, such as water, some of the water vapor is present in the gas and contributes to the total pressure. Assuming the gas is saturated with water vapor for that temperature, you can find the partial pressure due to the water vapor in a table of such water vapor values. This value depends only on the temperature. This vapor pressure must be subtracted from the total pressure to find the partial pressure of the gas being measured. See Chart Ⓝ in the Reference Tables section at the back of the book.

Correction of difference in the height of the fluid. When gases are collected in eudiometers (glass tubes closed at one end), it is not always possible to get the level of the liquid inside the tube to equal the level on the outside. This deviation of levels must be taken into account when determining the pressure of the enclosed gas. There are then two possibilities: (1) When the level inside is higher than the level outside the tube, the pressure on the inside is less, by the height of fluid in excess, than the outside pressure. If the fluid is mercury, you simply subtract the difference from the outside pressure reading (also in height of mercury and in the same units) to get the corrected pressure of the gas. If the fluid is water, you must first convert the difference to an equivalent height of mercury by dividing the difference by 13.6 (since mercury is 13.6 times as heavy as water,

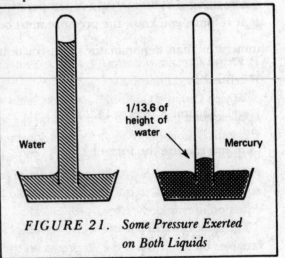

FIGURE 21. *Some Pressure Exerted on Both Liquids*

the height expressed in Hg will be 1/13.6 the height of water). Again, care must be taken that this equivalent height of Hg is in the same units as the expression for the outside pressure before it is subtracted. This gives the corrected pressure for the gas in the eudiometer. (2) When the level inside is lower than the level outside the tube, a correction must be added to the outside pressure. If the difference in height between the inside and the outside is expressed in height of water, you must take 1/13.6 of this quantity to correct it to mm of mercury units. This quantity is then added to the expression of the outside pressure, which must also be in mm of mercury. If the tube contains mercury, then the difference between the inside and outside levels is merely added to the outside pressure to get the corrected pressure for the enclosed gas.

• TYPE PROBLEM:

Hydrogen gas is collected in a eudiometer tube over water. It was impossible to level the outside water level with that in the tube, and the water level inside the tube was 40.8 mm higher than that outside. The barometric pressure was 730 mm of Hg. The water vapor pressure at the room temperature of 29°C was found in a handbook to be 30.0 mm. What is the pressure of the dry hydrogen?

Step 1: To find the true pressure of the gas, we must first subtract the water level difference expressed in mm of Hg:

$$\frac{40.8}{13.6} = 3 \text{ mm of Hg} \quad \text{So } 730 \text{ mm} - 3 \text{ mm} = 727 \text{ mm total pressure of gases}$$

in the eudiometer

Step 2: Correcting for the partial pressure due to water vapor in the hydrogen, we subtract the vapor pressure (30.0 mm) from 727 mm and get our answer: 697 mm.

Ideal Gas Law

The previous laws discussed do not include the relationship of number of moles of a gas to the pressure, volume, and temperature of the gas. A law derived from the kinetic-molecular theory relates these variables. It is called the Ideal Gas Law and is expressed as

$$PV = nRT$$

P, V, and T retain their usual meanings, but n stands for the number of moles of the gas and R represents the ideal gas constant.

Boyle's Law and Charles's Law are actually derived from the Ideal Gas Law. Boyle's Law applies when the number of moles and the temperature of the gas are constant. Then in $PV = nRT$, the number of moles, n, is constant; the gas constant (R) remains the same; and by definition T is constant. Therefore, $PV = k$. At the initial set of conditions of a problem, $P_1V_1 = $ a constant (k). At the second set of conditions, the terms on the right side of the equation are equal to the same constant, so $P_1V_1 = P_2V_2$. This matches the Boyle's Law equation introduced earlier.

The same can also be done with Charles's Law, since $PV = nRT$ can be expressed with the variables on the left and the constants on the right.

$$\frac{V}{T} = \frac{nR}{P}$$

In Charles's Law the number of moles and the pressure are constant. Substituting k for the constant term, $\frac{nR}{P}$, we have

$$\frac{V}{T} = k.$$

The expression relating two sets of conditions can be written as

$$\frac{V_1}{T_1} = \frac{V_2}{T_2}$$

To use the Ideal Gas Law in the form $PV = nRT$, the gas constant, R, must be determined. This can be done mathematically as shown in the following example.

One mole of oxygen gas was collected in the laboratory at a temperature of 24.0°C and a pressure of exactly 1 atmosphere. The volume was 24.35 liters. Find the value of R.

$$PV = nRT$$

Rearranging the equation to solve for R gives

$$R = \frac{PV}{nT}$$

Substituting the known values on the right, we have

$$R = \frac{1 \text{ atmosphere} \times 24.35 \text{ liters}}{1 \text{ mole} \times 297 \text{ kelvins}}$$

Calculating R, we get

$$R = 0.0820 \frac{\text{L atm}}{\text{mol K}}$$

Once R is known, the Ideal Gas Law can be used to find any of the variables, given the other three.

For example, calculate the pressure of 1.00 gram of hydrogen gas at 16.0°C and occupying 2.54 liters.

Rearranging the equation to solve for P, we get

$$P = \frac{nRT}{V}$$

The molar mass of hydrogen is 2.00 g/mol, so the number of moles in this problem would be

$$\frac{1.00\text{g}}{2.00\text{g/mol}} = .500 \text{ mol}$$

Substituting the known values, we have

$$P = \frac{(.500 \text{ mol})(.820 \frac{\text{L atm}}{\text{mol K}}) \, 289\text{K}}{2.54 \text{ L}}$$

Calculating the value, we get

$$P = 4.66 \text{ atm}$$

Another use of the ideal gas law is to find the number of moles of a gas when P, T, and V are known.

EXAMPLE:

How many moles of nitrogen gas are there in 0.38 liter of gas at 0°C and 380 mm Hg pressure?

Rearranging the equation to solve for n gives

$$n = \frac{PV}{RT}$$

Changing temperature to kelvins and pressure into atmospheres gives

$$T = 0° + 273 = 273 \text{ K}$$

$$P = \frac{380 \text{ mm Hg}}{760 \text{ mm Hg/atm}} = .5 \text{ atm}$$

Substituting in the equation, we have

$$n = \frac{(.5 \text{ atm})(.38L)}{(0.220 \frac{\text{L atm}}{\text{mol K}})(273K)} = .0085 \text{ mole}$$

$$n = .0085 \text{ mole of nitrogen gas}$$

Ideal Gas Deviations

In the use of the gas laws, we have assumed that the gases involved were "ideal" gases. This means that the molecules of the gas were not taking up space in the gas volume and that no intermolecular forces of attraction were serving to pull the molecules closer together. You will find that a gas behaves like an ideal gas at low pressures and high temperatures which move the molecules as far as possible from conditions which would cause condensation. In general, pressures below a few atmospheres will cause most gases to exhibit sufficiently ideal properties for the application of the gas laws with a reliability of a few percent or better.

Chapter
5
Review

1. The most abundant element in the earth's crust is (1) sodium (2) oxygen (3) silicon (4) aluminum.

2. A compound which can be decomposed to produce oxygen gas in the lab is (1) MnO_2 (2) NaOH (3) CO_2 (4) $KClO_3$.

3. In the usual laboratory preparation equation for the reaction in question 2, what is the coefficient of the O_2? (1) 1 (2) 2 (3) 3 (4) 4

4. In the graphic representation of the energy content of reactants and the resulting products, which would have a higher energy content in an exothermic reaction? (1) the reactants   (2) the products   (3) both the same   (4) neither one

5. The process of separating components of a mixture by making use of the difference in their boiling points is called (1) destructive distillation   (2) displacement   (3) fractional distillation   (4) filtration.

6. When oxygen combines with an element to form a compound, the resulting compound is called a(n) (1) salt   (2) oxide   (3) oxidation   (4) oxalate.

7. According to the activity chart of metals, which metal would react most vigorously in a dilute acid solution? (1) zinc   (2) iron   (3) aluminum   (4) magnesium

8. Graham's Law refers to (1) boiling points of gases   (2) gaseous diffusion   (3) gas compression problems   (4) volume changes of gases when the temperature changes.

9. When 200 ml of a gas at constant pressure is heated, its volume (1) increases   (2) decreases   (3) remains unchanged.

10. When 200 ml of a gas at constant pressure is heated from 0°C to 100°C, the volume must be multiplied by (1) 0/100   (2) 100/0   (3) 273/373   (4) 373/273.

11. If you wish to find the corrected volume of a gas which was at 20°C and 760 mm pressure and conditions were changed to 0°C and 700 mm pressure, by what fractions would you multiply the original volume? (1) $\frac{293}{273} \times \frac{760}{700}$ (2) $\frac{273}{293} \times \frac{700}{760}$ (3) $\frac{273}{293} \times \frac{760}{700}$ (4) $\frac{293}{273} \times \frac{700}{760}$

12. When the level of mercury inside a gas tube is higher than the level in the reservoir, you find the correct pressure inside the tube by taking the outside pressure reading and __?__ the difference in the height of mercury. (1) subtracting   (2) adding   (3) dividing by 13.6   (4) doing both (3) and (1)

13. If water is the liquid in question 12 instead of mercury, you can change the height difference to an equivalent mercury expression by (1) dividing by 13.6   (2) multiplying by 13.6   (3) adding 13.6   (4) subtracting 13.6.

14. Standard conditions are (1) 0°C and 14.7 mm   (2) 32°F and 76 cm   (3) 273°C and 760 mm   (4) 4°C and 7.6 m.

15. When a gas is collected over water, the pressure is corrected by (1) adding the vapor pressure of water   (2) multiplying by the vapor pressure of water   (3) subtracting the vapor pressure of water at that temperature   (4) subtracting the temperature of the water from the vapor pressure.

16. At 5 atmospheres pressure and 70°C how many moles are present in 1.5 liters of O_2 gas? (1) .036   (2) .267   (3) .536   (4) 1.60

• ANSWERS:

1. (2)	7. (4)	12. (1)
2. (4)	8. (2)	13. (1)
3. (3)	9. (1)	14. (2)
4. (1)	10. (4)	15. (3)
5. (3)	11. (3)	16. (2)
6. (2)		

Terms You Should Know

atmospheric pressure	oxidation	standard pressure
manometer	ozone	standard temperature
mercury barometer	pascal	torr

Boyle's Law
Charles' Law
Combined Gas Law
Dalton's Law of Partial Pressures
Graham's Law
Ideal Gas Law
Kinetic-Molecular Theory

Chemical Calculations (Stoichiometry) and the Mole Concept

THE MOLE CONCEPT

Providing a name for a quantity of things taken as a whole is common in everyday life. Some examples are a dozen, a gross, and a ream. Each of these represents a specific number of items and is not dependent on the commodity. A dozen of eggs, oranges, or bananas will always represent 12 items.

In chemistry we have a unit that describes a quantity of particles. It is called the *mole*. A mole is 6.02×10^{23} particles. The particles can be atoms, molecules, ions, electrons, and so forth. Because particles are so small in chemistry, the mole is a very convenient unit. The number 6.02×10^{23} is often referred to as Avogadro's number in honor of the Italian scientist whose hypothesis led to its determination.

MOLAR MASS AND MOLES

When the formula weight of an ionic compound is determined by the addition of its component atomic weights and expressed in grams, it is called the *gram-formula weight* or *molar mass*. An example is $CaCO_3$:

$$
\begin{array}{rcl}
1\ Ca &=& 40 \\
1\ C &=& 12 \\
3\ O &=& 48 = (3 \times 16) \\
\hline
CaCo_3 &=& 100 \text{ formula weight}
\end{array}
$$

$$100\ g = \text{gram-formula weight}$$

The molar mass is equivalent to the mass of one *mole* of that material.

The term *molar mass* or *gram-molecular weight* (gmw) can also be used when it is known that the material is a molecular substance and not an ionic lattice like NaCl or NaOH.

• **ANOTHER EXAMPLE:**

Find the weight of 1 mole of $KAl(SO_4)_2 \cdot 12 H_2O$.

$$1 \text{ K} = 39$$
$$1 \text{ Al} = 27$$
$$2 (SO_4) = 192 = 2(32 + 16 \times 4)$$
$$\underline{12 \text{ } H_2O = 216 = 12(2 + 16)}$$
$$1 \text{ mole} = 474 \text{ g}$$

• **TRY THESE PROBLEMS:**

Find the molecular weight or formula weight of the following:

1. HNO_3 Ans. 63
2. $C_6H_{10}O_5$ Ans. 162
3. H_2SO_4 Ans. 98
4. KCl Ans. 74.5
5. $C_{12}H_{22}O_{11}$ Ans. 342

In some cases, there may be a question of the mass of 1 mole of an element if you are not told specifically whether you are referring to the single atoms or a molecular state. An example of this situation may arise with hydrogen and other elements that form diatomic molecules. If you are asked the mass of 1 mole of hydrogen, you could say it is 1 gram if you are dealing with single atoms of hydrogen. This 1 mole of hydrogen atoms would contain 6.02×10^{23} atoms. In a similar fashion, 1 mole of hydrogen molecules (H_2) would have a molar mass of 2 grams. This 1 mole of hydrogen molecules would contain 6.02×10^{23} molecules and, since each molecule is composed of two atoms, 12.04×10^{23} atoms.

The other elements that exist as diatomic molecules are oxygen, nitrogen, fluorine, chlorine, bromine, iodine, and astatine.

Try these questions.

What is the mass of 1 mole of oxygen?

Answer: Since the atomic mass is 16 grams and the molecule is diatomic, 2×16 grams = 32 grams.

What is the mass of one molecule of oxygen?

Answer: There are 6.02×10^{23} molecules in 1 mole of this gas. 32 grams $\div$ 6.02×10^{23} molecules = 5.32×10^{-23} gram.

What is the mass of 1 mole of oxygen atoms?

Answer: Without the diatomic molecule structure, the atomic mass of oxygen expressed in grams is the answer. That is 16 grams.

What is the mass of one atom of oxygen?

Answer: The atomic molar mass, 16 grams, is divided by 6.02×10^{23}. This gives 2.66×10^{-23} gram.

If you gave the same answer to the first two questions, you are confused about the mole concept. To say that a molecule of oxygen gas has the same mass as a mole of oxygen gas is the same as saying that an apple has the same mass as a dozen apples.

MOLE RELATIONSHIPS

Since the mole is used often in chemistry to quantify the number of atoms, molecules, and several other items, it is important to know the relationships that exist and how to move from one to another. The following summarizes a number of these relationships:

When dealing with elements—
Moles of an element × molar mass (atomic) = mass of the element
Mass of an element/molar mass (atomic) = moles of the element
When dealing with compounds—
Moles of a compound × molar mass (molecular) = mass of the compound
Mass of a compound / molar mass (molecular) = moles of the compound
When dealing with the molecules of a compound—
Moles of molecules × 6.02 × 10^{23} = number of molecules
Number of molecules / 6.02 × 10^{23} = moles of molecules
When dealing with the atoms of elements—
Moles of atoms × 6.02 × 10^{23} = number of atoms
Number of atoms / 6.02 × 10^{23} = moles of atoms

GAS VOLUMES AND MOLECULAR WEIGHT

Because the volume of a gas may vary depending on the conditions of temperature and pressure, a standard is set for comparing gases. The standard conditions of temperature and pressure (abbreviated STP) are 0°C and 760 mm of mercury pressure.

The molecular weight of a gas expressed in grams and under standard conditions occupies 22.4 liters. This is an important relationship to remember! The 22.4 liters is referred to as the gram-molecular volume (gmv). Two men are associated with this relationship. Gay-Lussac's Law states that, when only gases are involved in a reaction, the volumes of the reacting gases and the volumes of the gaseous products are in ratio to each other as small whole numbers. This law may be illustrated by the following cases:

1 vol. hydrogen + 1 vol. chlorine = 2 vols. hydrogen chloride
2 vols. hydrogen + 1 vol. oxygen = 2 vols. steam

Avogadro's Law, which explains Gay-Lussac's, states that equal volumes of gases under the same conditions of temperature and pressure contain equal number of molecules. This means that 1 mole of any gas at STP occupies 22.4 liters, so:

32 g O_2 at STP occupies 22.4 liters
2 g H_2 at STP occupies 22.4 liters
44 g CO_2 at STP occupies 22.4 liters
2 O_2 (2 moles O_2) = 64 g = 44.8 liters
3 H_2 (3 moles H_2) = 6 g = 67.2 liters

DENSITY AND MOLECULAR WEIGHT

Since the density of a gas is usually given in grams/liter of a gas at STP, we can use the gram-molecular volume to gram-molecular weight relationship to solve the following type problems:

- **EXAMPLE:**

    Find the gram-molecular weight of a gas when the density is given as 1.25 g/L.

Now that it is known that 1 mole of a gas occupies 22.4 liters at STP, we can solve this problem by multiplying the weight of 1 liter by 22.4 liters/mole.

$$\frac{1.25 \text{ g}}{\text{liter}} \times \frac{22.4 \text{ liters}}{1 \text{ mole*}} = 28 \text{ g/mole}$$

Even if the weight given is not for 1 liter, the same setup can be used.

- **PROBLEM:**

    If 3 liters of a gas weigh 2 grams, find the molecular weight.

    Solution:

$$\frac{2 \text{ grams}}{3 \text{ liters}} \times \frac{22.4 \text{ liters}}{1 \text{ mole}} = 14.9 \text{ g/mole}$$

You can also find the density of a gas if you know the molecular weight. Since the molecular weight occupies 22.4 liters at STP, dividing the molecular weight by 22.4 liters will give you the weight per liter or density.

- **EXAMPLE:**

    Find the density of oxygen at STP.

    Solution:

    Oxygen is diatomic in its molecular structure.
    O_2 = molecular weight of 2 × 16 or 32 g/mole
    32 g/mole occupies 22.4 liters

    Therefore

$$\frac{32 \text{ g}}{1 \text{ mole}} \div \frac{22.4 \text{ L}}{1 \text{ mole}} = \frac{32 \text{ g}}{1 \text{ mole}} \times \frac{1 \text{ mole}}{22.4 \text{ L}} = 1.43 \text{ g/L}$$

You can find the density of a gas, then, by dividing its molecular weight by 22.4 liters.

WEIGHT-VOLUME RELATIONSHIPS

A typical *weight-volume problem*: How many liters of oxygen (STP) can you prepare from the decomposition of 42.6 grams of sodium chlorate?

1st step.   Write the balanced equation for the reaction.

$$2 \text{ NaClO}_3 \overset{\Delta}{\rightarrow} 2 \text{ NaCl} + 3 \text{ O}_2 \uparrow$$

*Mole may be abbreviated as mol.

2nd step. Write the given quantities and the unknown quantity above the appropriate substances.

42.6 grams x liters

$$2 \text{ NaClO}_3 \xrightarrow{\Delta} 2 \text{ NaCl} + 3 \text{ O}_2$$

3rd step. Calculate the equation weight or volume under the substances that have something indicated above them. Note that the units above and below *must* match.

42.6 grams x liters

$2 \text{ NaClO}_3 \xrightarrow{\Delta} 2 \text{ NaCl} + 3 \text{ O}_2$

213 grams   67.2 L

(2 × molecular wt.   (3 × 22.4 L)

     of NaClO_3)

4th step. Form the proportion:

$$\frac{42.6 \text{ g}}{213 \text{ g}} = \frac{x \text{ L}}{67.2 \text{ L}}$$

5th step. Solve for x:

$$x = 13.4 \text{ L O}_2$$

This problem can also be solved using methods other than the proportion method shown above.

Another method to proceed from the 3rd step is called the *factor-label method*. The reasoning is this: Since the equation shows that 213 grams of reactant produces 67.2 liters of the required product, multiplying the given amount by this equality (so that the units of the answer are correct) will give the same answer as above. So step 4 would be:

4th step. $42.6 \text{ grams NaClO}_3 \times \dfrac{67.2 \text{ L O}_2}{213 \text{ grams NaClO}_3} = 13.4 \text{ L O}_2$

Still another method of solving this problem is called the *mole method*. Steps 1 and 2 are the same. Then step 3 is as follows:

3rd step. Determine how many moles of substance are given.

$$42.6 \text{ g} \div \frac{106.5 \text{ grams}}{1 \text{ mole NaClO}_3} = .4 \text{ mole NaClO}_3$$

The equation shows that 2 moles NaClO_3 makes 3 moles O_2. So .4 mole NaClO_3 will yield

$$.4 \text{ mole NaClO}_3 \times \frac{3 \text{ moles O}_2}{2 \text{ moles NaClO}_3} = .6 \text{ mole O}_2$$

4th step. Convert the moles of O_2 to liters.

$$.6 \text{ mole O}_2 \times \frac{22.4 \text{ L O}_2}{1 \text{ mole O}_2} = 13.4 \text{ liters O}_2$$

• ANOTHER EXAMPLE:

Find the weight of $CaCO_3$ needed to produce 11.2 liters of CO_2 when the calcium carbonate is reacted with hydrochloric acid.

1st step. $CaCO_3 + 2\ HCl \rightarrow CaCl_2 + H_2O + CO_2$

$\quad\quad\quad\quad x\ \text{g}\quad\quad\quad\quad\quad\quad\quad\quad\quad\quad\quad\quad 11.2\ \text{L}$

2nd step. $CaCO_3 + 2\ HCl \rightarrow CaCl_2 + H_2O + CO_2$

$\quad\quad\quad\quad x\ \text{g}\quad\quad\quad\quad\quad\quad\quad\quad\quad\quad\quad\quad 11.2\ \text{L}$

3rd step. $CaCO_3 + 2\ HCl \rightarrow CaCl_2 + H_2O + CO_2$

$\quad\quad\quad\quad 100\ \text{g}\quad\quad\quad\quad\quad\quad\quad\quad\quad\quad\quad 22.4\ \text{L}$

• PROPORTION METHOD:

4th step. $\dfrac{x\ \text{g}}{100\ \text{g}} = \dfrac{11.2\ \text{L}}{22.4\ \text{L}}$

5th step. $x = 50\ \text{g}\ CaCO_3$

• FACTOR-LABEL METHOD:

4th step. $11.2\ \text{L}\ CO_2 \times \dfrac{100\ \text{g}\ CaCO_3}{22.4\ \text{L}\ CO_2} = 50\ \text{g}\ CaCO_3$

• MOLE METHOD

$$11.2\ \text{L}\ CO_2 \div \dfrac{22.4\ \text{L}}{1\ \text{mole}} = \tfrac{1}{2}\ \text{mole}\ Co_2$$

Equation shows 1 mole $CaCO_3$ yields 1 mole CO_2, so

$$\tfrac{1}{2}\ \text{mole}\ CO_2 \times \dfrac{1\ \text{mole}\ CaCO_3}{1\ \text{mole}\ CO_2} = \tfrac{1}{2}\ \text{mole}\ CaCO_3$$

Converting moles to grams:

$$\tfrac{1}{2}\ \text{mole}\ CaCO_3 \times \dfrac{100\ \text{g}\ CaCO_3}{1\ \text{mole}\ CaCO_3} = 50\ \text{g}\ CaCO_3$$

• TRY THESE PROBLEMS:

1. What weight of water must be electrolyzed to obtain 20 liters of oxygen?

Ans. 32.2 g*

2. How many grams of aluminum will be completely oxidized by 44.8 liters of oxygen?

Ans. 72 g Al*

WEIGHT-WEIGHT PROBLEMS

A typical problem concerning just weight relationships is as follows:

What weight of oxygen can be obtained from heating 100 grams of potassium chlorate?

1st step.   Write the balanced equation for the reaction.

$$2\ KClO_3 \rightarrow 2\ KCl + 3\ O_2$$

*Answer explained in section at the end of the book.

2nd step. Write the given quantities and the unknown quantity above the appropriate substances.

$$100 \text{ g} \qquad x \text{ g}$$
$$2 \text{ KClO}_3 \rightarrow 2 \text{ KCl} + 3 \text{ O}_2$$

3rd step. Calculate the equation weight under the substances that have something indicated above them. Note the units above and below *must* match.

$$100 \text{ g} \qquad x \text{ g}$$
$$2 \text{ KClO}_3 \rightarrow 2 \text{ KCl} + 3 \text{ O}_2$$
$$245 \text{ g} \qquad 96 \text{ g}$$

• USING THE PROPORTION METHOD:

4th step. Form the proportion.

$$\frac{100 \text{ g}}{245 \text{ g}} = \frac{x \text{ g}}{96 \text{ g}}$$

5th step. Solve for x.

$$x = 39.3 \text{ g of O}_2$$

• USING THE FACTOR-LABEL METHOD:

From the 3rd step on you would proceed as follows:

4th step. The equation indicates that 245 g $KClO_3$ yields 96 g of O_2. Therefore multiplying the given quantity by a factor made up of these two quantities arranged appropriately so that the units of the answer remain uncancelled will give the answer to the problem.

$$100 \text{ g KClO}_3 \times \frac{96 \text{ g O}_2}{245 \text{ g KClO}_3} = 39.3 \text{ g O}_2$$

• USING THE MOLE METHOD:

To solve this problem you would proceed as follows:

1st and
2nd steps. Same as above.

3rd step. Determine how many moles of substance are given.

$$100 \text{ g} \times \frac{1 \text{ mole KClO}_3}{122.5 \text{ g KClO}_3} = .815 \text{ mole KClO}_3$$

The equation shows that 2 moles of $KClO_3$ yields 3 moles of O_2. So .815 mole $KClO_3$ will yield

$$.815 \text{ mole KClO}_3 \times \frac{3 \text{ moles O}_2}{2 \text{ moles KClO}_3} = 1.22 \text{ moles O}_2$$

4th step.   Convert the moles of O_2 to grams of O_2.

$$1.22 \text{ moles } O_2 \times \frac{32 \text{ g } O_2}{1 \text{ mole } O_2} = 39 \text{ g } O_2$$

• ANOTHER EXAMPLE:

What weight of potassium hydroxide is needed to neutralize 20 grams of sulfuric acid?

1st step. $2 \text{ KOH} + H_2SO_4 \rightarrow K_2SO_4 + 2 H_2O$

$\quad\quad\quad\quad\quad x$ g $\quad$ 20 g

2nd step. $2 \text{ KOH} + H_2SO_4 \rightarrow K_2SO_4 + 2 H_2O$

$\quad\quad\quad\quad\quad x$ g $\quad$ 20 g

3rd step. $2 \text{ KOH} + H_2SO_4 \rightarrow K_2SO_4 + 2 H_2O$

$\quad\quad\quad\quad$ 112 g $\quad$ 98 g

• USING THE PROPORTION METHOD:

4th step. $\dfrac{x \text{ g}}{112 \text{ g}} = \dfrac{20 \text{ g}}{98 \text{ g}}$

5th step. $x = 22.8$ g KOH

• USING THE FACTOR-LABEL METHOD:

4th step. $20 \text{ g } H_2SO_4 \times \dfrac{112 \text{ g KOH}}{98 \text{ g } H_2SO_4} = 22.8$ g KOH

3rd step. $20 \text{ g } H_2SO_4 \times \dfrac{1 \text{ mole } H_2SO_4}{98 \text{ g } H_2SO_4} = .204 \text{ mole } H_2SO_4$

$.204 \text{ mole } H_2SO_4 \times \dfrac{2 \text{ mole KOH}}{1 \text{ mole } H_2SO_4} = .408 \text{ mole KOH}.$

4th step. $.408 \text{ mole KOH} \times \dfrac{56 \text{ g KOH}}{1 \text{ mole KOH}} = 22.8$ g KOH

• TRY THESE PROBLEMS:

1. What weight of manganese dioxide is needed to react with an excess of hydrochloric acid so that 100 g of chlorine is liberated?   Ans.: 122.5 g*
2. A 20-g sample of Mg is burned in 20 g of O_2. How much MgO is formed? (Hint: Determine which is in excess.)   Ans.: 33.3 g MgO*

VOLUME-VOLUME PROBLEMS

This type of problem involves only volume units and therefore can make use of Gay-Lussac's Law: "When gases combine, they combine in simple whole number ratios." These simple numbers are the coefficients of the equation.

*Answer explained in section at the end of the book.

A typical problem concerning just volume relationships is as follows:

What volume of NH_3 is produced when 22.4 L of nitrogen are made to combine with a sufficient quantity of hydrogen under the appropriate conditions?

1st step. Write the balanced equation for the reaction.

$$N_2 + 3 H_2 \rightarrow 2 NH_3$$

2nd step. Write the given quantities and the unknown quantity above the respective substance.

$$\begin{array}{cc} 22.4 \text{ L} & x \text{ L} \\ N_2 + 3 H_2 & \rightarrow 2 NH_3 \end{array}$$

3rd step. Set up a proportion using the coefficient of the substances that have something indicated above them as denominators.

$$\frac{22.4 \text{ L}}{1 \text{ L}} = \frac{x \text{ L}}{2 \text{ L}}$$

4th step. Solve for x.

$$x = 44.8 \text{ L}$$

• **USING THE FACTOR-LABEL METHOD:**

From the 2nd step on you would proceed as follows:

3rd step. The equation shows that 1 volume of N_2 will yield 2 volumes of NH_3. Therefore multiplying the given quantity by a factor made up of these two quantities appropriately arranged so that the units of the answer remain uncancelled will solve the problem.

$$22.4 \; \cancel{\text{L N}_2} \times \frac{2 \text{ L NH}_3}{1 \cancel{\text{L N}_2}} = 44.8 \text{ L NH}_3$$

• **USING THE MOLE METHOD:**

To solve this problem you would proceed as follows:

3rd step. The given quantity is converted to moles.

$$22.4 \cancel{\text{L N}_2} \times \frac{1 \text{ mole N}_2}{22.4 \cancel{\text{L N}_2}} = 1 \text{ mole N}_2$$

The equation shows that 1 mole of N_2 will yield 2 moles of NH_3. Therefore using this relationship will yield

$$1 \cancel{\text{mole N}_2} \times \frac{2 \text{ moles NH}_3}{1 \cancel{\text{mole N}_2}} = 2 \text{ moles NH}_3$$

Convert the moles of NH_3 to liters of NH_3

$$2 \cancel{\text{moles NH}_3} \times \frac{22.4 \text{ L NH}_3}{1 \cancel{\text{mole NH}_3}} = 44.8 \text{ L NH}_3$$

• ANOTHER EXAMPLE:

What volume of SO_2 will result from the complete burning of pure sulfur in 8 liters of oxygen until all the oxygen is used?

1st step. $S + O_2 \rightarrow SO_2$

2nd step. $\underset{S + O_2 \rightarrow SO_2}{\overset{8\text{ L} \quad x\text{ L}}{}}$

3rd step. $\dfrac{8\text{ L}}{1\text{ L}} = \dfrac{x\text{ L}}{1\text{ L}}$

4th step. $x = 8\text{ L of } SO_2$

• USING THE FACTOR-LABEL METHOD

3rd step. $8\ \cancel{L\ O_2} \times \dfrac{1\text{ L } SO_2}{1\ \cancel{L\ O_2}} = 8\text{ L } SO_2$

• USING THE MOLE METHOD (Not Really Practical in This Case):

3rd step. $8\ \cancel{L\ O_2} \times \dfrac{1\text{ mole } O_2}{22.4\ \cancel{L\ O_2}} = .358\text{ mole } O_2$

$.358\ \cancel{\text{mole } O_2} \times \dfrac{1\text{ mole } SO_2}{1\ \cancel{\text{mole } O_2}} = .358\text{ mole } SO_2$

4th step. $.358\ \cancel{\text{mole } SO_2} \times \dfrac{22.4\text{ L } SO_2}{1\ \cancel{\text{mole } SO_2}} = 8\text{ L } SO_2$

• TRY THESE PROBLEMS:

1. How many liters of hydrogen are necessary to react with sufficient chlorine to produce 12 liters of hydrogen chloride?
   Ans. 6 L*
2. How many liters of oxygen will be needed to burn 100 liters of carbon monoxide?
   Ans. 50 L*

PROBLEMS WITH AN EXCESS OF ONE REACTANT

It will not always be true that the amounts given in a particular problem are exactly in the proportion required for the reaction to use up all of the reactants. In other words, at times there will be some of one reactant left over after the other has been used up. This is similar to the situation when two eggs are required to mix with one cup of flour in a particular recipe, and you have four eggs and four cups of flour. Since the recipe will use two eggs for every one cup of flour, four eggs can use only two cups of flour and two cups of flour will be left over.

A chemical equation is very much like a recipe. Consider the following equation:

$$Zn + 2\ HCl \rightarrow ZnCl_2 + H_2$$

If you are given 65 grams of zinc and 65 grams of HCl, how many grams of hydrogen gas can be produced? Which reactant will be left over? How many grams of this reactant will not be consumed?

*Answer explained in section at the end of the book.

1st step. Set up the problem.

$$\begin{array}{ccc} 65\text{ g} & 65\text{ g} & x\text{ g} \\ \text{Zn} + 2\text{ HCl} &\rightarrow \text{ZnCl}_2 + & \text{H}_2 \\ 65\text{ g} & 73\text{ g} & 2\text{g} \end{array}$$

The given quantities are above the equation, and the equation weights are given beneath. To solve for the grams of hydrogen gas, the limiting reactant must be determined.

2nd step. Compare the given quantities with the equation requirements.

Knowing that it takes 65 grams of zinc to react with 73 grams of hydrochloric acid, it is reasonable to surmise that, since there are only 65 grams of hydrochloric acid, not all of the 65 grams of zinc can be used. The limiting factor will be the amount of hydrochloric acid.

3rd step. Solve for the quantity of product that will be produced using the amount of the limiting factor.

Now that we know that the limiting factor is the amount of hydrochloric acid (65 g), the equation can be solved. This means that the proportion is set up ignoring the 65 grams of zinc.

$$\frac{65\text{ g}}{73\text{ g}} = \frac{x\text{ g}}{2\text{ g}}$$

Solving for x, we get $x = 1.78$ g or 1.8 g of hydrogen produced.

4th step. To find the number of grams of zinc that were not consumed, a separate problem is necessary. The equation would be set up as follows:

$$\begin{array}{cc} y\text{ g} & 65\text{ g} \\ \text{Zn} + 2\text{ HCl} &\rightarrow \text{ZnCl}_2 + \text{H}_2 \\ 65\text{ g} & 73\text{ g} \end{array}$$

This gives the proportion:

$$\frac{y\text{ g}}{65\text{ g}} = \frac{65\text{ g}}{73\text{ g}}$$

$y = 65 \times 65 / 73 = 57.8$ or 58 g of zinc will be used.

Subtracting 58 from the original 65 g leaves 7 g of zinc that were not consumed.

Note: After each type of problem in this chapter there are several problems to try to solve; therefore a list of review questions is not included here.

Terms You Should Know

gram-formula weight (molar mass)	mole
gram-molecular volume	STP
Avogadro's Law	
Gay-Lussac's Law	

Liquids, Solids, and Phase Changes

LIQUIDS

In a liquid, the volume of the molecules and the intermolecular forces between them are much more important than in a gas. When you consider that in a gas the molecules constitute far less than 1% of the total volume, while in the liquid state the molecules constitute 70% of the total volume, it is clear that in a liquid the forces between molecules are more important. Because of this decreased volume and increased intermolecular interaction, a liquid expands and contracts only very slightly with a change in temperature and lacks the compressibility typical of gases.

Kinetics of Liquids

Even though the volume of space between molecules has decreased and the mutual attraction forces between neighboring molecules can have great effects on the molecules, they are still in motion. This motion can be verified under a microscope when colloidal particles are suspended in a liquid. The particle's zigzag path indicates molecular motion and supports the *Kinetic Molecular Theory*.

Increases in temperature increase the average kinetic energy of molecules and the rapidity of their movement. This is shown graphically in Figure 22. The molecules in the sample of cold liquid have, on the average, less kinetic energy than those in the warmer sample. Hence the temperature reading T_1 will be less than the temperature reading T_2. If a particular molecule gains enough kinetic energy when it is near the surface of a liquid, it can overcome the attractive forces of the liquid phase and escape into the gaseous phase. This is called a *change of phase*. When fast-moving molecules with high kinetic energy escape, the average energy of the remaining molecules is lower; hence the temperature is lowered.

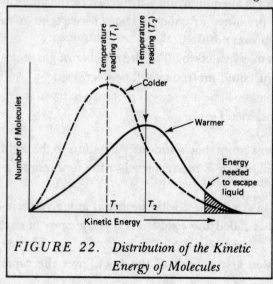

FIGURE 22. *Distribution of the Kinetic Energy of Molecules*

PHASE EQUILIBRIUM

Figure 23 shows water in a container enclosed by a bell jar. Observation of this closed system would show an initial small drop in the water level, but after some time the level would become constant. The explanation is that, at first, more energetic molecules near the surface are escaping into the gaseous phase faster than some of the gaseous water molecules are returning to the surface and possibly being caught by the attractive forces that will retain them in the liquid phase. After some time the rates of evaporation and condensation equalize. This is known as *phase equilibrium*.

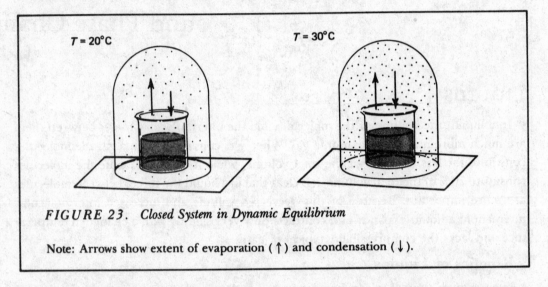

FIGURE 23. Closed System in Dynamic Equilibrium

Note: Arrows show extent of evaporation (↑) and condensation (↓).

In a closed system like this, when opposing changes are taking place at equal rates, the system is said to have *dynamic equilibrium*. At higher temperatures, since the number of molecules at higher energies increases, the number of molecules in the liquid phase will be reduced and the number of molecules in the gaseous phase will be increased. The rates of evaporation and condensation, however, will again become equal.

The behavior of the system described above illustrates what is known as *Le Châtelier's Principle*. It is stated as follows: When a system at equilibrium is disturbed by the application of a stress (change in temperature, pressure, or concentration) it reacts so as to minimize the stress, and attain a new equilibrium position.

In the discussion above, if the 20°C system is heated to 30°C, the number of gas molecules will be increased while the number of liquid molecules will be decreased:

$$\text{Heat} + H_2O(l) \rightleftarrows H_2O(g)$$

The equation shifts to the right (any similar system that is endothermic shifts to the right when temperature is increased) until equilibrium is reestablished at the new temperature.

The molecules in the vapor which are in equilibrium with the liquid at a given temperature exert a constant pressure. This is called the *equilibrium vapor pressure* at that temperature.

Chart Ⓝ in the Reference Tables section at the back of this book gives the vapor pressure of water at various temperatures.

BOILING POINT

The vapor pressure-temperature relation can be plotted on a graph for a closed system. (See Figure 24.) When a liquid is heated in an open container, the liquid and vapor are not in equilibrium and the vapor pressure increases until it becomes equal to the pressure above the liquid. At this point the average kinetic energy of the molecules is such that they are rapidly converted from the liquid to the vapor phase within the liquid as well as at the surface. The temperature at which this occurs is known as the *boiling point*.

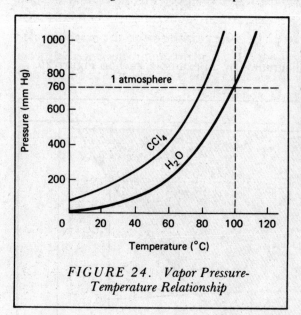

FIGURE 24. *Vapor Pressure-Temperature Relationship*

CRITICAL TEMPERATURE AND PRESSURE

There are conditions for particular substances when it is impossible for the liquid or gaseous phase to exist. Since the kinetic energy of a molecular system is directly proportional to the Kelvin temperature, it is logical to assume that there is a temperature at which the kinetic energy of the molecules is so great that the attractive forces between molecules are insufficient for the liquid phase to remain. The temperature above which the liquid phase of a substance cannot exist is called its *critical temperature*. Above its critical temperature, no gas can be liquefied regardless of the pressure applied. The minimum pressure required to liquefy a gas at its critical temperature is called its *critical pressure*.

SOLIDS

Whereas particles in gases have the highest degree of disorder, the solid state has the most ordered system. Particles are fixed in a rather definite position and maintain a definite shape. Particles in solids do vibrate in position, however, and may even diffuse through the solid. (Example: Gold clamped to lead shows diffusion of some gold atoms into the lead over long periods of time.) Other solids do not show diffusion because of strong ionic or covalent bonds in network solids. (Examples: NaCl and diamond, respectively.)

When heated at certain pressures, some solids vaporize directly without passing through the liquid phase. This is called *sublimation*. Solids like solid carbon dioxide and solid iodine exhibit this property because of unusually high vapor pressure.

The temperature at which atomic or molecular vibrations of a solid become so great that the particles break free from fixed positions and begin to slide freely over each other in a liquid state is called the *melting point*. The amount of energy required at the melting point temperature to cause the change of phase to occur is called the *heat of fusion*. The amount of this energy depends on the nature of the solid and the type of bonds present.

PHASE DIAGRAMS

The simplest way to discuss a phase diagram is by an example, such as Figure 25.

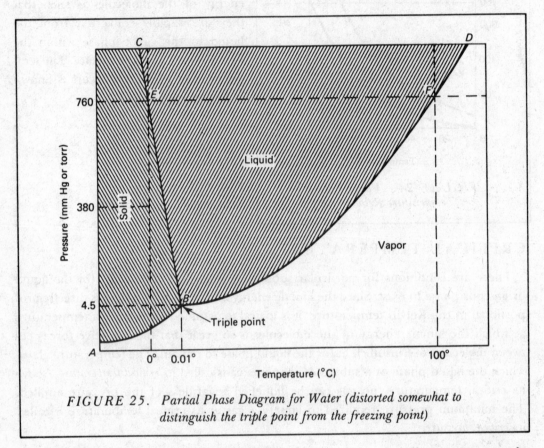

FIGURE 25. *Partial Phase Diagram for Water (distorted somewhat to distinguish the triple point from the freezing point)*

A phase diagram ties together the effects of temperature and also pressure on the phase changes of a substance. In Figure 25 the line *BD* is essentially the vapor-pressure curve for the liquid phase. Notice that at 760 mm (the pressure for 1 atmosphere) the water will boil (change to the vapor phase) at 100°C (point *F*). However, if the pressure is raised, the boiling point temperature increases; and, if the pressure is less than 760 mm, the boiling point decreases along the *BD* curve down to point *B*.

At 0°C the freezing point of water is found along the line *BC* at point *E* for pressure at 1 atmosphere or 760 mm. Again, this point is affected by pressure along the line *BC* so that, if pressure is decreased, the freezing point is slightly higher up to point *B* or .01°C.

Point *B* represents the point at which the solid, liquid, and vapor phases may all exist at equilibrium. This point is known as the *triple point*. It is the only temperature and pressure at which three phases of a pure substance can exist in equilibrium with one another in a system containing only the pure substance.

WATER

Purification of Water

Water is so often involved in chemistry that it is important to have a rather complete understanding of this compound and its properties. Pure water has become a matter of national concern. Although commercial methods of purification will not be discussed here, the usual laboratory method of obtaining pure water, distillation, will be covered.

The process of distillation involves the evaporation and condensation of the water molecules. The usual apparatus for the distillation of any liquid is shown in Figure 26.

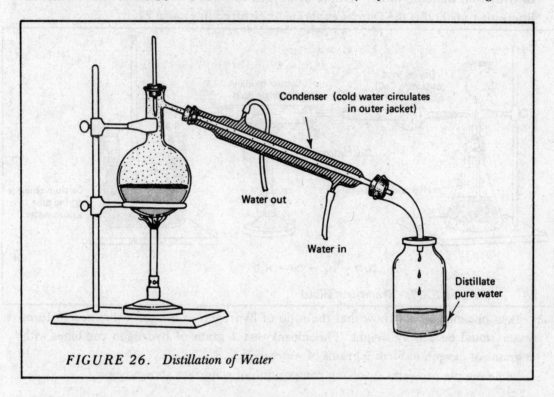

FIGURE 26. Distillation of Water

This method of purification will remove any substance that has a boiling point higher than that of water. It cannot remove dissolved gases or liquids that boil off before water. These substances will be carried over into the condenser and subsequently into the distillate.

When water contains bicarbonates of Ca, Mg, or Fe it is said to have carbonate hardness, formerly called "temporary" hardness. Such hardness may be removed by boiling since heat causes the bicarbonate to decompose to a carbonate precipitate. This can cause the detrimental boiler scales in industrial boilers if not removed.

Natural waters are also apt to contain varying amounts of calcium sulfate, magnesium sulfate, calcium chloride, and magnesium chloride. Water in which these chemicals are dissolved is said to have noncarbonate hardness, formerly known as "permanent" hardness. One way of softening water is to pass it through an artificial zeolite which replaces the "hard" positive ions (called cations) with "soft" ions of sodium. It is the "hard" cations that are responsible for the bathtub ring by reacting with the soap to form an insoluble compound. Another effective way which gives pure water like distillation is the use of organic deionizers (like Dowex, Amberlite, and Zeocarb). These are organic resins that exchange H ions for the cations, and OH ions for the anions.

Composition of Water

Water can be analyzed, that is, broken into its components, by electrolysis. This experiment shows that its composition by volume is two parts of hydrogen for one part of oxygen. Water composition can also be arrived at by synthesis. Synthesis is the formation of a compound by uniting its components to make the compound. Water can be formed by mixing hydrogen and oxygen in a eudiometer over mercury and passing an igniting spark through the mixture. Again the ratio of combination is found to be 2 parts hydrogen: 1 part oxygen. When these volumes combine in a steam-jacketed eudiometer, which keeps the water formed in the gas phase, it has been found that 2 volumes of hydrogen combine with 1 volume of oxygen to form 2 volumes of steam. Another interesting method is the Dumas experiment pictured in Figure 27.

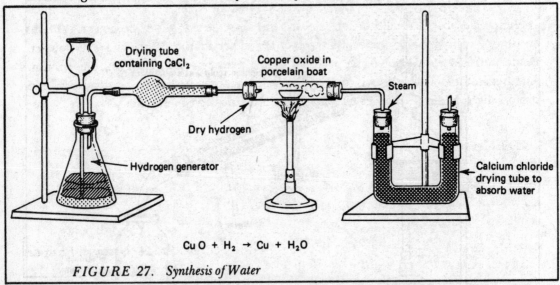

$$CuO + H_2 \rightarrow Cu + H_2O$$

FIGURE 27. *Synthesis of Water*

Data obtained would show that the ratio of hydrogen to oxygen combined to form water would be 1:8 by weight. This means that 1 gram of hydrogen combines with 8 grams of oxygen to form 9 grams of water.

Some sample problems involving composition of water are shown below.

• TYPE PROBLEM (BY WEIGHT):

An electric spark is passed through a mixture of 12 grams of hydrogen and 24 grams of oxygen in the eudiometer setup shown. Find the number of grams of water formed and the number of grams of gas left uncombined.

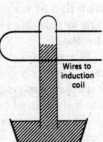

Since water forms in a ratio of 1:8 by weight, to use up the oxygen (which by inspection will be the limiting factor since it has enough hydrogen present to react completely) we need only 3 grams of hydrogen.

3 g of hydrogen + 24 g of oxygen = 27 g of water

This leaves (12 g − 3 g = 9 g) 9 g of hydrogen uncombined.

• TYPE PROBLEM (BY VOLUME):

A mixture of 8 mL of hydrogen and 200 mL of air is placed in a steam-jacketed eudiometer, and a spark is passed through the mixture. What will be the total volume of gases in the eudiometer?

Since this is a combination by volume, 8 mL of hydrogen require 4 mL of oxygen. (Ratio $H_2:O_2$ by volume = 2:1.)

The 200 mL of air is approximately 21% oxygen. This will more than supply the needed oxygen and leave 196 mL of the air uncombined.

The 8 ml of hydrogen and 4 mL of oxygen will form 8 mL of steam since the eudiometer is steam-jacketed and keeps the water formed in the gaseous state.

> (Ratio by volume of hydrogen:oxygen:steam = 2:1:2)
> TOTAL VOLUME = 196 mL of air + 8 mL of steam = 204 mL

Heavy water. A small portion of water is called "heavy" water because it contains an isotope of hydrogen, deuterium (symbol D), rather than ordinary hydrogen nuclei. Deuterium has a nucleus of one proton and one neutron rather than just one proton. Another isotope of hydrogen is tritium. Its nucleus is composed of two neutrons and one proton. Both of these isotopes have had use in the nuclear energy field.

Hydrogen peroxide. The prefix per- indicates that this compound contains more than the usual oxide. Its formula is H_2O_2. It is a well-known bleaching and oxidizing agent. Its electron-dot formula is shown in Figure 28.

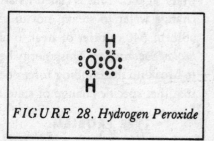

FIGURE 28. Hydrogen Peroxide

Properties and Uses of Water

Water has been used in the definition of various standards.
1. For weight—1 mL (or cc) of water at 4°C is 1 gram
2. For heat—(a) the heat needed to raise one gram of water
   one degree on the Celsius scale = 1 calorie (cal)
   1000 calories = 1 kilocalorie (kcal)
   (b) the heat needed to raise one pound of water one
   degree on the Fahrenheit scale = 1 British thermal
   unit (BTU)
3. Degree of heat—the freezing point of water = 0°C, 32°F
   —the boiling point of water = 100°C, 212°F

Water Calorimetry Problems

A calorimeter is a container well insulated from outside sources of heat or cold so that most of its heat is contained in the vessel. If a very hot object were placed in a calorimeter containing some ice crystals, we could find the final temperature of the mixture mathematically and check it experimentally. To do this, certain behaviors must be understood. Ice changing to water and then to steam is not a continuous and constant change of temperature as time progresses. In fact, the chart would look as shown in Figure 29.

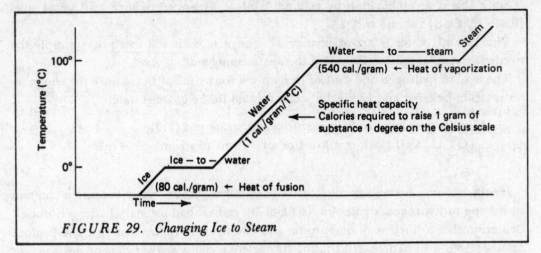

FIGURE 29. *Changing Ice to Steam*

From this graph, you see that heat is being used at 0°C and 100°C to change the state of water, but not its temperature. One gram of ice at 0°C needs 80 calories to change to water at 0°C. This is called its *heat of fusion*. Likewise, energy is being used at 100°C to change water to steam, not to change the temperature. One gram of water at 100°C absorbs 540 calories of heat to change to 1 gram of steam at 100°C. This is called its *heat of vaporization*. This energy being absorbed at the plateaus in the curve is being used to break up the bonding forces between molecules by increasing their energy content so that that specific change of state can occur.

• TYPE PROBLEM:

How much heat is needed to change 100 grams of ice at 0°C to steam at 100°C? To melt 100 g of ice at 0°C:

$$100 \text{ g} \times \frac{80 \text{ cal.}}{1 \text{ g}} = 8000 \text{ cal.} = 8 \text{ kcal.}$$

To heat 100 g of water from 0°C to 100°C:

$$100 \text{ g} \times (\overbrace{100° - 0°}^{\text{temperature change}})°C \times \frac{1 \text{ cal.}}{1 \text{ g} \times 1°C} = 10,000 \text{ cal.} = 10 \text{ kcal.}$$

To vaporize 100 g of water at 100°C to steam at 100°C:

$$100 \text{ g} \times \frac{540 \text{ cal.}}{1 \text{ g}} = 54,000 \text{ cal.} = 54 \text{ kcal.}$$

Total heat = 8 + 10 + 54 = 72 kcal.

Water's Reactions with Anhydrides

Anhydrides are certain oxides which react with water to form two classes of compounds—acids and bases.

Many metal oxides react with water to form bases such as sodium hydroxide, potassium hydroxide, and calcium hydroxide. For this reason, they are called *basic anhydrides*.

The common bases are water solutions which contain the hydroxyl (OH⁻) ion. Some common examples are:

$$Na_2O + H_2O \rightarrow 2NaOH$$
$$CaO + H_2O \rightarrow Ca(OH)_2$$

In general then: Metal oxide + H_2O → Metal hydroxide

In a similar manner, nonmetallic oxides react with water to form an acid such as sulfuric acid, carbonic acid, or phosphoric acid. For this reason, they are referred to as acid anhydrides. The common acids are water solutions containing hydrogen ions (H^+). Some common examples are:

$$CO_2 + H_2O \rightarrow H_2CO_3, \text{ carbonic acid}$$
$$SO_3 + H_2O \rightarrow H_2SO_4, \text{ sulfuric acid}$$
$$P_2O_5 + 3\ H_2O \rightarrow 2\ H_3PO_4, \text{ phosphoric acid}$$

In general, then: Nonmetallic oxides + H_2O → Acids

POLARITY AND HYDROGEN BONDING

Water is different from most liquids in that it reaches its greatest density at 4°C and then begins to expand from 4°C to 0°C (which is its freezing point). When it freezes at 0°C, its volume expands by about 9 percent. Most liquids contract as they cool and change state to a solid because their molecules have less energy, move more slowly, and are closer together. This abnormal behavior can be explained as follows. X-ray studies of ice crystal show that H_2O molecules are bound into large molecules in which each oxygen atom is connected through *hydrogen bonds* to four other oxygen atoms as shown in Figure 30.

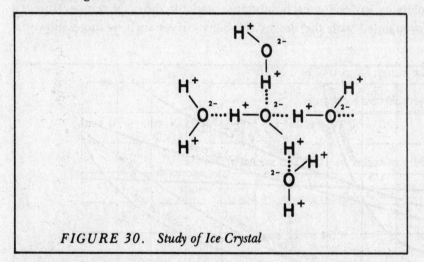

FIGURE 30. Study of Ice Crystal

This is a rather wide open structure which accounts for the low density of ice. As heat is applied and melting begins, this structure begins to collapse but not all the hydrogen bonds are broken. The collapsing increases the density but the remaining bonds keep the structure from completely collapsing. As the heat is absorbed, the kinetic energy of the molecules breaks more of these bonds as the temperature goes from 0° to 4°C. At the same time this added kinetic energy is tending to distribute the molecules farther apart. At 4°C these opposing forces are in balance—thus the greatest density. Above 4°C the increasing molecular motion again causes a decrease in density since it is the dominant force and offsets the breaking of any more hydrogen bonds.

This behavior of water can be explained by studying the water molecule itself. The water molecule is composed of two hydrogen atoms bonded by a polar covalent bond to one oxygen atom.

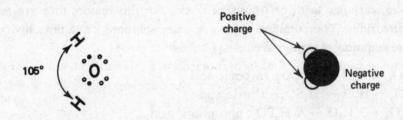

Because of the polar nature of the bonds, the molecule exhibits the charges as shown in the above drawing. It is this polar charge that causes the polar bonding previously discussed as the hydrogen bond. This bonding is stronger than the usual molecular attraction called van der Waals forces.

SOLUBILITY

Water is often referred to as "the universal solvent" because of the number of common substances that dissolve in water. When substances are dissolved in water to the extent that no more will dissolve at that temperature, the solution is said to be *saturated*. The substance dissolved is called the *solute* and the dissolving medium is called a *solvent*. To give an accurate statement of a substance's solubility, three conditions are mentioned: the amount of solute, the amount of solvent, and the temperature of the solution. Since the solubility varies for each substance and for different temperatures, a student must be acquainted with the use of solubility curves such as those shown in Figure 31.

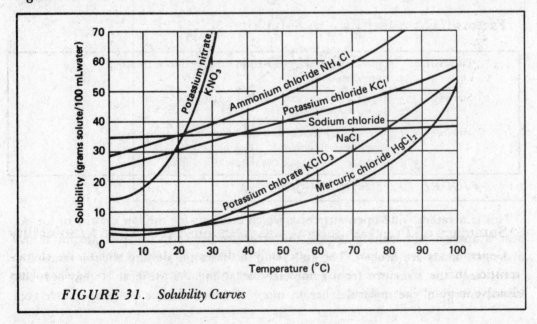

FIGURE 31. Solubility Curves

• TYPE PROBLEM USING THE SOLUBILITY CURVE:

A solution contains 20 grams of $KClO_3$ in 200 grams of H_2O at 80°C. How many more grams of $KClO_3$ can be dissolved to saturate the solution at 90°C?

Reading the graph at 90° and up to the graph line for $KClO_3$, you find that 100 grams of H_2O can dissolve 48 grams. So 200 grams could hold (2×48) grams or 96 grams. Therefore 96 g − 20 g = 76 g $KClO_3$ can be added to the solution.

General Rules of Solubility

All nitrates, acetates, and chlorates are soluble.

All common compounds of sodium, potassium, and ammonium are soluble.

All chlorides are soluble except those of silver, mercury(ious), and lead. (Lead chloride is noticeably soluble in hot water.)

All sulfates are soluble except those of lead, barium, strontium, and calcium. (Calcium sulfate is slightly soluble.)

The normal carbonates, phosphates, silicates, and sulfides are insoluble except those of sodium, potassium, and ammonium.

All hydroxides are insoluble except those of sodium, potassium, ammonium, calcium, barium, and strontium.

Some general trends of solubility are shown in the chart below.

	Temperature Effect	*Pressure Effect*
Solid	Solubility usually increases with temperature increase	Little effect
Gas	Solubility usually decreases with temperature increase	Solubility varies in direct proportion to the pressure applied to it. *Henry's Law*

Factors That Affect Rate of Solubility

Pulverizing	increases surface exposed to solvent, thus increasing rate of solubility
Stirring	brings more solvent which is unsaturated into contact with solute
Heating	increases molecular action and gives rise to mixing by convection currents. (This heating affects the solubility as well as the rate of solubility.)

Summary of Types of Solutes and Relationship of Type to Solubility

Generally speaking, solutes are most likely to dissolve in solvents with similar characteristics, that is, ionic and polar solutes dissolve in polar solvents and nonpolar solutes dissolve in nonpolar solvents.

It should also be mentioned that polar molecules that do not ionize in aqueous solution (e.g., sugar, alcohol, glycerol) have molecules as solute particles; polar molecules that partially ionize in aqueous solution (e.g., ammonia, acetic acid) have a mixture of molecules and ions as solute particles; and polar molecules that completely ionize in aqueous solution (e.g., hydrogen chloride, hydrogen iodide) have ions as solute particles.

WATER SOLUTIONS

Water molecules must overcome the forces which hold molecules or ions together to make them go into solution. This mechanism of the actual process is complex. To make sugar molecules go into solution, the water molecules must overcome the forces holding the solid sugar molecule together. The water molecules cluster around the sugar molecules, pull them off, and disperse, forming the solution.

For an ionic crystal such as salt, the water molecules orient themselves around the ions (which are charged particles) and again must overcome the forces holding the ions together. Since the water molecule is polar, this orientation around the ion is an attraction of the polar ends of the water molecule. For example:

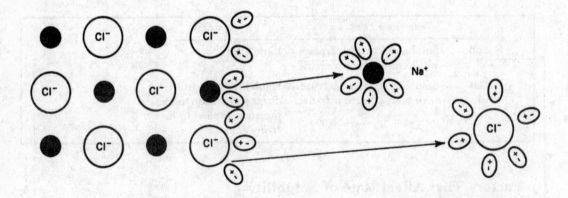

Once surrounded, the ion is insulated to an extent from other ions in solution because of the *dipole* property of water. These water molecules which surround the ion differ in number for various ions, and the whole group is called a *hydrated ion*. In general, polar substances and ions dissolve in polar solvents and nonpolar substances like fats dissolve in nonpolar solvents like gasoline. The process of going into solution may be *exothermic* if energy is released in the process, or *endothermic* if energy from the water is used up to a greater extent than the energy released in freeing the particle.

When two liquids are mixed and they dissolve in each other, they are said to be completely *miscible*. If they separate and do not mix, they are said to be *immiscible*.

Two molten metals may be mixed and allowed to cool. This gives a "solid solution" called an *alloy*.

CONTINUUM OF WATER MIXTURES

Figure 32 shows the general sizes of the particles found in a water mixture.

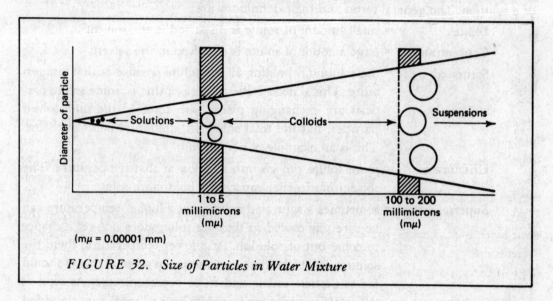

FIGURE 32. *Size of Particles in Water Mixture*

The basic difference between a colloid and a suspension is the diameter of the particles dispersed. All the boundaries marked are only the general ranges in which the distinctions between solutions, colloids, and suspensions are generally made.

The characteristics of water mixtures are as follows:

Solutions	Colloids	Suspensions
. . . . 1 mμ. . . . . 100 mμ. . . . . 1000 mμ. . . . . . . 10,000 mμ		
Clear; may have color		Cloudy; opaque color
Particles do not settle		Particles settle on standing
Pass through ordinary filter paper		Do not pass through ordinary filter paper
Pass through membranes	*Do not pass through semipermeable membranes like animal bladders, cellophane, and parchment, which have very small pores	
Not visible	Visible in ultramicroscope	Visible with microscope or naked eye
	Show Brownian movement	No Brownian movement

*Separation of a solution from a colloidal dispersion through a semipermeable membrane is called *dialysis*.

When a bright light is directed at right angles to the stage of the ultramicroscope, the individual reflections of colloidal particles can be seen. They can be observed to be following a random zigzag path. This is explained as follows: the molecules in the dispersing medium are in motion and continuously bumping into the colloidal particles causing them to change direction in a random fashion. This observation is called *Brownian movement* after the man who first observed it.

EXPRESSIONS OF CONCENTRATION

There are general terms and very specific terms used to express the concentration of a solution. The general terms and their definitions are:

Dilute	= small amount of solute is dispersed in the solvent
Concentrated	= large amount of solute is dissolved in the solvent
Saturated	= the solution is holding all the solute possible at that temperature. This is not a static condition; that is, some solute particles are exchanging places with some of the undissolved particles, but the total solute in solution remains the same. This is an example of equilibrium.
Unsaturated	= more solute can go into solution at that temperature. The solvent has further capacity to hold more solute.
Supersaturated	= sometimes a saturated solution at a higher temperature can be carefully cooled so that the solute does not get a chance to come out of solution. At a lower temperature, then, the solution will be holding more solute in solution than it should for saturation and is said to be supersaturated. As soon as the solute particles are jarred or have a "seed" particle added to the solution to act as a nucleus, they rapidly come out of the solution so that the solution reverts to the saturated state.

It is interesting to note that the words *saturated* and *concentrated* are NOT synonymous. In fact, it is possible to have a saturated dilute solution when the solute is only slightly soluble and a small amount of it makes the solution saturated, but its concentration is still dilute.

The more specific terms used to describe concentration are mathematically calculated. The term *percentage concentration* is based on the % of solute in the solution by weight. The general formula is:

$$\frac{\text{No. of grams of solute}}{\text{No. of grams of solution}} \times 100\% = \%\text{ concentration}$$

• TYPE PROBLEM:

How many grams of NaCl are needed to prepare 200 grams of a 10% salt solution?

10% of 200 grams = *20 grams* of salt

You could also solve the problem using the above formula and solving for the unknown quantity.

$$\frac{x \text{ g solute}}{200 \text{ g solution}} \times 100\% = 10\%$$

$$x \text{ g solute} = \frac{10}{100} \times 200 = 20 \text{ grams of solute}$$

The next two expressions depend on the fact that if the formula weight of a substance is expressed in grams, it is called a *gram-formula weight* (gfw) or one *mole. Gram-*

molecular weight can be used in place of gram-formula weight when the substance is really of molecular composition and not ionic like NaCl or NaOH. The definitions and examples are:

Molarity (abbreviated *M*) is defined as the number of moles of a substance dissolved in 1 liter of solution.

- **EXAMPLE:**

    A 1 molar H_2SO_4 solution would have 98 grams of H_2SO_4 (its gram-formula weight) in 1 liter of the solution.

Molality (abbreviated *m*) is defined as the number of moles of the solute dissolved in 1000 grams of solvent.

- **EXAMPLE:**

    A 1 molal solution of H_2SO_4 would have 98 grams of H_2SO_4 dissolved in 1000 grams of water. This, you will notice, gives you a total volume greater than the 1 liter, whereas the molar solution had 98 grams in 1 liter of solution.

The next expression of concentration is used less frequently and depends on the knowledge of what a *gram-equivalent weight* is. It can be defined as the amount of a substance which reacts with or displaces 1 gram of hydrogen or 8 grams of oxygen. A simple method of determining the number of equivalents in a formula is to count the number of hydrogens or find the total positive oxidation numbers since each +1 charge can be replaced by a hydrogen.

In H_2SO_4 (gram-formula weight = 98 g), there are 2 hydrogens so the gram-equivalent weight will be:

$$\frac{98 \text{ g}}{2} = 49 \text{ g}$$

In Al_2SO_4 (gfw = 150 g), there are 2 Al's, each of which has a +3 oxidation number, making a total of +6. The gram-equivalent weight will be:

$$\frac{150 \text{ g}}{6} = 25 \text{ g}$$

This idea of gram-equivalent weight is used in the definition of the next means of expressing concentration, namely the *normality* of a solution.

The *normality* (abbreviated *N*) of a solution is defined as the number of gram-equivalent weights of solute in a liter of solution.

- **EXAMPLE:**

    If 49 grams of H_2SO_4 is mixed with enough H_2O to make 500 mL of solution, what is the normality?

    Since normality is expressed for a liter, we must double the expression to 98 g of H_2SO_4 in 1000 mL of solution. To find the normality we find the number of gram-equivalents in a liter of solution. So

    $$\frac{98 \text{ g (no. of grams in 1 liter of solution)}}{49 \text{ g (gram-equivalent wt.)}} = 2 \text{ } N \text{ (normal)}$$

DILUTION

Since the expression of molarity gives the quantity of solute per volume of solution, the amount of solute dissolved in a given volume of solution is equal to the product of the concentration times the volume. Hence .5 liter of 2 M solution contains

$M \times V =$ amount of solute (in moles)

$$\frac{2 \text{ moles}}{\text{liter}} \times .5 \text{ liter} = 1 \text{ mole (of solute in .5 liter)}$$

Notice that volume units must be identical.

If you dilute a solution with water, the amount or number of moles of solute present remains the same, but the concentration changes. So you can use the expression:

Before   After
$$M_1V_1 = M_2V_2$$

This expression is useful in solving problems involving dilution.

• TYPE PROBLEM:

If you wish to make 1 liter of solution that is 6 M into a 3 M solution, how much water must be added?

$$M_1V_1 = M_2V_2$$
$$6 \, M \times 1 \, L = 3 \, M \times ? \, L$$

Solving this expression:

? L = 2 liters. This is the total volume of the solution after dilution. This means that 1 liter of water had to be added to the original volume of 1 liter to get a total of 2 liters for the dilute solution volume.

An important use of the molarity concept is in the solution of *titration* problems, which are covered in Chapter 11, along with pH expressions of concentration for acids.

COLLIGATIVE PROPERTIES OF SOLUTIONS

Colligative properties are properties that depend primarily on the concentration of particles and not the type of particle. There is usually a direct relationship between the concentration of particles and the effect recorded.

The vapor pressure of an aqueous solution is always lowered by the addition of more solute. From the molecular standpoint, it is easy to see that there are fewer molecules of water per unit volume in the liquid, and therefore fewer molecules of water in the vapor phase are required to maintain equilibrium. The concentration in the vapor drops and so does the pressure that molecules exert. This is shown graphically on the next page.

Notice that the effect of this change in vapor pressure is registered in the freezing point and the boiling point. The freezing point is lowered, and the boiling point is raised, in direct proportion to the number of particles of solute present. For water solutions, the concentration expression that tells this relationship is molality (m), that is, the number of moles of solute per kilogram of solvent. For molecules that do not

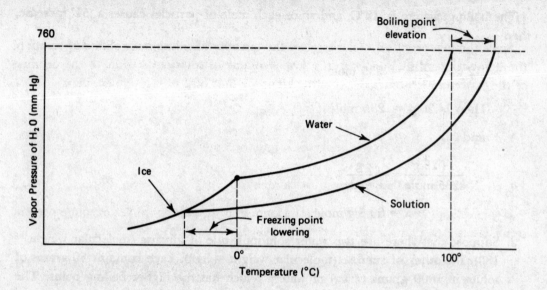

dissociate, it has been found that a 1 m solution freezes at −1.86°C and boils at 100.51°C. A 2 m solution would then freeze at twice this lowering, or −3.72°C, and boil at twice the 1 molal increase of .51°C, or 101.02°C.

The chart below summarizes the colligative effect for aqueous solutions.

Type	Concen-tration (m)	Examples	Moles of Particles	Freezing Point (°C)	Boiling Point (°C)
Molecular nonioniz-ing	1	Sugar, urea	1	−1.86	100.51
Molecular completely ionized	1	HCl	2	−3.72	101.02
Ionic com-pletely dissociated	1	NaCl	2	−3.72	101.02
Ionic com-pletely dissociated	1	$CaCl_2$ $Cu(NO_3)_2$	3	−5.58	101.53

• SAMPLE PROBLEMS:

1. A 1.50-g sample of urea is dissolved in 105.0 g of water and produces a solution that boils at 100.12°C. What is the molecular mass of urea from the data?

Since this property is related to the molality, then

$$\frac{1.5 \text{ g}}{105.0 \text{ g}} = \frac{x \text{ g}}{1000 \text{ g}}$$

$$x = 14.28 \text{ g in } 1000 \text{ g of water}$$

The boiling change is .12°C, and since each mole of particles causes a .51° increase, then

$$.12 \, g \div \frac{.51 \, g}{mole} = .235 \text{ mole}$$

Then 14.28 g = .235 mole

and

$$\frac{14.28 \text{ g}}{.235 \text{ mole}} = \frac{x \text{ g}}{1 \text{ mole}}$$

$$x = 60.8 \text{ g/mole}$$

2. Suppose that there are two water solutions, one of glucose (molecular weight = 180), the other of sucrose (molecular weight = 342). Each contains 50 grams of solute in 1000 grams (1 kg) of water. Which has the higher boiling point? The lower freezing point?

The molality of both these nonionizing substances is found by dividing the number of grams of solute by the molecular weight.

$$\text{Glucose:} \quad \frac{50 \text{ g}}{180 \text{ g/mole}} = .278 \text{ mole}$$

$$\text{Sucrose:} \quad \frac{50 \text{ g}}{342 \text{ g/mole}} = .146 \text{ mole}$$

Since the freezing point and boiling point are colligative properties, the effect depends only on the concentration. Because the glucose has a higher concentration, it will have a higher boiling point and a lower freezing point.

CRYSTALLIZATION

Many substances form a repeated pattern structure as they come out of solution. The structure is bounded by plane surfaces that make definite angles with each other to form a geometric form. This is called a crystal. The smallest portion of the crystal lattice that is repeated throughout the crystal is called the unit cell. The kinds of unit cells are shown in Figure 33.

The crystal structure can also be classified by its internal axis, as shown in Figure 34.

Crystals that hold a definite proportion of water in their crystal structure are called *hydrates*. Their formulas indicate this water in the following manner: $CuSO_4 \cdot 5 \, H_2O$; $CaSO_4 \cdot 2 \, H_2O$; and $Na_2CO_3 \cdot 10 \, H_2O$. (The · is read as "with.") When these crystals are heated gently, the water of hydration can be forced out of the crystal and the structure collapses into an anhydrous (without water) powder. The dehydration of hydrated $CuSO_4$ serves as a good example of this since the hydrated crystals are a deep blue because of the water molecules present with the copper ions. When this water is removed, the structure crumbles into the anhydrous white powder. Some hydrated crystals, such as magnesium sulfate (Epsom salt), lose the water of hydration on exposure to air at

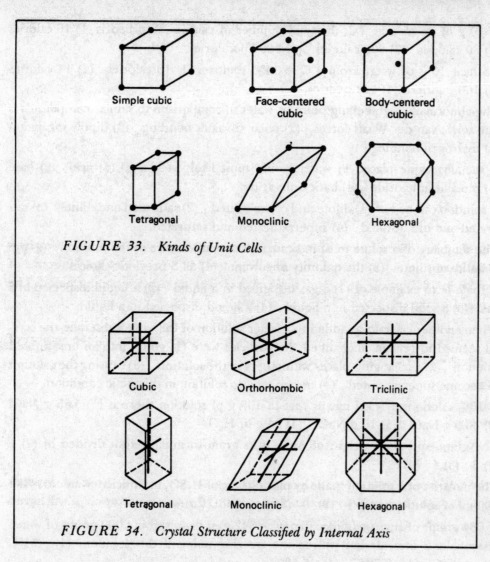

FIGURE 33. *Kinds of Unit Cells*

FIGURE 34. *Crystal Structure Classified by Internal Axis*

ordinary temperatures. They are said to be *efflorescent*. Other hydrates, such as magnesium chloride and calcium chloride, absorb water from the air and become wet. They are said to be *deliquescent*, or *hygroscopic*. This is why calcium chloride is often used as a drying agent in laboratory experiments.

Chapter

7

Review

1. Distillation of water cannot remove (1) volatile liquids like alcohol (2) dissolved salts (3) suspensions (4) precipitates.

2. The ratio in water of hydrogen to oxygen by weight is (1) 1:9 (2) 2:1 (3) 1:2 (4) 1:8.

3. Decomposing water by an electric current will give a ratio of hydrogen to oxygen by volume equal to (1) 1:9 (2) 2:1 (3) 1:2 (4) 1:8.

4. If 10 g of ice melt at 0°C, the total number of calories absorbed is (1) 10 calories (2) 80 calories (3) 800 calories (4) 800 kilocalories.

5. To heat 10 g of water from 4°C to 14°C requires (1) 10 calories (2) 14 calories (3) 100 calories (4) 140 calories.

6. The abnormally high boiling point of water in comparison to similar compounds is due to (1) van der Waals forces (2) polar covalent bonding (3) dipole insulation (4) hydrogen bonding.

7. A metallic oxide placed in water would most likely yield a(n) (1) acid (2) base (3) metallic anhydride (4) basic anhydride.

8. A solution can be *both* (1) dilute and concentrated (2) saturated and dilute (3) saturated and unsaturated (4) supersaturated and saturated.

9. The solubility of a solute must indicate (1) the temperature of the solution (2) the quantity of solute (3) the quantity of solvent (4) all 3 previous choices.

10. A foam is an example of (1) a gas dispersed in a liquid (2) a liquid dispersed in a gas (3) a solid dispersed in a liquid (4) a liquid dispersed in a liquid.

11. When another crystal was added to a water solution of the same substance, the crystal seemed to remain unchanged. Its particles were (1) going into an unsaturated solution (2) exchanging places with others in the solution (3) causing the solution to become supersaturated (4) not going into solution in this static condition.

12. A 10% solution of NaCl means that in 100 g of solution there is (1) 5.85 g NaCl (2) 58.5 g NaCl (3) 10 g NaCl (4) 94 g of H_2O.

13. The gram-equivalent weight of $AlCl_3$ is its gram-formula weight divided by (1) 1 (2) 3 (3) 4 (4) 6.

14. The molarity of a solution made by placing 98 g of H_2SO_4 in sufficient water to make 500 ml of solution is (1) $\frac{1}{2}$ (2) 1 (3) 2 (4) 3 (5) 4.

15. If 684 grams of sucrose (molecular wt. = 342 g) is dissolved in 2000 grams of water (essentially 2 liters), what will be the freezing point of this solution? (1) −.51°C (2) −1.86°C (3) −3.72°C (4) −6.58°C

• **ANSWERS:**

1. (1)	*4. (3)	7. (2)	10. (1)	13. (2)
2. (4)	*5. (3)	8. (2)	11. (2)	*14. (3)
3. (2)	6. (4)	9. (4)	*12. (3)	*15. (2)

Terms You Should Know

anhydride	endothermic	molarity
boiling point	exothermic	normality
Brownian movement	heat of fusion	phase equilibrium
colligative property	heat of vaporization	polarity
critical pressure	"heavy" water	solute
critical temperature	hydrate	solvent
crystal	melting point	sublimation
dynamic equilibrium	molality	van der Waals forces

*Answer explained in section at the end of the book.

Chemical Reactions

TYPES OF REACTIONS

The many kinds of reactions you may encounter can be placed into categories. This leads to four basic kinds of reactions: combination, decomposition, single replacement, and double replacement.

The first type, *combination*, can also be called *synthesis*. This means the formation of a compound from the union of its elements. Some examples of this type are:

$$Zn + S \rightarrow ZnS$$
$$2\,H_2 + O_2 \rightarrow 2\,H_2O$$
$$C + O_2 \rightarrow CO_2$$

The second type of reaction, *decomposition*, can also be referred to as *analysis*. This means the breakdown of a compound to release its components as individual elements or compound. Some examples of this type are:

$$2\,H_2O \rightarrow 2\,H_2 + O_2 \text{ (electrolysis of water)}$$
$$C_{12}H_{22}O_{11} \rightarrow 12\,C + 11\,H_2O$$
$$2\,HgO \rightarrow 2\,Hg + O_2$$

The third type is called *single replacement* or *single displacement*. This type can best be shown by some examples where one substance is displacing another. Some examples are:

$$Fe + CuSO_4 \rightarrow FeSO_4 + Cu$$
$$Zn + H_2SO_4 \rightarrow ZnSO_4 + H_2\uparrow$$
$$Cu + 2\,AgNO_3 \rightarrow Cu(NO_3)_2 + 2\,Ag$$

The last type of reaction is called *double replacement* or *double displacement* because there is an actual exchange of "partners" to form new compounds. Some examples of this are:

$$AgNO_3 + NaCl \rightarrow AgCl + NaNO_3$$
$$H_2SO_4 + 2\,NaOH \rightarrow Na_2SO_4 + 2\,H_2O \text{ (neutralization)}$$
$$CaCO_3 + 2\,HCl \rightarrow H_2CO_3 + CaCl_2$$
$$\qquad\qquad\quad \big| \text{ (unstable)}$$
$$\qquad\qquad\quad \rightarrow H_2O + CO_2\uparrow$$

PREDICTING REACTIONS

One of the most important topics of chemistry deals with the reasons why reactions take place. Taking each of the above types of reactions, let us see how a prediction can be made concerning how the reaction gets the driving force to make it occur.

1. Combination

The best source of information to predict a chemical combination is the heat of formation table. A heat of formation table gives the number of calories evolved or absorbed when a mole (gram-formula weight) of that compound is formed by the direct union of its elements. In this book a positive number indicates heat is absorbed and a negative number indicates that heat is evolved. (The opposite of this may be true in some texts.) It makes some difference whether the compounds formed are in the solid, liquid, or gaseous state. Unless otherwise indicated (g = gas, l = liquid), the compounds are in the solid state. The values given are in kilocalories (1 kilocalorie is equal to 1000 calories). One calorie is the amount of heat needed to raise the temperature of 1 gram of water 1 degree on the Celsius scale. The symbol ΔH is used to indicate the heat of formation.

If the heat of formation is a large number preceded by a minus sign, the combination is likely to occur spontaneously and the reaction is exothermic. If, on the other hand, the number is small and negative or a positive number, heat will be needed to get the reaction to go at any noticeable rate. Some examples are:

$$Zn + S \rightarrow ZnS + 48.5 \text{ kcal} \qquad \Delta H = -48.5 \text{ kcal}$$

This means that 1 mole of zinc (65 grams) reacts with 1 mole of sulfur (32 grams) to form 1 mole of zinc sulfide (97 grams) and releases 48.5 kilocalories of heat.

• SIMILARLY:

$$Mg + \tfrac{1}{2}O_2 \rightarrow MgO + 143.84 \text{ kcal} \qquad \Delta H = -143.84 \text{ kcal}$$

indicates that the formation of 1 mole of magnesium oxide requires 1 mole of magnesium and $\tfrac{1}{2}$ mole of oxygen with the release of 143.84 kilocalories of heat. Notice the use of the fractional coefficient for oxygen. If the equation had been written with the usual whole-number coefficients, 2 moles of magnesium oxide would have been released.

$$2\,Mg + O_2 \rightarrow 2\,MgO + 2(+143.84) \text{ kcal}$$

Since, by definition, the heat of formation is given for the formation of *one* mole, this latter thermal equation shows $2 \times (-143.84)$ kcal released.

• ANOTHER EXAMPLE:

$$H_2(g) + \tfrac{1}{2}O_2(g) \rightarrow H_2O(l) + 68.32 \text{ kcal} \qquad \Delta H = -68.32 \text{ kcal}$$

In combustion reactions the heat evolved when 1 mole of a substance is completely oxidized is called the *heat of combustion* of the substance. So, in this equation:

$$C(g) + O_2(g) \rightarrow CO_2(g) + 94.05 \text{ kcal} \qquad \Delta H = -94.05 \text{ kcal}$$

the ΔH is the heat of combustion of carbon. Because the energy of a system is conserved during chemical activity, this same equation could be arrived at by adding the following equations:

$$
\begin{aligned}
C(s) \; + \tfrac{1}{2}O_2(g) &\rightarrow CO(g) & \Delta H &= -26.41 \text{ kcal} \\
CO(g) + \tfrac{1}{2}O_2(g) &\rightarrow CO_2(g) & \Delta H &= -67.64 \text{ kcal} \\
\hline
C(s) \; + \; O_2(g) &\rightarrow CO_2(g) & \Delta H &= -94.05 \text{ kcal}
\end{aligned}
$$

2. Decomposition

The prediction of decomposition reactions uses the same source of information, the heat of formation table. If the heat of formation is a high exothermic (ΔH is negative) value, the compound will be difficult to decompose since this same quantity of energy must be returned to the compound. A low heat of formation indicates decomposition would not be difficult, such as the decomposition of mercuric oxide with $\Delta H = -21.68$ kcal/mole:

$$2 \; HgO \rightarrow 2 \; Hg + O_2 \; \text{(Priestley's method of preparation)}$$

A high positive heat of formation indicates extreme instability of a compound, and it can explosively decompose.

3. Single Replacement

A prediction of the feasibility of this type of reaction can be based on a comparison of the heat of formation of the original compound and that of the compound to be formed. For example, in a reaction of zinc with hydrochloric acid, the 2 moles of HCl have $\Delta H = 2 \times -22.06$ kcal.

$$Zn + 2 \; HCl \rightarrow ZnCl_2 + H_2 \uparrow \qquad \textit{Note:} \; \Delta H = 0 \; \text{for elements}$$
$$2 \times -22.06$$
$$-44.12 \; \text{kcal} \neq -99.40 \; \text{kcal}$$

and the zinc chloride has $\Delta H = -99.40$. This comparison leaves an excess of 55.28 kcal of heat given off so the reaction would occur.

In this next example, $-220.5 - (-184.0) = -36.5$ excess kcal to be given off as the reaction occurs:

$$Fe + \quad CuSO_4 \quad \rightarrow \quad FeSO_4 \quad + Cu$$
$$-184.0 \; \text{kcal} \qquad -220.5 \; \text{kcal}$$

Another simple way of predicting single replacement reactions is to check the relative positions of the two elements in the electromotive chart (Table 10, p. 140). If the element that is to replace the other in the compound is higher on the chart, the reaction will occur. If it is below, there will be no reaction.

4. Double Replacement

For double replacement reactions to go to completion, that is, proceed until the supply of one of the reactants is exhausted, one of the following conditions must be present: an insoluble precipitate is formed, a nonionizing substance is formed, or a gaseous product is given off.

To predict the formation of an insoluble precipitate, you should have some knowledge of the solubility of compounds. Table 9 gives some general solubility rules.

SOLUBLE	EXCEPT
Na⁺ NH₄⁺ K⁺ } compounds	
Acetates	
Bicarbonates	
Chlorates	
Chlorides . . . . .Ag⁺, Hg⁺, Pb (PbCl₂ sol. in hot water)	
Nitrates	
Sulfates . . . . .Ba, Ca (slight) Pb	
INSOLUBLE	
Carbonates, phosphates . . . . .Na, NH₄, K compounds	
Sulfides, hydroxides . . . . .Na, NH₄, K, Ba, Ca	

TABLE 9. Solubility of Compounds

(A table of solubilities could also be used as reference.)

An example of this type of reaction is given in its complete ionic form.

$$(K^+ + Cl^-) + (Ag^+ + NO_3^-) \rightarrow AgCl \downarrow + (K^+ + NO_3^-)$$

The silver ions combine with the chloride ions to form the insoluble precipitate, silver chloride. If it had been a reaction like this, there would merely have been a mixture of the ions

$$(K^+ + Cl^-) + (Na^+ + NO_3^-) \rightarrow K^+ + NO_3^- + Na^+ + Cl^-$$

shown in the final solution.

Another reason for a reaction of this type to go to completion is the formation of a nonionizing product such as water. This weak electrolyte keeps its component ions in molecular form and thus removes the possibility of reversing the reaction. All neutralization reactions are of this type.

$$(H^+ + Cl^-) + (Na^+ + OH^-) \rightarrow H_2O + Na^+ + Cl^-$$

This example shows the ions of the reactants, hydrochloric acid and sodium hydroxide, and the non-electrolyte product water with sodium and chloride ions in solution. Since the water does not ionize to any extent, the reverse reaction cannot occur.

The third reason for double displacement to occur is the evolution of a gaseous product. An example of this is calcium carbonate reacting with hydrochloric acid:

$$CaCO_3 + 2\,HCl \rightarrow CaCl_2 + H_2O + CO_2 \uparrow$$

Another example of a compound that evolves a gas in sodium sulfite with an acid:

$$Na_2SO_3 + 2\,HCl \rightarrow 2\,NaCl + H_2O + SO_2 \uparrow$$

In general, acids with carbonates or sulfites are good examples of this type of equation.

Hydrolysis Reactions

Hydrolysis reactions are the opposite of neutralization reactions. In hydrolysis the salt and water react to form an acid and a base. For example, if sodium chloride is placed in solution, this reaction occurs to some degree:

$$(Na^+ + Cl^-) + H_2O \rightarrow (Na^+ + OH^-) + (H^+ + Cl^-)$$

In this hydrolysis reaction the same number of hydrogen ions and of hydroxide ions is released so that the solution is neutral. However, if Na_2CO_3 is dissolved, we have:

$$(2\ Na^+ + CO_3^{2-}) + 2\ H_2O \rightarrow (2\ Na^+ + 2\ OH^-) + H_2CO_3$$

The H_2CO_3 is written together because it is a slightly ionized acid or, in other words, a weak acid. Since the hydroxide ions are free in the solution, the solution is basic. Notice that this was the salt of a strong base and a weak acid that formed a basic solution. This generalization is true for this type of salt.

If we use the salt of a strong acid and a weak base, the reaction would be:

$$(Zn^{2+} + 2\ Cl^-) + H_2O \rightarrow (2\ H^+ + 2\ Cl^-) + Zn(OH)_2$$

In this case the hydroxide ions are held in the weakly ionizing compound while the hydrogen ions are free to make the solution acidic. In general then, the salt of a strong acid and a weak base forms an acid solution by hydrolysis.

A salt of a weak acid and a weak base forms a neutral solution since neither hydrogen ion nor hydroxide ion will be present in excess.

Entropy

In many of the preceding predictions of reactions, we used the concept that reactions will occur when they result in the lowest possible energy state.

There is, however, another driving force to reactions that relates to their state of disorder or of randomness. This state of disorder is called *entropy*. A reaction is also driven, then, by this need for the greater degree of disorder. An example of this is the intermixing of gases in two connected flasks when a valve is opened to allow the two isolated gases to travel between the two flasks. Since temperature remains constant throughout the process, the total heat content cannot have changed to a lower energy level and yet the gases will become evenly distributed in the two flasks. The system has thus reached a higher degree of disorder or entropy.

ORIGIN OF CHEMICAL ENERGY

In general, all chemical reactions either liberate or absorb heat. The origin of chemical energy lies in the position and motion of atoms, molecules, and subatomic particles. The total energy possessed by a molecule is the sum of all the forms of potential and kinetic energy associated with it.

The energy changes in a reaction are due, to a large extent, to the changes in potential energy that accompany the breaking of chemical bonds in reactants to form new bonds in products.

The molecule may also have rotational, vibrational, and translational energy, along with some nuclear energy sources. All these make up the total energy of molecules. In beginning chemistry, the greatest concern in reactions is the electronic energy involved in the making and breaking of chemical bonds.

Since it is virtually impossible to measure the total energy of molecules, the energy change is usually the experimental data that we deal with in reactions. This change in quantity of energy is known as the change in *enthalpy* (heat content) of the chemical system and is symbolized by ΔH.

CHANGES IN ENTHALPY

Changes in enthalpy for exothermic and endothermic reactions can be shown graphically.

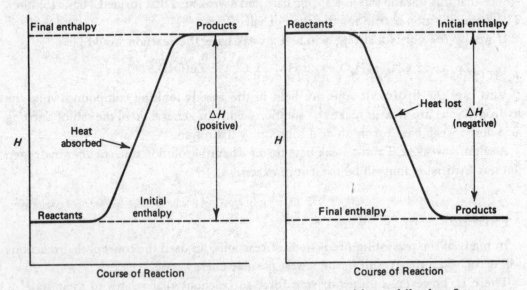

Notice that the ΔH for an endothermic reaction is positive, while that for an exothermic reaction is negative. It should be noted also that changes in enthalpy are always independent of the path taken to change a system from the initial state to the final state.

Since the quantity of heat absorbed or liberated during a reaction varies with the temperature, scientists have adopted 25°C and 1 atmosphere pressure as the *standard state* condition for reporting heat data.

To calculate the enthalpy of a reaction, it is necessary to write an equation for the reaction. The standard enthalpy change, designated by ΔH, for a given reaction is usually expressed in kilocalories and depends on how the equation is written. For example, here is the equation for the reaction of hydrogen with oxygen in two ways:

$$H_2(g) + \tfrac{1}{2} O_2(g) \rightarrow \quad H_2O(g) \quad \Delta H° = -57.8 \text{ kcal}$$
$$2 H_2(g) + O_2(g) \rightarrow \quad 2 H_2O(g) \quad \Delta H° = -115.6 \text{ kcal}$$

Experimentally, the ΔH for the formation of 1 mole of $H_2O(g)$ is −57.8 kcal. Since the second equation represents the formation of 2 moles of $H_2O(g)$, the quantity is twice −57.8 or −115.6 kcal. It is assumed that the initial and final states are measured at 25°C and 1 atmosphere, although the reaction occurs at a higher temperature.

• PROBLEM:

How much heat is liberated when 40.0 g of $H_2(g)$ reacts with excess oxygen (g)? The reaction equation is:

$$H_2(g) + \tfrac{1}{2} O_2(g) \rightarrow H_2O(g) \qquad \Delta H° = -57.8 \text{ kcal}$$

This represents 1 mole or 2 g of $H(g)$ forming 1 mole of $H_2O(g)$:

$$40 \text{ g} \times \frac{1 \text{ mole}}{2 \text{ g}} = 20 \text{ moles of hydrogen}$$

Since each mole gives off -57.8 kcal, then

$$20 \text{ moles} \times \frac{-57.8 \text{ kcal}}{1 \text{ mole}} = -1156 \text{ kcal}$$

Notice that the physical state of each participant must be given since the phase changes involve energy changes.

ADDITIVITY OF REACTION HEATS

Chemical equations and $\Delta H°$ values may be manipulated algebraically. Finding the ΔH for the formation of vapor from liquid water shows how this can be done.

$$H_2(g) + \tfrac{1}{2} O_2(g) \rightarrow H_2O(g) \qquad \Delta H° = -57.8 \text{ kcal}$$
$$H_2(g) + \tfrac{1}{2} O_2(g) \rightarrow H_2O(l) \qquad \Delta H° = -68.3 \text{ kcal}$$

Since we want the equation for $H_2O(l) \rightarrow H_2O(g)$, we can reverse the second equation. This changes the sign of ΔH.

$$H_2O(l) \rightarrow H_2(g) + \tfrac{1}{2} O_2(g) \qquad \Delta H° = 68.3 \text{ kcal}$$
Adding
$$H_2(g) + \tfrac{1}{2} O_2(g) \rightarrow H_2O(g) \qquad \Delta H° = -57.8 \text{ kcal}$$
yields
$$H_2O(l) + H_2(g) + \tfrac{1}{2} O_2(g) \rightarrow$$
$$H_2(g) + \tfrac{1}{2} O_2(g) + H_2O(g) \qquad \Delta H = 10.5 \text{ kcal}$$

Simplification gives a net equation of

$$H_2O(l) \rightarrow H_2O(g) \qquad \Delta H = 10.5 \text{ kcal}$$

The principle underlying the preceding calculations is known as *Hess's Law of Heat Summation*. This principle states that, when a reaction can be expressed as the algebraic sum of two or more other reactions, the heat of the reaction is the algebraic sum of the heats of these reactions. This is based upon the *First Law of Thermodynamics*, which, simply stated, says that the total energy of the universe is constant and cannot be created or destroyed.

This allows calculations of ΔH's that cannot be easily determined experimentally. An example is the determination of the ΔH of CO from the ΔH_r of CO_2.

$$C(s) + O_2(g) \rightarrow CO_2(g) \qquad \Delta H_r° = -94.0 \text{ kcal}$$
$$CO(g) + \tfrac{1}{2} O_2(g) \rightarrow CO_2(g) \qquad \Delta H_r° = -67.6 \text{ kcal}$$

The equation wanted is

$$C(s) + \tfrac{1}{2} O_2(g) \rightarrow CO(g)$$

To get this, we reverse the second equation and add it to the first:

$$C(s) + O_2(g) \rightarrow CO_2(g) \qquad \Delta H_r^\circ = -94.0 \text{ kcal}$$
$$CO_2(g) \rightarrow \tfrac{1}{2} O_2(g) + CO(g) \qquad \Delta H_r^\circ = +67.6 \text{ kcal}$$

Addition yields

$$C(s) + \tfrac{1}{2} O_2(g) \rightarrow CO(g) \qquad \Delta H = -26.4 \text{ kcal}$$

This relationship can be shown schematically as done below:

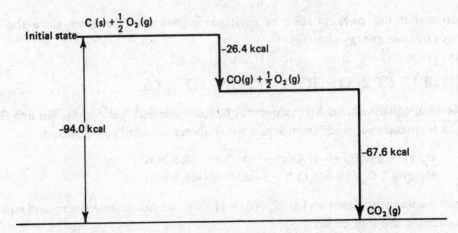

Standard heats of formation (enthalpy) are designated as ΔH_f. Some commonly used ones are listed in Table Ⓔ in the Reference Tables section.

An alternative (and easier) method of calculating enthalpies is based on the concept that ΔH is equal to the difference between the total enthalpies of the reactants and products. This can be expressed as follows:

$$\Delta H_r^\circ = \Sigma \text{ (sum of) } \Delta H_f^\circ \text{ (products)} - \Sigma \text{ (sum of) } \Delta H_f^\circ \text{ (reactants)}$$

• **PROBLEM:**

Calculate ΔH_r° for the decomposition of sodium chlorate.

$$NaClO_3(s) \rightarrow NaCl(s) + \tfrac{3}{2} O_2(g)$$

Step 1. Obtain ΔH_f° for all substances.
$NaClO_3(s) = -85.7$ kcal/mole
$NaCl(s) \quad = -98.2$ kcal/mole
$O_2(g) \quad = \quad 0$ kcal/mole (all elements = 0)

Step 2. Substitute these in the equation
$$\Delta H_r^\circ = \Sigma \Delta H_f^\circ \text{ (products)} - \Sigma \Delta H_f^\circ \text{ (reactants}$$
$$\Delta H_r = -98.2 + 85.7 \text{ kcal}$$
$$H_r = -12.5 \text{ kcal}$$

• ANOTHER EXAMPLE:

Calculate ΔH_r for this oxidation of ammonia:

$$4 NH_3(g) + 5 O_2(g) \rightarrow 6 H_2O(g) + 4 NO(g)$$

The individual ΔH_f°'s are

$$
\begin{aligned}
4 NH_3 &= 4(-11 \text{ kcal}) &&= -44.9 \text{ kcal} \\
5 O_2(g) &= 0 \\
6 H_2O(g) &= 6(-57.8 \text{ kcal}) &&= -346.8 \text{ kcal} \\
4 NO(g) &= 4(21.6 \text{ kcal}) &&= 86.4 \text{ kcal}
\end{aligned}
$$

Substituting these in the ΔH_r equation gives

$$\Delta H_r = [-346.8 \ (6 \ H_2O) + 86.4 \ (4 \ NO)] - [-44.0 \ (4 \ NH_3) + 0 \ (5 \ O_2)]$$
$$\Delta H_r = -2.16.4 \text{ kcal}$$

BOND DISSOCIATION ENERGY

The same principle of additivity applies to bond energies. Experimentation has found average bond energies for particular bonds. Some of the more common are:

Bond	Energy (kcal/mole)	Bond	Energy (kcal/mole)
H—H	104	O=O	119
C—H	99	O—H	111
C—C	83	C—O	86
C=C	146	C=O	177
C≡C	200	H—F	135
O—O	35	H—Cl	103
N—N	39	H—Br	87
Cl—Cl	58	H—I	71

• SAMPLE CALCULATION:

Find the bond energy, H_{be}, for ethane.

$$
\text{Ethane} = \begin{array}{c} \ \ \ H \ \ \ H \\ \ \ \ | \ \ \ \ | \\ H-C-C-H \\ \ \ \ | \ \ \ \ | \\ \ \ \ H \ \ \ H \end{array} \quad \text{and has 6 C-H bonds and 1 C-C bond.}
$$

$$H_{be} = 6 \ E_{C-H} + 1 \ E_{C-C} = 6(99 \text{ kcal}) + 1(83 \text{ kcal}) = 677 \text{ kcal}$$

ENTHALPY FROM BOND ENERGIES

A reaction's enthalpy can be approximated through the summation of bond energies. An example is this reaction:

$$H_2(g) + Cl_2(g) = 2 HCl(g)$$

Bonds Broken	Energy Absorbed (kcal)	Bonds Formed	Energy Evolved (kcal)
H—H	104.2		
Cl—Cl	57.8	2 H—Cl	206.0
Totals	162.0 kcal		206.0 kcal

The difference between heat evolved and heat absorbed is $206.0 - 162.0 = 44.0$ kcal. This is for 2 moles of HCl, or -44 kcal divided by $2 = -22$ kcal mole. This answer is very close to the experimentally determined ΔH_f.

Chapter 8

Review

1. A synthesis reaction will occur spontaneously if the heat of formation of the product is (1) large and negative (2) small and negative (3) large and positive (4) small and positive.

2. The reaction of aluminum with dilute H_2SO_4 can be classified as (1) synthesis (2) decomposition (3) single replacement (4) double replacement.

3. For a metal atom to replace another kind of metallic ion in a solution, the metal atom must be (1) a good oxidizing agent (2) higher in the electromotive chart than the metal in solution (3) lower in the electromotive chart than the metal in solution (4) equal in activity to the metal in solution.

4. One reason for a double displacement reaction to go to completion is that (1) a product is soluble (2) a product is given off as a gas (3) the products can react with each other (4) the products are miscible.

5. Hydrolysis will give an acid reaction when which of these is placed in solution with water? (1) Na_2SO_4 (2) K_2SO_4 (3) $NaNO_3$ (4) $Cu(NO_3)_2$

6. A salt derived from a strong base and a weak acid will undergo hydrolysis and give a solution that will be (1) basic (2) acid (3) neutral (4) volatile.

7. Enthalpy is an expression for the (1) heat content (2) energy state (3) reaction rate (4) activation energy.

8. The $\Delta H°$ of a reaction is recorded for (1) 0°C (2) 25°C (3) 100°C (4) 200°C.

9. The property of being able to add enthalpies is based on the (1) Law of Conservation of Heat (2) First Law of Thermodynamics (3) Law of Constants (4) Law of $E = mc^2$.

10. If the ΔH_r is -100 kcal/mole, it indicates the reaction is (1) endothermic (2) unstable (3) in need of a catalyst (4) exothermic.

• ANSWERS:

1. (1)	3. (2)	5. (4)	7. (1)	9. (2)
2. (3)	4. (2)	6. (1)	8. (2)	10. (4)

Terms You Should Know

combination reaction
decomposition reaction
double replacement reaction
enthalpy
First Law of Thermodynamics
Hess's Law of Heat Summation

entropy
heat of combustion
hydrolysis reaction
single replacement reaction

Rates of Chemical Reactions

MEASUREMENTS OF REACTION RATES

The measurement of reaction rate is based on the rate of appearance of a product or disappearance of a reactant. It is usually expressed in terms of a change in concentration of one of the participants per unit time.

Experiments have shown that for most reactions the concentrations of all participants change most rapidly at the beginning of the reaction, that is, the concentration of the product shows the greatest rate of increase, and the concentrations of the reactants the highest rate of decrease, at this point. This means that the rate of a reaction changes with time. Therefore a rate must be identified with a specific time.

FACTORS AFFECTING REACTION RATES

There are five important factors that control the rate of chemical reactions. These are summarized below.

1. The nature of the reactants. Some elements and compounds, because of the bonds broken or formed, react more rapidly with each other.
2. The surface area exposed. Since most reactions depend on the reactants coming into contact, the surface exposed proportionally affects the rate of the reaction.
3. The concentrations. The reaction rate is usually proportional to the concentrations of the reactants.
4. The temperature. A temperature increase of 10°C above room temperature usually causes the reaction rate to double or triple.
5. The catalyst. Catalysts can either speed up or slow down the rate of a reaction.

COLLISION THEORY OF REACTION RATES

This theory makes the assumption that, for a reaction to occur, there must be collisions between the reacting species. This means that the rate of reaction depends on two factors: the number of collisions per unit time, and the fraction of these collisions that are successful because enough energy is involved.

There is a definite relationship between the concentrations of the reactants and the number of collisions. Graphically, the reaction of

$$A + B \rightarrow AB$$

can be examined as the concentrations are changed.

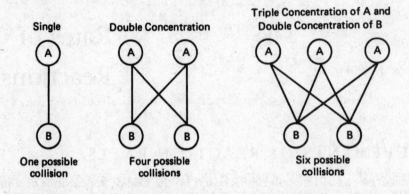

This shows that the number of collisions, and consequently the rate of reaction, are proportional to the product of the concentrations. Simply stated, then, the rate of the reaction is directly proportional to the concentrations.

ACTIVATION ENERGY

Often a reaction rate may be increased or decreased by affecting the activation energy, that is, the energy necessary to cause a reaction to occur. This is shown graphically below.

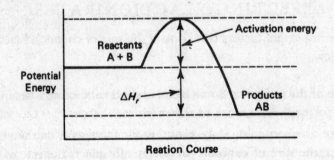

A *catalyst* is a substance that is introduced into a reaction to either speed up or slow down the reaction. This is accomplished by changing the amount of activation energy needed. If a catalyst were used to speed up a reaction, its effect could be shown like this:

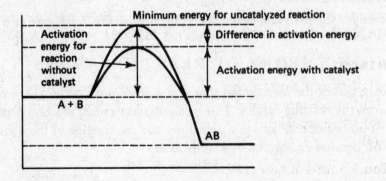

REACTION RATE LAW

The relationship between the rate of a reaction and the masses (expressed as concentrations) of the reacting substances is summarized in the *Law of Mass Action*. It states that the rate of a chemical reaction is proportional to the product of the concentrations of the reactants. For a general reaction between A and B, represented by

$$aA + bB \rightarrow \cdots$$

the rate law expression is

$$r \propto [A]^a[B]^b$$

or, inserting a constant of proportionality that mathematically changes the expression to an equality, we have

$$r = k[A]^a[B]^b$$

Here k is called the *specific rate constant* for the reaction at the temperature of the reaction.

The exponents a and b may be added to give the total reaction order. For example:

$$H_2(g) + I_2(g) \rightarrow 2\ HI(g)$$
$$r = k[H_2]^1[I_2]^1$$

The sum of the exponents is $1 + 1 = 2$, and therefore we have a second-order reaction.

Chapter

9

Review

1. List the five factors that affect the rate of a reaction:
   1.   4.
   2.   5.
   3.

2. The addition of a catalyst to a reaction (1) changes the enthalpy   (2) changes the entropy   (3) changes the nature of the products   (4) changes the activation energy.

3. An increase in concentration (1) is related to the number of collisions directly   (2) is related to the number of collisions inversely   (3) has no effect on the number of collisions.

4. At the beginning of a reaction, the reaction rate for the reactants is (1) largest, then decreasing   (2) largest and remains constant   (3) smallest and increasing   (4) smallest and remains constant.

5. The reaction rate law applied to the reaction $aA + bB \rightarrow AB$ gives the expression
   (1) $r \propto [A]^b[B]^a$ (2) $r \propto [AB]^a[A]^b$ (3) $r \propto [B]^a[AB]^b$ (4) $r \propto [A]^a[B]^b$

   • **ANSWERS:**
   1. (1) Nature of the reactants   (4) Temperature
      (2) Surface area exposed   (5) Presence of a catalyst
      (3) Concentrations
   2. (4)   3. (1)   4. (1)   5. (4)

Terms You Should Know

catalyst
Law of Mass Action

Chemical Equilibrium

REVERSIBLE REACTIONS AND EQUILIBRIUM

In some reactions no product is formed to allow the reaction to go to completion, that is, the reactants and products can still interact in both directions. This can be shown as follows:

$$A + B \rightleftarrows C + D$$

The double arrow indicates that C and D can react to form A and B, while A and B react to form C and D.

The reaction is said to have reached *equilibrium* when the forward reaction rate is equal to the reverse reaction rate. Notice this is *not* a static condition but a dynamic one. This must be pointed out because in appearance the reaction *seems* to have stopped. An example of an equilibrium is a crystal of copper sulfate in a saturated solution of copper sulfate. Although the crystal seems to remain unchanged to the observer, there is actually an equal exchange of crystal material with the copper sulfate in solution. As some solute comes out of solution, an equal amount is going into solution.

To express the rate of reaction in numerical terms, we can use the *Law of Mass Action* which states that: the rate of a chemical reaction is proportional to the product of the concentrations of the reacting substances. The concentration is expressed in moles of a gas per liter of volume or moles of solute per liter of solution. Suppose, for example, that 1 mole/liter of gas A_2 (diatomic molecule) is allowed to react with 1 mole/liter of another diatomic gas, B_2, and they form a gas AB; let R be the rate for the forward reaction forming AB. The bracketed symbols $[A_2]$ and $[B_2]$ represent the concentrations in moles per liter for these diatomic molecules. Then $A_2 + B_2 \rightarrow 2AB$ has the rate expression

$$R \propto [A_2] \times [B_2]$$

where $\propto$ is the symbol for "proportional to." When $[A_2]$ and $[B_2]$ are both 1 mole/liter, the reaction rate is a certain constant value (k_1) at a fixed temperature.

$$R = k_1 \ (k_1 \text{ is called the rate constant})$$

For any concentration of A and B, the reaction rate is

$$R = k_1 \times [A_2] \times [B_2]$$

If $[A_2]$ is 3 moles/liter and $[B_2]$ is 2 moles/liter, the equation becomes

$$R = k_1 \times 3 \times 2 = 6\,k_1$$

The reaction rate is six times the value for 1 mole/liter concentration of both reactants.

At the fixed temperature of the forward reaction, AB molecules are also decomposing. If we designate this reverse reaction as R', then, since

$$2\text{ AB (or AB + AB)} \rightarrow A_2 + B_2$$

2 molecules of AB must decompose to form a molecule of A_2 and one of B_2. Thus the reverse reaction in this equation is proportional to the square of the molecular concentration of AB.

$$R' \propto [AB] \times [AB]$$
$$\text{or}\quad R' \propto [AB]^2$$
$$\text{and}\quad R' \propto k_2 \times [AB]^2$$

where k_2 represents the rate of decomposition of AB at the fixed temperature. Both reactions can be shown in this manner:

$$A_2 + B_2 \rightleftarrows 2\text{ AB}\qquad \text{(note double arrows)}$$

When the first reaction begins to produce AB, some AB is available for the reverse reaction. If the initial condition is only the presence of A_2 and B_2 gases, then the forward reaction will occur rapidly to produce AB. As the concentration of AB increases, the reverse reaction will increase. At the same time, the concentrations of A_2 and B_2 will be decreasing and consequently the forward reaction rate will decrease. Eventually the two rates will be equal, that is, $R = R'$. At this point, equilibrium has been established, and

$$k_1[A_2] \times [B_2] = k_2[AB]^2$$

or

$$\frac{k_1}{k_2} = \frac{[AB]^2}{[A_2] \times [B_2]} = K_e$$

The convention is that k_1 (forward reaction) is placed over k_2 (reverse reaction) to get this expression. Then k_1/k_2 can be replaced by K_e, which is called the *equilibrium constant* for this reaction under the particular conditions.

In another general example, like this:

$$aA + bB \rightleftarrows cC + dD$$

the reaction rates can be expressed as

$$R = k_1[A]^a \times [B]^b$$
$$R' = k_2[C]^c \times [D]^d$$

Note that the values of k_1 and k_2 are different, but that each is a constant for the conditions of the reaction. At the start of the reaction, [A] and [B] will be at their greatest values, and R will be large; [C], [D], and R' will be zero. Gradually R will decrease and R'

will become equal. At this point the reverse reaction is forming the original reactants just as rapidly as they are being used up by the forward reaction. Therefore no further change in R, R', or any of the concentrations will occur.

If we set R' equal to R, we have:

$$k_2 \times [C]^c \times [D]^d = k_1 \times [A]^a \times [B]^b$$

or

$$\frac{[C]^c \times [D]^d}{[A]^a \times [B]^b} = \frac{k_1}{k_2} = K_e$$

We see that, for the given reaction and the given conditions, K_e is a constant. K_e is called the *equilibrium constant*. If K_e is large, it means that equilibrium does not occur until the concentrations of the original reactants are small and those of the products large. If K_e is small, it means that equilibrium occurs almost at once and relatively little product is produced.

The equilibrium constant, K_e, has been determined experimentally for many reactions and is printed in chemical handbooks. (See Tables Ⓙ and Ⓚ in the Reference Tables section.)

Suppose we find the K_e for reacting H_2 and I_2 at 490°C to be equal to 45.9. Then the equilibrium constant for this reaction

$$H_2 + I_2 \rightleftarrows 2\,HI \text{ at } 490°C$$

is

$$K_e = \frac{[HI]^2}{[H_2][I_2]} = 45.9$$

• SAMPLE PROBLEM:

Three moles of H_2 and 3 moles of I_2 are introduced into a 1-liter box at a temperature of 490°C. Find the concentration of each substance in the box when equilibrium is established.

Initial conditions:

$[H_2] = 3$ moles/liter

$[I_2]\ = 3$ moles/liter

$[HI] = 0$ mole/liter

The reaction proceeds to equilibrium and

$$K_e = \frac{[HI]^2}{[H_2][I_2]} = 45.9$$

At equilibrium, then,

$[H_2] = (3 - x)$ moles/liter (where x is the number of moles of H_2 that are in the form of HI at equilibrium)

$[I_2]\ = (3 - x)$ moles/liter (the same x is used since 1 mole of H_2 requires 1 mole of I_2 to react to form 2 moles of HI)

$[HI] = 2x$ moles/liter

so

$$K_e = \frac{(2x)^2}{(3-x)(3-x)} = 45.9$$

If

$$\frac{(2x)^2}{(3-x)^2} = 45.9$$

then taking the square root of each side gives

$$\frac{2x}{3-x} = 6.77$$

Solving for x:

$$x = 2.32$$

Substituting this x value into the concentration expressions at equilibrium we have

$[H_2] = (3 - x) = .68$ mole/liter
$[I_2] = (3 - x) = .68$ mole/liter
$[HI] = 2x = 4.64$ moles/liter

The crucial step in this type of problem is setting up your concentration expressions from your knowledge of the equation. Suppose that this problem had been:

Find the concentrations at equilibrium for the same conditions as in the preceding example except that only 2 moles of HI are injected into the box.

$[H_2] = 0$ mole/liter
$[I_2] = 0$ mole/liter
$[HI] = 2$ moles/liter

At equilibrium

$[HI] = (2 - x)$ moles/liter

(For every mole of HI that decomposes, only $\frac{1}{2}$ mole of H_2 and $\frac{1}{2}$ mole of I_2 are formed.)

$[H_2] = \frac{1}{2}x$
$[I_2] = \frac{1}{2}x$
$K_e = \dfrac{(2-x)^2}{(x/2)^2} = 45.9$

Solving for x gives:

$$x = .456$$

Then, substituting into the equilibrium conditions,

$[HI] = 2 - x = 1.54$ moles/liter
$[I_2] = \frac{1}{2}x = .228$ mole/liter
$[H_2] = \frac{1}{2}x = .228$ mole/liter

LE CHÂTELIER'S PRINCIPLE

A general law, *Le Châtelier's Principle,* can be used to explain the results of applying any change of condition (stress) on a system in equilibrium. It states that if a stress is placed upon a system in equilibrium, the equilibrium is displaced in the direction which counteracts the effect of the stress. An increase in concentration of a substance favors the reaction which uses up that substance and lowers its concentration. A rise in temperature favors the reaction which absorbs heat and so tends to lower the temperature. These ideas are further developed below.

EFFECTS OF CHANGING CONDITIONS

Effect of Changing the Concentrations

When a system at equilibrium is disturbed by adding or removing one of the substances (thus changing its concentration), all the concentrations will change until a new equilibrium point is reached with the same value of K_e.

If the concentration of a reactant in the forward action is increased, the equilibrium is displaced to the right, favoring the forward reaction. If the concentration of a reactant in the reverse reaction is increased, the equilibrium is displaced to the left. Decreases in concentration will produce effects opposite to those produced by increases.

Effect of Temperature on Equilibrium

If the temperature of a given equilibrium reaction is changed, the reaction will shift to a new equilibrium point. If the temperature of a system in equilibrium is raised, the equilibrium is shifted in the direction that absorbs heat. Note that the shift in equilibrium as a result of temperature change is actually a change in the value of the equilibrium constant. This is different from the effect of changing the concentration of a reactant; when concentrations are changed, the equilibrium shifts to a condition that maintains the same equilibrium constant.

Effect of Pressure on Equilibrium

A change in pressure affects only equilibria in which a gas or gases are reactants or products. Le Châtelier's Law can be used to predict the direction of displacement. If it is assumed that the total space in which the reaction occurs is constant, the pressure will depend on the total number of molecules in that space. An increase in number of molecules will increase pressure; a decrease in number of molecules will decrease pressure. If the pressure is increased, that reaction will be favored which lowers the pressure, i.e., which decreases the number of molecules.

An example of using these principles is the Haber process of making ammonia. The reaction is

$$N_2 + 3 H_2 \rightleftarrows 2 NH_3 + \text{heat (at equilibrium)}$$

If the concentrations of the nitrogen and hydrogen are increased, the forward reaction is increased. At the same time, if the ammonia produced is removed by dissolving it into water, the forward reaction is again favored.

Being an exothermic reaction, the addition of heat must be considered with care. Increasing the temperature causes an increase of molecular motion and collisions, thus allowing the product to form more readily. At the same time, the equilibrium equation shows that the reverse reaction is favored by this increased temperature so a compromise temperature of about 500°C is used to get the best yield.

An increase of pressure will cause the forward reaction to be favored since the equation shows that 4 molecules of reactants are forming 2 molecules of products. This tends to reduce the increase in pressure by forming more ammonia.

EQUILIBRIA IN HETEROGENEOUS SYSTEMS

The examples so far have been of systems made up of only gaseous substances. The expression of the K_e of systems is changed with the presence of other phases.

Equilibrium Constant for Systems Involving Solids

If the experimental data for this reaction are studied:

$$CaCO_3(s) \rightleftarrows CaO(s) + CO_2(g)$$

it is found that at a given temperature an equilibrium is established in which the concentration of CO_2 is constant. It is also true that the concentrations of the solids have no effect on the CO_2 concentration as long as both solids are present. Therefore the K_e, which would conventionally be written like this:

$$K_e = \frac{[CaO][CO_2]}{[CaCO_3]}$$

can be modified by incorporating the concentrations of the two solids. This can be done since the concentration of solids is fixed by the density. The K_e becomes a new constant, K:

$$K = [CO_2]$$

Any heterogeneous reaction involving gases does not include the concentrations of pure solids. As another example, K for the reaction

$$NH_4Cl(s) \rightleftarrows NH_3(g) + HCl(g)$$

is

$$K = [NH_3][HCl]$$

Ionization Constants

When substances that do not ionize completely are placed in solution, an equilibrium is reached between the substance and its ions. The mass action expression can be used to derive an equilibrium constant for this condition. It is called the *ionization constant*. For example, an acetic acid solution ionizing is shown as

$$HC_2H_3O_2 + H_2O \rightleftarrows H_3O^+ + C_2H_3O_2^-$$

$$K = \frac{[H_3O^+][C_2H_3O_2^-]}{[HC_2H_3O_2][H_2O]}$$

The concentration of water in moles/liter is found by dividing the weight of 1 liter of water (which is 1000 grams at 4°C) by its gram-molecular weight, 18 g, giving H_2O a value of 55.6 moles/liter. Since this number is so large compared to the other numbers involved in the equilibrium constant, it is practically constant and is incorporated into a new equilibrium constant, designated as K_i. Then the new expression is

$$K_i = \frac{[H_3O^+][C_2H_3O_2^-]}{[HC_2H_3O_2]}$$

Ionization constants have been found experimentally for many substances and listed in chemical tables. The ionization constants of ammonia and acetic acid are about 1.8×10^{-5}. For boric acid $K_i = 5.8 \times 10^{-10}$ and for carbonic acid $K_i = 4.3 \times 10^{-7}$.

If the concentrations of the ions present in the solution of a weak electrolyte are known, the value of the ionization constant can be calculated. Also, if the value of K_i is known, the concentrations of the ions can be calculated.

A small value for K_i means that the concentration of the un-ionized molecule must be relatively large compared to the ion concentrations. A large value for K_i means that the concentration of ions is relatively high. Therefore the smaller the ionization constant of an acid, the weaker the acid. Thus, in the cases of the three acids referred to above, the ionization constants show that the weakest of these is the boric acid, and the strongest, the acetic acid. It must be remembered that in all cases where ionization constants are used, the electrolytes must be weak in order to be involved in ionic equilibria.

Ionization Constant of Water

Since water is a very weak electrolyte, its ionization constant can be expressed as follows:

$$2\ H_2O \rightleftarrows H_3O^+ + OH^-$$

(Equilibrium constant) $K = \dfrac{[H_3O^+][OH^-]}{[H_2O]^2}$ (since $[H_2O]^2$ remains relatively constant, it is incorporated into K_w)

(Ionization constant) $K_w = [H_3O^+][OH^-] = 1 \times 10^{-14}$ at 25°C

From this expression, we see that for distilled water $[H_3O^+] = [OH^-] = 1 \times 10^{-7}$. Therefore the pH, which is $-\log [H_3O^+]$, is

pH $= -\log [1 \times 10^{-7}]$
pH $= -[-7] = 7$ for a neutral solution

The pH range of 1 to 6 is acid, and the pH range of 8 to 14 is basic.

• SAMPLE PROBLEM:

(This sample incorporates the entire discussion of ionization constants, including finding the pH.)

Calculate (a) the $[H_3O^+]$, (b) the pH, and (c) the percentage dissociation for .100 M acetic acid at 25°C. The symbol K_a is used for the ionization of acids. K_a for $HC_2H_3O_2$ is 1.8×10^{-5}.

(a) For this reaction

$$H_2O(l) + HC_2H_3O_2(l) \rightleftarrows H_3O^+(aq) + C_2H_3O_2^-(aq)$$

and

$$K_a = \frac{[H_3O^+][C_2H_3O_2^-]}{[HC_2H_3O_2]} = 1.8 \times 10^{-5}$$

Let x = number of moles/liter of $HC_2H_3O_2$ that dissociate and reach equilibrium. Then

$$[H_3O^+] = x, [C_2H_3O_2^-] = x, [HC_2H_3O_2] = .1 - x$$

Substituting in the expression for K_a gives

$$K_a = 1.8 \times 10^{-5} = \frac{(x)(x)}{.10 - x}$$

Since weak acids, like acetic, at concentrations of .01 M or greater dissociate very little, the equilibrium concentration of the acid is very nearly equal to the original concentration, that is,

$$.10 - x \cong .10.$$

Because of this, the expression can be changed to

$$1.8 \times 10^{-5} = \frac{(x)(x)}{.10}$$
$$x^2 = 1.8 \times 10^{-6}$$
$$x = 1.3 \times 10^{-3} = [H_3O^+]$$

(b) Substituting this in the pH expression gives

$$pH = -\log [H_3O^+] = -\log [1.3 \times 10^{-3}]$$
$$pH = 3 - \log 1.3$$
$$pH = 2.9$$

(c) The percentage of dissociation of the original acid may be expressed as

$$\% \text{ dissociation} = \frac{\text{moles/liter that dissociate}}{\text{original concentration}} (100)$$

$$\% \text{ dissociation} \times \frac{1.3 \times 10^{-3}}{1.0 \times 10^{-1}} (100) = 1.3\%$$

Solubility Products

A saturated solution of a substance has been defined as an equilibrium condition between the solute and its ions. For example:

$$AgCl \rightleftarrows Ag^+ + Cl^-$$

The equilibrium constant would be:

$$K = \frac{[Ag^+][Cl^-]}{[AgCl]}$$

Since the concentration of the solute remains constant for that temperature, the [AgCl] is incorporated into the K to give the K_{sp}, called the *solubility constant*:

$$K_{sp} = [Ag^+][Cl^-] = 1.2 \times 10^{-10} \text{ at } 25°C$$

This setup can be used to solve problems in which the ionic concentrations are given and the K_{sp} is to be found or the K_{sp} is given and the ionic concentrations are to be determined.

• SAMPLE PROBLEM:

By experimentation it is found that a saturated solution of $BaSO_4$ at 25°C contains 3.9×10^{-5} mole/liter of Ba^{2+} ions. Find the K_{sp} of this salt.

Since $BaSO_4$ ionizes into equal numbers of Ba^{2+} and SO_4^{2-}, the barium ion concentration will equal the sulfate ion concentration. So the solution is

$$BaSO_4 \rightleftarrows Ba^{2+} + SO_4^{2-}$$
$$K_{sp} = [Ba^{2+}][SO_4^{2-}] .$$

Therefore

$$K_{sp} = (3.9 \times 10^{-5})(3.9 \times 10^{-5}) = 1.5 \times 10^{-9}$$

• ANOTHER TYPE PROBLEM:

If the K_{sp} of radium sulfate, $RaSO_4$, is 4×10^{-11}, calculate its solubility in pure water.

Let x = moles of $RaSO_4$ that dissolve per liter of water. Then, in the saturated solution,

$$[Ra^{2+}] = x \text{ moles/liter}$$
$$[SO_4^{2-}] = x \text{ moles/liter}$$
$$RaSO_4(s) \rightleftarrows Ra^{2+} + SO_4^{2-}$$
$$[Ra^{2+}][SO_4^{2-}] = K_{sp} = 4 \times 10^{-11}$$

Let $x = [Ra^{2+}]$ and $[SO_4^{2-}]$. Then

$$(x)(x) = 4 \times 10^{-11}$$
$$x = 6 \times 10^{-6} \text{ mole/liter}$$

Thus the solubility of $RaSO_4$ is 6×10^{-6} mole per liter of water, giving a solution $6 \times 10^{-6} M$ in Ra^{2+} and $6 \times 10^{-6} M$ in SO_4^{2-}.

COMMON ION EFFECT

When a reaction has reached equilibrium, and an outside source adds more of one of the ions that is already in solution, the result is to cause the reverse reaction to occur at a faster rate and reestablish the equilibrium. This is called the *common ion effect*. For example, in this equilibrium reaction:

$$NaCl(s) \rightleftarrows Na^+ + Cl^-$$

the addition of concentrated HCl (12 M) adds H^+ and Cl^- both at a concentration of 12 M. This increases the concentration of the Cl^- and disturbs the equilibrium. The reaction will shift to the left and cause some solid NaCl to come out of solution.

The "common" ion is the one already present in an equilibrium before a substance is added that increases the concentration of the ion present. This consequently reverses the solution reaction and decreases the solubility of the original substance, as shown in the above example.

FACTORS RELATED TO THE MAGNITUDE OF K_e

Relation of Minimum Energy (Enthalpy) to Maximum Disorder (Entropy)

Some reactions are said to go to completion because the equilibrium condition is achieved when practically all the reactants have been converted to products. At the other extreme, some reactions reach equilibrium immediately with very little product being formed. These two examples are representative of very large K_e values and very small K_e values, respectively. There are essentially two driving forces that control the extent of a reaction and determine when equilibrium will be established. These are the drive to the lowest heat content, or *enthalpy*, and the drive to the greatest randomness or disorder, which is called *entropy*. Reactions with negative ΔH's (enthalpy or heat content) are exothermic, and reactions with positive ΔS's (entropy or randomness) are proceeding to greater randomness.

The Second Law of Thermodynamics states that the entropy of the universe increases for any spontaneous process. This means that the entropy of a system may increase or decrease but that, if it decreases, then the entropy of the surroundings must increase to a greater extent so that the overall change in the universe is positive. In other words,

$$\Delta S_{universe} = \Delta S_{system} + \Delta S_{surroundings}$$

The following is a list of conditions in which ΔS is positive for the system:

1. When a gas is formed from a solid, for example,

    $CaCO_3(s) \rightarrow CaO(s) + CO_2(g)$.

2. When a gas is evolved from a solution, for example,

    $Zn(s) + 2H^+(aq) \rightarrow H_2(g) + Zn^{2+}(aq)$.

3. When the number of moles of gaseous product exceeds the moles of gaseous reactant, for example,

    $2 C_2H_6(g) + 7 O_2(g) \rightarrow 4 CO_2(g) + 6 H_2O(g)$.

4. When crystals dissolve in water, for example,

    $NaCl(s) \rightarrow Na^+(aq) + Cl^-(aq)$.

Looking at specific examples, we find that in some cases endothermic reactions occur when the products provide greater randomness or positive entropy. This reaction is an example:

$$CaCO_3(s) \rightleftarrows CaO(s) + (CO_2(g)$$

The production of the gas and thus greater entropy might be expected to take this reaction almost to completion. However, this is not true because another force is hampering this reaction. It is the absorption of energy, and thus the increase in enthalpy, as the $CaCO_3$ is heated.

The equilibrium condition, then, at a particular temperature, is a compromise between the increase in entropy and the increase in enthalpy of the system.

The Haber process of making ammonia is another example of this compromise of driving forces which affect the establishment of an equilibrium. In this reaction

$$N_2(g) + 3\ H_2(g) \rightleftharpoons 2\ NH_3(g) + heat$$

the forward reaction to reach the lowest heat content and thus release energy cannot go to completion because the force to maximum randomness is driving the reverse reaction.

These factors can be combined in an equation that summarizes the change of *free energy* in a system. This is designated as ΔG. The relationship is

$$\Delta G = \Delta H - T \Delta S \qquad (T \text{ is temperature in kelvins})$$

The sign of ΔG can be used to predict the spontaneity of a reaction at constant temperature and pressure. If ΔG is negative, the reaction is (probably) spontaneous; if ΔG is positive, the reaction is improbable; and if ΔG is 0, the system is at equilibrium and there is no net reaction.

The ways in which the factors in the equation affect ΔG are shown in this table:

ΔH	ΔS	Will It Happen?	Comment
Exothermic (−)	+	Yes	No exceptions
Exothermic (−)	−	Probably	At low temperature
Endothermic (+)	+	Probably	At high temperature
Endothermic (+)	−	No	No exceptions

This drive to achieve a minimum of free energy may be interpreted as the driving force of a chemical reaction.

Chapter

10

Review

1. For the reaction $A + B \rightleftharpoons C + D$, the equilibrium constant can be expressed as:

(1) $K = \dfrac{[A][B]}{[C][D]}$ (2) $K = \dfrac{[C][B]}{[A][D]}$ (3) $K = \dfrac{[C][D]}{[A][B]}$ (4) $K = \dfrac{C \cdot D}{A \cdot B}$

2. The concentrations in an expression of the equilibrium constant are given in
(1) moles/ml (2) g/liter (3) gram-equivalents/liter (4) moles/liter.

3. In the equilibrium expression for the reaction of

$$BaSO_4 \rightleftarrows Ba^{2+} + SO_4^{2-}$$

the K_{sp} is equal to   (1) $[Ba^{2+}][SO_4^{2-}]$ (2) $\dfrac{[Ba^{2+}][SO_4^{2-}]}{BaSO_4}$

(3) $\dfrac{[Ba^{2+}][SO_4^{2-}]}{[BaSO_4]}$ (4) $\dfrac{[BaSO_4]}{[Ba^{2+}][SO_4^{2-}]}$.

4. The K_w of water is equal to (1) 1×10^{-7} (2) 1×10^{-17} (3) 1×10^{-14} (4) 1×10^{-1}.

5. The pH of a solution that has a hydrogen ion concentration of 1×10^{-4} mole/liter is (1) 4   (2) -4 (3) 10   (4) -10.*

6. The pH of a solution that has a hydroxide ion concentration of 1×10^{-4} mole/liter is (1) 4   (2) -4 (3) 10   (4) -10.*

7. A small value for K, the equilibrium constant, indicates that (1) the concentration of the un-ionized molecules must be relatively small compared to the ion concentrations   (2) the concentration of the ionized molecules must be larger than the ion concentrations   (3) the substance ionizes to a large degree   (4) the concentration of the un-ionized molecules must be relatively large compared to the ion concentrations.

8. In the Haber process for making ammonia, an increase in pressure favors (1) the forward reaction   (2) the reverse reaction   (3) neither reaction.

9. A change in which of these conditions will change the K of an equilibrium? (1) temperature   (2) pressure   (3) concentration of reactants   (3) concentration of products.

10. If $Ca(OH)_2$ is dissolved in a solution of NaOH, its solubility, compared to that in pure water, is (1) increased   (2) decreased   (3) unaffected.

• ANSWERS:

1. (3)	5. (1)	8. (1)
2. (4)	6. (3)	9. (1)
3. (1)	7. (4)	10. (2)
4. (3)		

Terms You Should Know

common ion effect	equilibrium	ionization constant
enthalpy	equilibrium constant	solubility constant
entropy	Haber process	

Le Châtelier's Principle

*Answer explained in section at the end of the book.

Acids, Bases, and Salts

DEFINITIONS AND PROPERTIES

Acids

There are some characteristic properties by which an acid may be defined. The most important are:

1. Water solutions of acids conduct electricity. This conduction depends on their degree of ionization. A few acids ionize almost completely, while others ionize only to a slight degree. The table below indicates some common acids and their degrees of ionization.

ACIDS		
Completely or Nearly Completely Ionized	Moderately Ionized	Slightly Ionized
Nitric	Oxalic	Hydrofluoric
Hydrochloric	Phosphoric	Acetic
Sulfuric	Sulfurous	Carbonic
Hydriodic		Hydrosulfuric
Hydrobromic		(Most others)

2. Acids will react with metals that are more active than hydrogen ions (see chart on page 59) to liberate hydrogen. (Some acids are also strong oxidizing agents and will not release hydrogen. Somewhat concentrated nitric acid is such an acid.)

3. Acids have the ability to change the color of indicators. Some common indicators are litmus and phenolphthalein. Litmus is a dyestuff obtained from plant life. When litmus is added to an acidic solution or paper impregnated with litmus is dipped into an acid, the neutral purple color changes to a pink-red color. The phenolphthalein is red in a basic solution and becomes colorless in a neutral or acid solution.

130

4. Acids react with bases so that the properties of both are lost to form water and a salt. This is called *neutralization*. The general equation is:

   Acid + Base → Salt + Water

An example is:

$$Mg(OH)_2 + H_2SO_4 \rightarrow MgSO_4 + 2\ H_2O$$

5. If an acid is known to be a weak solution, you might taste it and note the sour taste.

6. Acids react with carbonates to release carbon dioxide. An example:

$$CaCO_3 + 2\ HCl \rightarrow CaCl_2 + H_2CO_3 \text{ (unstable and decomposes)}$$
$$\rightarrow H_2O + CO_2 \uparrow$$

The most common theory used in first-year chemistry is the *Arhennius Theory*, which states that an acid is a substance that yields hydrogen ions in an aqueous solution. Although we speak of these hydrogen ions in the solution, they are really not separate ions but become attached to the oxygen of the polar water molecule to form the H_3O^+ ion (the hydronium ion). So it is really this hydronium ion we are concerned with in an acid solution.

A list of common acids and their formulas is given in Chapter 4, Table 7, with an explanation of the naming procedures for acids preceding Table 7.

Bases

Bases may also be defined by some operational definitions that are based on experimental observations. Some of the important ones are as follows:

1. Bases are conductors of electricity in an aqueous solution. Their degree of con-

BASES	
Completely or Nearly Completely Ionized	Slightly Ionized
Potassium hydroxide	Ammonium hydroxide
Sodium hydroxide	(All others)
Barium hydroxide	
Strontium hydroxide	
Calcium hydroxide	

duction depends on their degree of ionization. See the table of the common bases above and their degrees of ionization.

2. Bases cause a color change in indicators. Litmus changes from red to blue in a basic solution and phenolphthalein becomes a pink color from its colorless state.

3. Bases react with acids to neutralize each other and form a salt and water.

4. Bases react with fats to form a class of compounds called soaps. The earlier generations used this method to make their own soap.

5. Aqueous solutions of bases feel slippery and the stronger bases are very caustic to the skin.

The Arhennius Theory defines a base as a substance which yields hydroxide ions (OH^-) in an aqueous solution.

Some of our common bases have common names. The list below gives some of these:

Sodium hydroxide   = lye, caustic soda
Potassium hydroxide   = caustic potash
Calcium hydroxide   = slaked lime, hydrated lime, limewater
Ammonium hydroxide = ammonia water, household ammonia

Much of the sodium hydroxide produced today comes from the Hooker cell electrolysis apparatus. We have already discussed the electrolysis apparatus for the decomposition of water. When an electric current is passed through a salt water solution, hydrogen, chlorine, and sodium hydroxide are the products. The formula for this equation is:

$$2\ NaCl + 2\ HOH \xrightarrow[\text{energy}]{\text{electrical}} H_2\uparrow + Cl_2\uparrow + 2\ NaOH$$

Broader Acid-Base Theories

Besides the common Arhennius Theory of acids and bases discussed for aqueous solutions, there are two other theories which are widely used. They are known as the Brönsted Theory and the Lewis Theory.

The Brönsted Theory (1923) defines acids as proton donors and bases as proton acceptors. This definition agrees with the aqueous solution definition of an acid giving up hydrogen ions in solution, but goes beyond to other cases as well.

An example of this is dry HCl gas reacting with ammonia gas to form the white solid NH_4Cl.

$$HCl(g) + NH_3(g) \rightarrow NH_4^+Cl^-(s)$$

The HCl is the proton donor or acid and the ammonia is a Brönsted base which accepts the proton.

Conjugate Acids and Bases

In an acid-base reaction, the original acid gives up its proton to become a conjugate base. In other words, after losing its proton, it is capable of gaining a proton, thus qualifying as a base. The original base accepts a proton, so it now is classified as a conjugate acid since it could release this newly acquired proton and behave like an acid.

Some examples of this are given below:

$$HCl \quad + \quad NH_3 \longrightarrow NH_4^+ \quad + \quad Cl^-$$

Acid   Base   Conjugate acid   Conjugate base

*   *   *   *

$$HC_2H_3O_2 + H_2O \longrightarrow H_3O^+ \quad + \quad C_2H_3O_2^-$$

Acid   Base   Conjugate acid   Conjugate base

The Lewis Theory (1916) of acids and bases is defined in terms of the electron-pair concept. It is probably the most generally useful concept of acids and bases. According to the Lewis definition, an acid is an electron-pair acceptor; and a base is an electron-pair donor. An example of this is the formation of ammonium ions from ammonia gas and hydrogen ions.

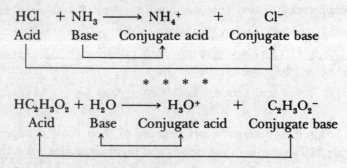

× hydrogen electron

○ nitrogen electrons

Notice that the hydrogen ion is in fact accepting the electron pair of the ammonia, so it is a Lewis acid. The ammonia is donating its electron pair, so it is a Lewis base.

Acid Concentration Expressed as pH

Frequently, acid and base concentrations are expressed in a system called the pH system. The pH can be defined as $-\log [H^+]$, where $[H^+]$ is the concentration of hydrogen ions expressed in moles per liter. The logarithm is the exponent of 10 when the number is written in the base 10. For example:

$$100 = 10^2 \text{ so logarithm of } 100, \text{ base } 10 = 2$$
$$10,000 = 10^4 \text{ so logarithm of } 10,000, \text{ base } 10 = 4$$
$$.01 = 10^{-2} \text{ so logarithm of } .01, \text{ base } 10 = -2$$

The logarithm of more complex numbers can be found in a logarithm table.

An example of a pH problem is:

Find the pH of a .1 molar solution of HCl.

1st step.   Since HCl ionizes almost completely into H^+ and Cl^-, the $[H^+] = .1$ mole/liter.

2nd step.   By definition

$$pH = -\log [H^+]$$

So

$$pH = -\log [10^{-1}]$$

3rd step.   The logarithm of 10^{-1} is -1

So

$$pH = -(-1)$$

4th step.   The pH then is $= 1$.

Since water has a normal H^+ concentration of 10^{-7} mole/liter because of the slight ionization of water molecules, the water pH is 7 when it is neither acid nor base. The normal pH range is from 1 to 14.

	Acid	Neutral	Base	
1 ←———————————→		7 ←———————————————→		14

INDICATORS

Some indicators can be used to determine pH because of their color change somewhere along this pH scale. Some common indicators and their respective color changes are given below.

Indicator	pH *Range* *of Color Change*	*Color below* *Lower* pH	*Color above* *Higher* pH
Methyl orange	3.1–4.4	Red	Yellow
Bromthymol blue	6.0–7.6	Yellow	Blue
Litmus	4.5–8.3	Red	Blue
Phenolphthalein	8.3–10.0	Colorless	Red

An example of how to read this chart is: At pH values below 4.5, litmus is red; above 8.3, it is blue. Between these values, it is a mixture of the two colors.

VOLUMETRIC ANALYSIS

Knowledge of the concentrations of solutions and the reactions they take part in can be used to determine the concentrations of "unknown" solutions or solids. The use of volume measurement in solving these problems is called *volumetric analysis.*

A common example of a volumetric analysis uses acid-base reactions. If you are given a base of known concentration, let us say .10 *M* NaOH, and you want to determine the concentration of an HCl solution, you could *titrate* the solutions in the following manner.

First, introduce a measured quantity, 25.0 mL, of the NaOH into a flask by using a pipet or buret. Next, introduce 2 drops of a suitable indicator. Since NaOH and HCl are considered a strong base and a strong acid, respectively, an indicator that changes color in the middle pH range would be appropriate. Litmus solution would be one choice. It is blue in basic solution but changes to red when the solution becomes acidic. Slowly introduce the HCl until the color change occurs. This is called the *endpoint.*

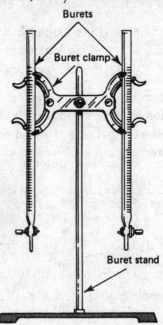

Burets

Buret clamp

Buret stand

Suppose 21.5 mL was needed to produce the color change; the reaction that occurred was

$$H^+(aq) + OH^-(aq) \rightarrow H_2O$$

until all the OH^- were neutralized, and then the excess H^+ caused the litmus paper to change color. To solve the question of the concentration of the NaOH this equation is used:

$$M_{acid} \times V_{acid} = M_{base} \times V_{base}$$

Substituting the known amounts in this equation gives

$$x\, M_{acid} \times 21.5\, mL = .1\, M \times 25.0\, mL$$
$$x = .116\, M$$

• SAMPLE PROBLEMS:

1. Find the concentration of acetic acid in vinegar if 21.6 mL of .20 M NaOH is needed to titrate a 25-mL sample of the vinegar.

   Solution
   Using the equation $M_{acid} \times V_{acid} = M_{base} \times V_{base}$ we have

   $$x\, M \times 25\, mL = .2 \times 21.6\, mL$$
   $$x = .173\, M_{acid}$$

   Another type of titration problem involves a solid and a titrated solution.

2. A solid mixture contains NaOH and NaCl. If a .100-g sample of this mixture required 10 mL of .100 M HCl to titrate the sample to its endpoint, what is the percent of NaOH in the sample?

   Solution
   Since

   $$Molarity = \frac{No.\ of\ moles}{liter\ of\ solution}$$

   then

   $$M \times V = \frac{No.\ of\ moles}{liter\ of\ solution} \times liters\ of\ solution = No.\ of\ moles$$

   Substituting the HCl information in the equation, we have

   $$\frac{.100\ mole}{liter\ of\ solution} \times .010\ liter\ of\ solution = .001\ mole$$
   (Note: this is 10 mL expressed in liters)

   Since 1 mole of HCl neutralizes 1 mole of NaOH, .001 mole of NaOH must be present in the mixture.

   1 mole NaOH = 40.0 g

So

.001 mole $\times$ 40.0 g/mole = .04 g NaOH

Therefore .04 g of NaOH was in the .1-g sample of the solid mixture. The percent is .04 g/.1 g $\times$ 100 = 40%.

BUFFER SOLUTIONS

Buffer solutions are equilibrium systems that resist changes in acidity and maintain constant pH when acids or bases are added to them. A typical laboratory buffer can be prepared by mixing equal molar quantities of a weak acid such as $HC_2H_3O_2$ and its salt, $NaC_2H_3O_2$. When a strong base such as NaOH is added to the buffer, the acetic acid reacts (and consumes) most of the excess OH^- ion. The OH^- ion reacts with the H^+ ion from the acetic acid, thus reducing the H^+ ion concentration in this equilibrium:

$$HC_2H_3O_2 \rightleftharpoons H^+ + C_2H_3O_2^-$$

This reduction of H^+ causes a shift to the right, forming additional $C_2H_3O_2^-$ ions and H^+ ions. For practical purposes, each mole of OH^- added consumes 1 mole of $HC_2H_3O_2$ and produces 1 mole of $C_2H_3O_2^-$ ions.

When a strong acid such as HCl is added to the buffer, the H^+ ions react with the $C_2H_3O_2^-$ ions of the salt and form more undissociated $HC_2H_3O_2$. This does not alter the H^+ ion concentration. Proportional increases and decreases in the concentrations of $C_2H_3O_2^-$ and $HC_2H_3O_2$ do not significantly affect the acidity of the solution.

SALTS

A salt is an ionic compound containing positive ions other than hydrogen and negative ions other than hydroxide ions. The usual method of preparing a particular salt is by neutralizing the appropriate acid and base to form the salt and water.

The naming of salts is discussed on page 49.

Chapter

11

Review

1. The difference between HCl and $HC_2H_3O_2$ as acids is (1) the first has less hydrogen in solution   (2) the second has more ionized hydrogen   (3) the first is highly ionized   (4) the second is highly ionized.

2. The hydronium ion is represented as (1) H_2O^+ (2) H_3O^+ (3) HOH^+ (4) H^-

3. H_2SO_4 is a strong acid because it is (1) slightly ionized   (2) unstable   (3) an organic compound   (4) highly ionized.

4. In the reaction of an acid with a base the common ionic reaction involves ions of (1) hydrogen and hydroxide   (2) sodium and chloride   (3) hydrogen and hydronium   (4) hydroxide and nitrate.

5. The pH of an acid solution is (1) 3   (2) 7   (3) 9   (4) 10.

6. The pH of a solution with an hydrogen ion concentration of 1×10^{-3} is (1) +3 (2) −3   (3) ±3   (4) 1 + 3.

7. According to the Brönsted Theory, an acid is (1) a proton donor   (2) a proton acceptor   (3) an electron donor   (4) an electron acceptor.

8. A buffer solution (1) changes pH rapidly with the addition of an acid   (2) does not change pH at all   (3) resists changes in pH   (4) changes pH only with the addition of a strong base.

9. The point at which a titration is complete is called the (1) endpoint   (2) equilibrium point   (3) calibrated point   (4) chemical point.

10. If 10 mL of 1 M HCl was required to titrate a 20-mL NaOH solution of unknown concentration to its endpoint, what was the concentration of the NaOH? (1) .5 M (2) 1.5 M (3) 2 M (4) 2.5 M

• **ANSWERS:**

1. (3)	5. (1)	8. (3)
2. (2)	6. (1)	9. (1)
3. (4)	7. (1)	*10. (1)
4. (1)		

Terms You Should Know

acid	conjugate base	pH
base	endpoint	salt
buffer solution	litmus	volumetric analysis
conjugate acid	neutralization	

Arhennius Theory
Brönsted Theory
Lewis Theory

*Answer explained in section at the end of the book.

Oxidation-Reduction and Electrochemistry

IONIZATION

In the early 1830's, Michael Faraday discovered that the water solutions of certain substances conduct an electric current. He called these substances *electrolytes*. Our definition today of an electrolyte is much the same. It is a substance which dissolves in water to form a solution which will conduct an electric current.

The usual apparatus to test for this conductivity is a light bulb placed in series with two prongs that are immersed in the solution tested.

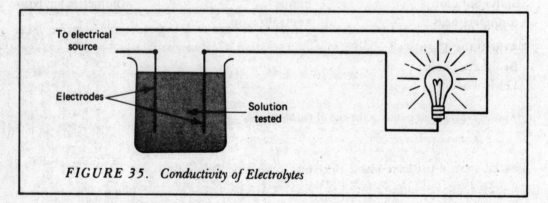

To electrical source

Electrodes

Solution tested

FIGURE 35. *Conductivity of Electrolytes*

Using this type of apparatus, we can classify solutions as good, moderate, or poor electrolytes. If they do not conduct at all, they are called nonelectrolytes. The table below gives a classification of some common substances.

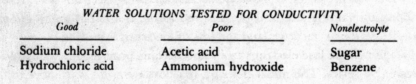

WATER SOLUTIONS TESTED FOR CONDUCTIVITY

Good	Poor	Nonelectrolyte
Sodium chloride	Acetic acid	Sugar
Hydrochloric acid	Ammonium hydroxide	Benzene

The reason that these substances conduct with varying degrees is related to the number of ions in solution.

The ionic lattice substances like sodium chloride are dissociated by the water molecules so that the individual positive and negative ions are dispersed throughout the solution. In the case of a covalent bonded substance the degree of polarity determines the extent to which it will be ionized. The water molecules, which are polar themselves, can help weaken and finally break the polar covalent bonds by clustering around the substance. When the ions are formed in this manner, the process is called *ionization;* when the ionic lattice comes apart, it is called *dissociation.* Substances which are non-electrolytes are usually bonded so that the molecule is a nonpolar molecule. The polar water molecule cannot orient itself around the molecule and cause its ionization.

Let us see how the current was carried through the apparatus in Figure 35. The electricity causes one electrode to become positively charged and one negatively charged. These are called the *anode* and *cathode,* respectively. If the solution contains ions, they will be attracted to the electrode with the charge opposite to their own. This means that the positive ions migrate to the cathode, so they are referred to as *cations.* The negative ions migrate toward the positive anode, so they are called *anions.* When these ions arrive at the respective electrodes, the negative ions give up electrons and the positive ions accept electrons, and we have a completed path for the electric current. The more highly ionized the substance, the more current flows and the brighter the light bulb glows. It was the Swedish chemist Arhennius who proposed a theory to explain the behavior of electrolytes in aqueous solutions.

OXIDATION-REDUCTION

In the discussion of acids and metals reacting to produce hydrogen (page 59), the activity of metals compared to hydrogen is shown in the form of a chart. A more complete chart of this activity of metals, called the *electromotive series,* is shown in Table 10.

From this chart you notice that zinc is above copper and, therefore, more active. This means zinc can displace copper ions in a solution of copper sulfate.

$$Zn^0 + Cu^{2+} + SO_4^{2-} \rightarrow Cu^0 + Zn^{2+} + SO_4^{2-}$$

The zero indi-   ⟶ The zinc ion
cates the zinc ————⌐   carries a +2
atom is neutral.   charge.

The Zn atom must have lost 2 electrons to become Zn^{2+} ions.

$$Zn^0 \rightarrow Zn^2 + 2e^- \text{ (electrons)}$$

At the same time, the Cu^{2+} must have gained 2 electrons to become the Cu^0 atom.

$$Cu^{2+} + 2e^- \text{ (electrons)} \rightarrow Cu^0$$

These two equations are called half-reactions. The Zn losing electrons is called oxidation. The copper ion gaining electrons is called reduction. It is important to remember that the gain of electrons is *reduction* and the loss of electrons is *oxidation.*

The metal elements that lose electrons easily and become positive ions are placed high in the electromotive series. The metal elements that lose electrons with more difficulty are placed lower on the chart.

The energy required to remove electrons from metallic atoms can be assigned numerical values called electrode potentials.

Element			STANDARD HALF-CELL VOLTAGES Oxidation Electron Reaction	Electrode Potential,* E^0_{ox}
Potassium			$K \rightarrow K^+ + e^-$	+2.93 V
Calcium			$Ca \rightarrow Ca^{2+} + 2e^-$	+2.87 V
Sodium	Increasing	Increasing	$Na \rightarrow Na^+ + e^-$	+2.71 V
Magnesium	Tendency	Tendency	$Mg \rightarrow Mg^{2+} + 2e^-$	+2.37 V
Aluminum	for	for	$Al \rightarrow Al^{3+} + 3e^-$	+1.67 V
Zinc	Atoms	Ions	$Zn \rightarrow Zn^{2+} + 2e^-$	+0.76 V
Iron	to	to Gain	$Fe \rightarrow Fe^{2+} + 2e^-$	+0.44 V
Tin	Lose	Electrons	$Sn \rightarrow Sn^{2+} + 2e^-$	+0.14 V
Lead	Electrons	and	$Pb \rightarrow Pb^{2+} + 2e^-$	+0.13 V
Hydrogen	and	Form	$H_2 \rightarrow 2H^+ + 2e^-$	0.00 V
Copper	Form	Atoms	$Cu \rightarrow Cu^{2+} + 2e^-$	−0.34 V
Mercury	Positive	of the	$2\,Hg \rightarrow Hg_2^{2+} + 2e^-$	−0.79 V
Silver	Ions	Metal	$Ag \rightarrow Ag^+ + e^-$	−0.80 V
Mercury			$Hg \rightarrow Hg^{2+} + 2e^-$	−0.85 V
Gold			$Au \rightarrow Au^{3+} + 3e^-$	−1.50 V

*A measure in volts of the tendency of atoms to gain or lose electrons.

TABLE 10. Electromotive Series

Element			Reduction Electron Reaction	E^0_{red}
Fluorine	Increasing Tendency	Increasing Tendency	$F_2 + 2e^- \rightarrow 2F^-$	+2.85 V
Chlorine	for Atoms to Gain	for Ions to Lose	$Cl_2 + 2e^- \rightarrow 2\,Cl^-$	+1.36 V
Bromine	Electrons and Form	Electrons and Form	$Br_2 + 2e^- \rightarrow 2\,Br^-$	+1.06 V
Iodine	Negative Ions	Atoms	$I_2 + 2e^- \rightarrow 2\,I^-$	+0.53 V

TABLE 11. Activity of Nonmetals

These voltages depend on the nature of the reaction, the concentrations of reactants and products, and the temperature. Throughout this discussion, we use standard concentrations, that is, all ions or molecules in aqueous solution are at a concentration of 1 molar. Furthermore, all gases taking part in the reactions are at 1 atmosphere pressure, and the temperature is 25°C. The voltage measured under these conditions is called *standard voltage*.

The electrode potentials are shown in the last column of Table 10. Notice that hydrogen is used as the standard with an electrode potential of zero. These values help you predict what reactions will occur and how readily they will occur. These reactions are called *voltaic cell reactions*.

The following examples will clarify the use of electrode potentials.

If magnesium reacts with chlorine, we can write the equation

$$Mg + Cl_2 \rightarrow MgCl_2$$

The two half-reactions with the electrode potentials would be:

Oxidation half-reaction: $Mg \rightarrow Mg^{2+} + 2e^-$	$E^0 = +2.37$ V
Reduction half-reaction: $Cl_2 + 2e^- \rightarrow 2\ Cl^-$	$E^0 = +1.36$ V
Net reaction: $Mg + Cl_2 \rightarrow Mg^{2+} + 2\ Cl^-$	$E^0 = +3.73$ V

In the net reaction, E^0 is a positive number. This indicates that the reaction occurs spontaneously.

You should also note that the total number of electrons lost in oxidation is equal to the total number of electrons gained so that the net reaction (arrived at by adding the two reactions) does not contain any electrons.

Another example is sodium reacting with chorine:

$$2\ Na + Cl_2 \rightarrow 2\ NaCl$$

The two half-reactions would be:

Oxidation half-reaction: $2\ Na \rightarrow 2\ Na^+ + 2e^-$	$E^0 = +2.71$ V
Reduction half-reaction: $Cl_2 + 2e^- \rightarrow 2\ Cl^-$	$E^0 = +1.36$ V
Net reaction: $2\ Na + Cl_2 \rightarrow 2\ Na^+ + 2\ Cl^-$	$E^0 = +4.07$ V

Note: The oxidation potentials (E^0) are *not* multiplied by the coefficients in calculating the E^0 for the reaction.

Again, the E^0 for this reaction is positive and the reaction is spontaneous.

The next example shows a negative E^0.

Copper metal placed in an acid solution would be shown as follows:

Oxidation: $Cu \rightarrow Cu^{2+} + 2e^-$	$E^0 = -0.34$ V
Reduction: $2\ H^+ + 2e^- \rightarrow H_2$	$E^0 = 0.00$ V
Net reaction: $Cu + 2\ H^+ \rightarrow Cu^{2+} + H_2$	$E^0 = -0.34$ V

Since E^0 is negative, we know the reaction will not take place.

You notice in the chart that the oxidation reaction with its E^0 value is shown for the metals. If you must use the reduction reaction of these metals, the equation must be reversed, and the sign of E^0 changed. An example of this is placing a piece of copper into a solution of silver ions.

Oxidation	$Cu \rightarrow Cu^{2+} + 2e^-$	$E^0 = -0.34$ V
Reversed to show reduction	$2\ Ag^+ + 2e^- \rightarrow 2\ Ag^0$	$E^0 = +.80$ V
	$Cu + 2\ Ag^+ \rightarrow Cu^{2+} + 2\ Ag^0$	$E^0 = +.46$ V

This reaction would occur spontaneously since E^0 is positive.

NON-STANDARD-STATE CELL POTENTIALS

You are undoubtedly aware that the voltage of a battery does not remain constant. After being used, the voltage begins to decrease. The reason for this is related to the changes in concentration of the reactants and products that are causing the electron flow. For the voltaic standard mentioned earlier, the assumed concentration of 1 molar solutions was the basis of the standard voltages given. A German scientist, Walter Nernst, developed a mathematical relationship that enables us to calculate the cell voltages and direction of a spontaneous reaction at other than standard-state concentrations. For a general oxidation-reduction reaction

$$a\text{A} + b\text{B} \rightleftarrows c\text{C} + d\text{D}$$

the *Nernst equation* has the form

$$E_{cell} = E^0_{cell} - \frac{.059}{n} \log \frac{[\text{C}]^c[\text{D}]^d}{[\text{A}]^a[\text{B}]^b}$$

where E^0_{cell} is the standard-state cell voltage, n is the number of electrons exchanged in the equation for the reaction, and .059 is a constant at 298°K. The concentrations are all in molarity. Concentrations of solids and liquid solvents are considered constant and are not included in the expression.

Notice from this mathematical expression that, if the concentrations of the reactants increase or if the product is decreased, the voltage will increase ($E > E^0$). If the concentrations of the products are increased or the concentrations of reactants decreased, the voltage will decrease ($E < E^0$).

• SAMPLE PROBLEM:

Calculate the cell voltage of

$$\text{Zn(s)} \mid \text{Zn}^{2+}(.001 \; M) \parallel \text{Ag}^+(.1 \; M) \mid \text{Ag(s)}$$

(Each side of the $\parallel$ represents the half-reaction characters, with the concentration when appropriate.)

The reactions would be

Cathode reduction reaction:	$2 \text{Ag}^+ + 2e^- \rightarrow 2 \text{Ag(s)}$	$E^0 = + .80 \text{ V}$
Anode oxidation reaction:	$\text{Zn(s)} \rightarrow \text{Zn}^{2+} + 2e^-$	$E^0 = + .76 \text{ V}$
Cell	$2 \text{Ag}^+ + \text{Zn(s)} \rightleftarrows \text{Zn}^{2+} + 2 \text{ Ag}$	$E^0 = +1.56 \text{ V}$

Substituting the known values in the Nernst equation, solve for the E_{cell}:

$$E_{cell} = E^0_{cell} - \frac{.059}{n} \log \frac{[\text{Zn}^{2+}]}{[\text{Ag}^+]^2}$$

$$E_{cell} = 1.56 - \frac{.059}{2} \log \frac{[10^{-3}]}{[10^{-1}]^2}$$

$$E_{cell} = 1.56 - .03 \log 10^{-1}$$

$$E_{cell} = 1.56 - .03$$

$$E_{cell} = 1.59 \text{ V}$$

ELECTROLYTIC CELLS

Reactions that do not occur spontaneously can be forced to take place by supplying energy with an external current. These reactions are called *electrolytic reactions*. Some examples of this type of reaction are electroplating, electrolysis of salt solution, electrolysis of water, and electrolysis of molten salts. An example of this setup is shown in Figure 36.

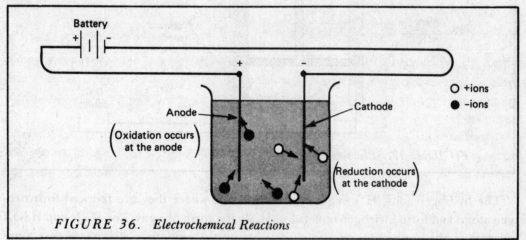

FIGURE 36. *Electrochemical Reactions*

If this solution contained Cu^{2+} ions and Cl^- ions, the half-reactions would be:

Anode half-reaction:

Oxidation	$2\,Cl^- \rightarrow Cl_2 \uparrow + 2e^-$	$E^0 = -1.36 \text{ V}$

Cathode half-reaction:

Reduction	$Cu^{2+} + 2e^- \rightarrow Cu^0$	$E^0 = +0.34 \text{ V}$
Net reaction:	$Cu^{2+} + 2\,Cl^- \rightarrow Cu^0 + Cl_2 \uparrow$	$E^0 = -1.02 \text{ V}$

(Notice that the E^0 for this reaction is negative so an outside source of energy must be used to make it occur.)

Another electrolysis example is the electrolysis of a water solution of sodium chloride. This water solution contains chloride ions which are attracted to the anode and set free as chlorine molecules. The cathode reaction is somewhat more complicated. Although the sodium ions are attracted to the cathode, they are not set free as atoms. Remember water can ionize to some extent and the electromotive series shows that the hydrogen ion is reduced more easily than the sodium ion. Therefore, hydrogen, and *not* sodium, is set free at the cathode. The reaction can be summarized like this:

Cathode reaction:	$2\,H_2O + 2e^- \rightarrow H_2 \uparrow + 2\,OH^-$
Anode reaction:	$2\,Cl^- \rightarrow 2e^- + Cl_2$
Net reaction:	$2\,H_2O + 2\,Cl^- \rightarrow H_2 \uparrow + Cl_2 \uparrow + 2\,OH^-$

Another example of electrolysis is the decomposition of water by using an apparatus like the one shown in Figure 37.

The solution in this apparatus contains distilled water and a small amount of H_2SO_4. The reason for adding H_2SO_4 is to make the solution an electrolyte since distilled water alone will not conduct an electric current. The solution, therefore, contains ions of H^+, HSO_4^-, and SO_4^{2-}.

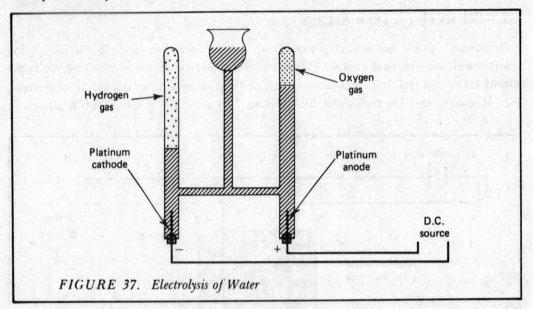

FIGURE 37. Electrolysis of Water

The hydrogen ions (H^+) migrate to the cathode where they are reduced to hydrogen atoms and form hydrogen molecules (H_2) in the form of a gas. The SO_4^{2-} and HSO_4^- migrate to the anode but are not oxidized since the oxidation of water occurs more readily. These ions then are merely spectator ions. The oxidized water reacts as shown in the half-reaction below.

The half-reactions can be shown as follows:

Cathode reaction:
Reduction $\qquad\qquad 4\,H^+ + 4e^- \rightarrow 2\,H_2 \uparrow$
Anode reaction:
Oxidation $\qquad\qquad 2\,H_2O \rightarrow O_2 \uparrow + 4\,H^+ + 4e^-$

Net reaction: $\qquad\quad 2\,HOH \rightarrow 2\,H_2 \uparrow + O_2 \uparrow$

Notice that the equation shows 2 volumes of hydrogen gas are released while only 1 volume of oxygen gas is liberated. This agrees with the discussion of the composition of water in Chapter 7.

APPLICATIONS OF ELECTROCHEMICAL CELLS

Commercial Voltaic Cells

One of the most common voltaic cells is the ordinary "dry cell" used in flashlights. Its makeup is shown in the drawing on page 145, along with the reactions.

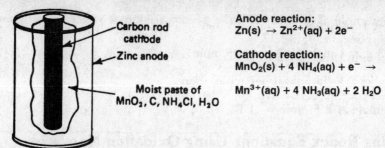

Anode reaction:
$Zn(s) \rightarrow Zn^{2+}(aq) + 2e^-$

Cathode reaction:
$MnO_2(s) + 4 NH_4(aq) + e^- \rightarrow$

$Mn^{3+}(aq) + 4 NH_3(aq) + 2 H_2O$

The automobile lead storage battery is also a voltaic cell. When it discharges, the reactions are:

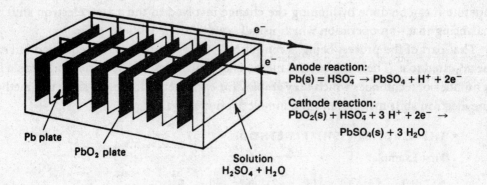

Anode reaction:
$Pb(s) = HSO_4^- \rightarrow PbSO_4 + H^+ + 2e^-$

Cathode reaction:
$PbO_2(s) + HSO_4^- + 3 H^+ + 2e^- \rightarrow$

$PbSO_4(s) + 3 H_2O$

QUANTITATIVE ASPECTS OF ELECTROLYSIS

The amounts of products iiberated at the electrodes of an electrolytic cell are related to the quantity of electricity passed through the cell and the electrode reactions.

In electrolysis, 1 mole of electrons is called a *faraday* of electric charge. In the reaction of electrolysis of molten NaCl, 1 faraday will liberate 1 mole of sodium atoms.

Cathode reactions: $\cdot Na^+ + e^- \rightarrow Na(s)$

At the same time, 1 mole of Cl^- ions at the anode will form 1 mole of chlorine atoms and thus $\frac{1}{2}$ mole of chlorine molecules.

If the metallic ion had been Ca^{2+}, 1 faraday would release only $\frac{1}{2}$ mole of calcium atoms since each calcium ion requires 2 electrons, as shown here:

$Ca^{2+} + 2e^- \rightarrow Ca(s)$

• SAMPLE PROBLEM:

How many faradays are required to reduce 2.93 g of nickel ions from melted $NiCl_2$?

Solution
The reaction is

$Ni^{2+} + 2e^- \rightarrow Ni(s)$

The 2.93 g represents

$$2.93 \text{ g} \times 1 \text{ mole}/58.7 \text{ g} = .05 \text{ mole}$$

Since it takes 2 F (faradays) per mole, then

$$.05 \text{ mole} \times 2 \text{ F/mole} = .1 \text{ F}$$

Balancing Redox Equations Using Oxidation Numbers

In general, many chemical equations are so simple that the process of writing them (once the reactants and products are known) can be carried out by mere inspection. Many redox reactions, however, are of such complexity that the process of writing the equations by methods of trial and error is a time-consuming one. In these cases the operation can be done by limiting the change involved to the actual electron shift, and balancing that—an operation which usually can be done without difficulty.

That part of the process being accomplished, the remainder of the equation can easily be adjusted to it. There are several methods by which this can be done, and people have a number of techniques which they employ for each method. We will show two methods: the electron shift method and the ion-electron method.

• THE ELECTRON SHIFT METHOD:

First Example:

$$\overset{+1}{\text{H}}\overset{-1}{\text{Cl}} + \overset{+4}{\text{Mn}}\overset{-2}{\text{O}_2} = \overset{+1}{\text{H}_2}\overset{-2}{\text{O}} + \overset{+2}{\text{Mn}}\overset{-1}{\text{Cl}_2} + \overset{0}{\text{Cl}_2}$$

1. The molecular formulas are written in a statement of the reaction, and the oxidation states are assigned to the elements. These appear in the above equation.

2. Inspection shows that the oxidation state of the manganese atoms has changed from +4 to +2, and that the oxidation state of the chlorine atoms which emerge from the action as a free element has changed from −1 to 0:

$$1 \ (\text{Mn}^{4+} + 2e^+ = \text{Mn}^{2+})$$
$$2 \ (\text{Cl}^{1-} \qquad = \text{Cl}^0 + 1e^-)$$

3. The reason for multiplying the two half-reactions by 1 and 2 is to balance the electrons gained with the electrons lost.

4. It appears that $1 \ \text{Mn}^{4+} + 2 \ \text{Cl}^{1-} = 1 \ \text{Mn}^{2+} + 2 \ \text{Cl}^0$

5. Placing these coefficients in the equation gives

$$2 \ \text{HCl} + 1 \ \text{MnO}_2 = {}^*\text{? } \text{H}_2\text{O} + 1 \ \text{MnCl}_2 + \text{Cl}_2$$

6. From the numbers thus established, the remaining coefficients can easily be deduced. Two *more* molecules of HCl are required to furnish the chlorine for MnCl_2, and the two atoms of oxygen in MnO_2 form 2 H_2O.

7. *The final equation is:* $4\text{HCl} + \text{MnO}_2 = 2 \ \text{H}_2\text{O} + \text{MnCl}_2 + \text{Cl}_2$

Now consider a more complicated reaction. Only the elements which show a cnange in oxidation state are indicated.

*? indicates it is not known as yet what this coefficient will be.

Second Example:

$$\overset{-1}{\text{HCl}} + \overset{+7}{\text{KMnO}_4} = \text{H}_2\text{O} + \text{KCl} + \overset{+2}{\text{MnCl}_2} + \overset{0}{\text{Cl}_2}$$

The electron change is: $5\ (\text{Cl}^{-1} \qquad\qquad = \text{Cl}^0 + 1e^-)$

$$\underline{1\ (\text{Mn}^{7+} + 5e^- \qquad = \text{Mn}^{2+})}$$

Therefore $\quad 5\text{HCl} + 1\ \text{KMnO}_4 = \tfrac{5}{2}\text{Cl}_2 + 1\ \text{MnCl}_2$

$\uparrow$

$\boxed{\text{5 atoms of chlorine}}$

equal $\tfrac{5}{2}$ Cl_2 molecules

Multiplying by 2 to avoid fractions gives:

$$10\ \text{HCl} + 2\ \text{KMnO}_4 = 5\ \text{Cl}_2 + 2\ \text{MnCl}_2$$

Substituting in the original equation:

$$10\ \text{HCl} + 2\ \text{KMnO}_4 = *?\ \text{H}_2\text{O} + *?\ \text{KCl} + 2\ \text{MnCl}_2 + 5\ \text{Cl}_2$$

For the remainder of the coefficients, 2 KMnO_4 produces 2 KCl and 8 H_2O, while 2 KCl + 2 MnCl_2 call for six additional molecules of HCl. So finally the expression becomes

$$16\ \text{HCl} + 2\ \text{KMnO}_4 = 8\ \text{H}_2\text{O} + 2\ \text{KCl} + 2\ \text{MnCl}_2 + 5\ \text{Cl}_2$$

Third Example:

$$\text{As}_2\text{S}_5 + \text{KClO}_3 + \text{H}_2\text{O} = \text{H}_3\text{AsO}_4 + \text{H}_2\text{SO}_4 + \text{KCl}$$

Your final coefficients should be 3, 20, 24, 6, 15, 20.†

• ION-ELECTRON METHOD:

The second method is more complex but seems to represent the true mechanism of such reactions more closely.

In this method, only units which actually have individual existence (atoms, molecules or ions) in the particular reaction being studied are taken into consideration. The principal oxidizing agent and the principal reducing agent are chosen from these (by a method to be indicated later) and the electron loss or gain of these two principal actors is then determined by taking into consideration the fact that, since electrons can neither be created nor destroyed, electrons lost by one of these actors must be gained by the other. This is accomplished by using two separate partial equations representing the changes undergone by each of the two principal actors.

Probably the best way to show how the method operates will be to follow in detail the steps taken in balancing the example of an actual oxidation-reduction reaction.

First Example:

Assume that the equation $\text{K}_2\text{CrO}_4 + \text{HCl} \rightarrow \text{KCl} + \text{CrCl}_3 + \text{H}_2\text{O} + \text{Cl}_2$ is to be balanced.

†Answer explained in section at the end of the book.

1. Determine which of the substances present are involved in the oxidation-reduction.

   This is done by listing all substances present on each side of the arrow and the crossing out those which appear on both sides of the arrow without any portion being changed in any way.

   $$K^+, \; \cancel{CrO_4^{2-}}, \; H^+, \; Cl^- \rightarrow K^+, \; Cl^-, \; Cr^{3+}, \; H_2O, \; Cl_2$$

   > Note that Cl^- is not crossed out although it appears on both sides because some of the Cl^- from the left side appears in a changed form, namely, Cl_2, on the right.

   The two substances on the left side which are *not* crossed out are the ones involved in the oxidation-reduction.

   If, as *in this case* there are *more* than two, disregard H^+, OH^-, or H_2O; this will leave the two principal actors.

2. Indicate in two as yet unbalanced partial equations the fate of each of the two active agents thus:

   $$CrO_4^{2-} \rightarrow Cr^{3+}$$
   $$Cl^- \quad \rightarrow Cl_2$$

3. Balance these equations chemically, inserting any substance necessary.

   $$CrO_4^{2-} + 8H^+ \rightarrow Cr^{3+} + 4H_2O$$
   $$2Cl^- \quad\quad\quad \rightarrow Cl_2$$

   In the upper partial equation, H^+ had to be added in order to remove the oxygen from the CrO_4^{2-} ion.

   H^+ always is used for this purpose in acid solutions. If the solution is basic, H_2O must be used for this purpose, thus:

   $$CrO_4^{2-} + 4H_2O \rightarrow Cr^{3+} + 8OH^-$$

4. Balance these equations electrically by adding electrons *on either side* so that the total electric charge is the same on the left and right sides, thus:

   $$CrO_4^{2-} + 8H^+ + 3e^- \rightarrow Cr^{3+} + 4H_2O$$
   $$2Cl^- \quad\quad\quad\quad \rightarrow Cl_2 + 2e^-$$

5. Add these partial equations.

   But before we add we must realize that electrons can neither be created nor destroyed. Electrons are gained in one of these equations and lost in the other. THOSE GAINED IN THE ONE MUST COME FROM THE OTHER. Therefore we must multiply each of these equations through by numbers so chosen that the number of electrons gained in the one will be the same as the number lost in the other, thus:

   $$2 \, [CrO_4^{2-} + 8H^+ + 3e^- \rightarrow Cr^{3+} + 4H_2O]$$
   $$3 \, [2Cl^- \quad\quad\quad\quad \rightarrow Cl_2 + 2e^-]$$

THIS IS THE KEY TO THE METHOD!

Adding the multiplied equations, we get:

$$2CrO_4^{2-} + 16H^+ + 6Cl^- \rightarrow 2Cr^{3+} + 8H_2O + 3Cl_2$$

This sum tells us all that really happens in the oxidation-reduction but if a conventionally balanced equation is desired for problem purposes, etc., it can be obtained by carrying the coefficient into the original skeleton equation, thus:

$$2K_2CrO_4 + 16HCl \rightarrow 4KCl + 2CrCl_3 + 8H_2O + 3Cl_2$$

(4 inserted by inspection)

Second Example:

$$2\,KMnO_4 + H_2O + KI \rightarrow 2\,MnO_2 + 2\,KOH + KIO_3$$
$$2[MnO_4^- + 2\,H_2O + 3e^- \rightarrow MnO_2 + 4\,OH^-]$$
$$I^- + 6\,OH^- \rightarrow IO_3^- + 3\,H_2O + 6e^-$$

Adding:

$$2\,MnO_4^- + I^- + H_2O \rightarrow 2\,MnO_2 + IO_3^- + 2\,OH^-$$

Note that, since this takes place in the basic solution, H_2O was used to remove oxygen in the upper partial.

Note also that in the second partial it was necessary to ADD oxygen and that this was done by means of OH^- ion.

If the solution had been *acid*, then H_2O would have been used for this purpose, thus:

$$I^- + 3H_2O \rightarrow IO_3^- + 6H^+ + 6e^-$$

In general, you will meet three types of partial equations:

1. Where electrons only are needed to balance, as in the second partial of the first example above.

2. Where oxygen must be removed from an ion.

   In acid solution, H^+ is used for this purpose as in the first partial of the first example above.

   In basic solution, H_2O is used for this as in the first partial of the second example above.

3. Where oxygen must be added to an ion.

   In basic solution, OH^- is used for this as in the second partial of the second example above.

   In acid solution, H_2O is used for this as in:

$$SO_3^{2-} + H_2O \rightarrow SO_4^{2-} + 2H^+ + 2e^-$$

Now, try to balance the following equation.

Third Example:

$$KMnO_4 + H_2SO_3 \rightarrow H_2SO_4 + MnSO_4 + K_2SO_4 + H_2O$$

Since this takes place in acid solution, you will have to use H_2O to add oxygen to SO_3^{2-}.

Coefficients are 2, 5, 2, 2, 1, 3.*

• **TRY THESE PROBLEMS:**

1. $Zn + HNO_3 \rightarrow Zn(NO_3)_2 + NH_4NO_3 + H_2O$ Ans. 4, 10, 4, 1, 3*
2. $Cu + HNO_3 \rightarrow Cu(NO_3)_2 + H_2O + NO$ Ans. 3, 8, 3, 4, 2*
3. $KMnO_4 + HCl \rightarrow KCl + MnCl_2 + H_2O + Cl_2$ Ans. 2, 16, 2, 2, 8, 5*

Chapter
12
Review

1. Which of the following when dissolved in water and placed in the conductivity apparatus would cause the light to glow? (1) table salt   (2) ethyl alcohol   (3) sugar   (4) glycerine

2. The reason that a current could flow was because (1) ions combine to form molecules   (2) molecules migrate to the charge plates   (3) ions migrate to the charge plates   (4) sparks cross the gap.

3. The extent of ionization depends on the (1) nature of the solvent   (2) nature of the solute   (3) concentration of the solution   (4) the temperature of the solution   (5) all the above.

4. Which of the following is TRUE? (1) The number of positive ions in solution equals the number of negative ions.   (2) The positive ions are called anions.   (3) The positive ions are called cathodes.   (4) The total positive charge equals the total negative charge.   (5) None of the above.

5. The hydronium ion is represented as (1) H_2O^+ (2) H_3O^+ (3) HOH^+ (4) H^-.

6. In the electrolysis of copper chloride, the substance liberated at the anode is (1) copper   (2) chlorine   (3) hydrogen   (4) copper chloride.

7. Ions are particles which exist (1) only in water solutions   (2) in some crystals   (3) in polar covalent compounds   (4) in covalent compounds which are not polar.

8. Ionic compounds will conduct an electric current when they are (1) solidified   (2) melted   (3) frozen   (4) dehydrated.

9. The cathode in a direct-current circuit is an electrode which is (1) always negative   (2) always positive   (3) always neutral   (4) sometimes negative and sometimes positive.

10. Electrolysis of a dilute solution of sodium chloride results in the cathode product (1) sodium   (2) chlorine   (3) hydrogen   (4) oxygen.

11. Electromotive series:

$Zn^0 = Zn^{2+} + 2e^-$ $E^0 = +.76$ V
$Au^0 = Au^{3+} + 3e^-$ $E^0 = -1.42$ V

*Answer explained in section at the end of the book.

If a gold foil was placed in a solution containing Zn ions, the reaction potential would be (1) −1.34 V   (2) −2.18 V   (3) −.66 V   (4) +2.18 V   (5) +1.34 V.

12. If the reaction potential is positive, this indicates that (1) the reaction will not occur (2) the reaction will occur and give off energy   (3) the reaction will occur if heat or energy is added   (4) the reaction will power an outside alternating electric current.

• **ANSWERS:**

1. (1)	5. (2)	9. (1)
2. (3)	6. (2)	10. (3)
3. (5)	7. (2)	11. (2)
4. (4)	8. (2)	12. (2)

Terms You Should Know

anions	electrode potential	ionization
anode	electrolyte	Nernst equation
cathode	electrolytic reaction	oxidation
cations	electromotive series	reduction
dissociation	faraday	standard voltage

Some Representative Groups and Families

In the following sections, a brief description is given of some of the important and representative groups of elements usually referred to in most first-year chemistry courses. The presentation of each group will follow along this outline: the important element or elements, their occurrence, their preparation both in laboratory and in industrial processes, their properties, and some of their important uses.

SULFUR FAMILY

Since we discussed oxygen separately in a preceding chapter, the most important element left to discuss is sulfur.

Sulfur is found free in the volcanic regions of Japan, Mexico, and Sicily. It was removed from the rock mixtures by heating in retorts or furnaces.

In 1867 sulfur was discovered in underground deposits in Louisiana and Texas. The overlying layer of earth was "quicksand" which prohibited the usual mining operations. An American chemist, Herman Frasch, devised an ingenious plan for its removal which bears his name, the Frasch process. It consists of drilling a hole and inserting three concentric pipes. Through the outer pipe superheated water (at 160°C) is forced into the sulfur bed. This melts the sulfur. Compressed air, which is forced down the innermost pipe, causes the liquid sulfur to rise up through the middle pipe to the surface. At the surface the liquid sulfur is pumped into vast molds in which it solidifies. See Figure 38.

The sulfur obtained in this process is usually greater than 99.5% pure.

Sulfur has three allotropic forms. Their differences can be summarized in chart form (Table 12).

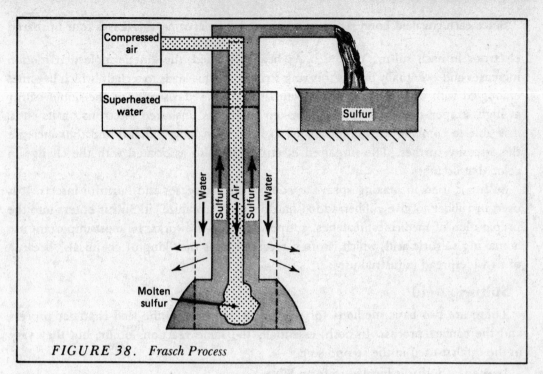

FIGURE 38. Frasch Process

Characteristic	ALLOTROPIC FORM		
	Rhombic	Monoclinic	Amorphous
Shape	Rhombic or octahedral crystals	Needle-shaped monoclinic crystals	Noncrystalline
Color	Pale-yellow opaque, brittle	Yellow, waxy, translucent, brittle	Dark, tough, elastic
Preparation	Crystallizes from CS_2 solution of sulfur	Crystallizes from molten sulfur as it cools	Obtained by quick-cooling very hot liquid sulfur
Stability	Stable below 96°C	Stable between 96° and 119°C	Unstable; slowly changes to rhombic

TABLE 12. Allotropic Forms of Sulfur

The solid sulfur is composed of S_8 molecules as shown below:

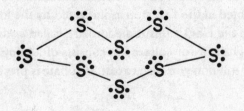

Since each covalent bond is a sharing of a pair of electrons, this leaves four unshared electrons in each sulfur: $\left(\overset{s^{s}{}^{s}}{\underset{\diagup}{S}} {}_{\diagdown} \right)$. As heat is applied, the kinetic molecular motion increases and eventually breaks this ring structure. This leads to a chain which becomes entangled with other chains and explains the increased viscosity of the liquid sulfur at high temperatures. Because the broken chain has unshared electrons on its ends, it is able to combine with other chains to form even longer chains and thus increase the viscosity further. The unpaired electrons are also associated with the change in color that occurs.

Sulfur is used in making sprays to control plant diseases and harmful insects. It is used in rubber to give rubber added hardness or "vulcanize" it. Sulfur enters into the preparation of medicines, matches, gunpowder, and fireworks. Its most important use is making sulfuric acid, which is often referred to as the "king of chemicals" because of its widespread industrial uses.

Sulfuric Acid

There are two basic methods for preparing *sulfuric acid:* the lead-chamber process and the contact process. In both, essentially the same reactions occur, but they vary in the catalyst used in the second step.

1st step.   Sulfur is burned to form SO_2:

$$S + O_2 \rightarrow SO_2$$

2nd step. SO_2 is oxidized with the aid of a catalyst to form SO_3. The contact process uses insulated steel towers containing layers of V_2O_3 pellets as the catalyst. The lead-chamber process uses a series of lead chambers wherein SO_2 is oxidized by means of NO_2, supplied for that purpose. The basic reaction in both cases is

$$2\ SO_2 + O_2 \rightarrow 2\ SO_3$$

3rd step. H_2O is added to SO_3 to form H_2SO_4:

$$SO_3 + H_2O \rightarrow H_2SO_4$$

In the contact process water can be added as fast as the SO_3 is absorbed to give 98 or 99% H_2SO_4. If less water is added, "oleum" or "fuming" sulfuric acid is formed. The older lead-chamber process generally makes a more dilute solution of H_2SO_4.

Important properties of sulfuric acid.   Sulfuric acid ionizes in two steps:

$$H_2SO_4 + H_2O \leftrightharpoons H_3O^+ + HSO_4^-$$
$$HSO_4^- + H_2O \leftrightharpoons H_3O^+ + SO_4^{2-}$$

to form a strong acid solution. The ionization is more extensive in a dilute solution. Most hydronium ions are formed in the first step as indicated by the longer forward reaction arrow. Salts formed with the HSO_4^- (bisulfate ion) are called *acid salts;* the SO_4^{2-} (sulfate ion) forms *normal salts.* The test for sulfate ion consists of adding a solution of HCl and $BaCl_2$ or $Ba(NO_3)_2$ to the solution you wish to test. If sulfate is present, a white precipitate of $BaSO_4$ will form.

Sulfuric acid reacts like other acids, as shown below:

With active metals: $Zn + H_2SO_4 \rightarrow ZnSO_4 + H_2\uparrow$ (for dilute H_2SO_4)

With bases: $2\ NaOH + H_2SO_4 \rightarrow Na_2SO_4 + 2\ H_2O$

With metal oxides: $MgO + H_2SO_4 \rightarrow MgSO_4 + H_2O$

With carbonates: $CaCO_3 + H_2SO_4 \rightarrow CaSO_4 + H_2O + CO_2\uparrow$

Sulfuric has other particular characteristics as shown below:

As an oxidizing agent:

$$Cu + 2\ H_2SO_4\ (concentrated) \rightarrow CuSO_4 + SO_2\uparrow + 2\ H_2O$$

As a dehydrating agent with carbohydrates:

$$C_{12}H_{22}O_{11}\ (sugar) \xrightarrow[H_2SO_4]{Conc.} 12\ C + 11\ H_2O$$

Important uses of H_2SO_4

1. Making other acids (because of its high boiling point, 338°C):

$$NaCl + H_2SO_4 \xrightarrow{heat} NaHSO_4 + HCl\uparrow$$

(With more heat and an excess of salt, the second hydrogen ion can also be removed to form the salt Na_2SO_4.)

2. Washing objectionable colors from gasoline made by the "cracking" process.
3. Dehydrating agent in making explosives, dyes, and drugs.
4. Freeing metals of iron or steel of scale and rust. This is called "pickling."
5. Electrolyte in ordinary lead storage batteries.
6. Used in precipitating bath in making rayon.

This is a list of just the more important uses of sulfuric acid. It is by no means complete.

Other Important Compounds of Sulfur

Hydrogen sulfide is a colorless gas having an odor of rotten eggs. It is fairly soluble in water and poisonous in rather small concentrations. It can be prepared by reacting ferrous sulfide with an acid, such as dilute HCl:

$$FeS + 2\ HCl \rightarrow FeCl_2 + H_2S\uparrow$$

It does burn in excess oxygen to form compounds of water and sulfur dioxide. If insufficient oxygen is available, some free sulfur will form. It is only a weak acid in a water solution. Hydrogen sulfide is used widely in qualitative laboratory tests since many metallic sulfides precipitate with recognizable colors. These sulfides are sometimes used as paint pigments. Some common sulfides and their colors are:

ZnS — White

CdS — Bright yellow

As_2S_3 — Lemon yellow

Sb_2S_3 — Orange
CuS — Black
HgS — Black
PbS — Brown-black

Another important compound of sulfur is *sulfur dioxide*. It is a colorless gas with a suffocating odor. It can be prepared by burning sulfur in air:

$$S + O_2 \rightarrow SO_2$$

or decomposing sulfites:

$$Na_2SO_3 + H_2SO_4 \rightarrow Na_2SO_4 + H_2O + SO_2\uparrow$$

Sulfur dioxide is the acid anhydride of sulfurous acid:

$$SO_2 + H_2O \rightarrow H_2SO_3$$

It is used as a bleach on moist straw, silk, wool, and paper because of its ability to reduce the coloring agent. This process is reversed by oxidation in sunlight as shown by the yellowing of straw and paper. Potassium permanganate can also be bleached by a water solution of SO_2 in the form of H_2SO_3:

$$5\ H_2SO_3 + 2\ KMnO_4 \rightarrow K_2SO_4 + 2\ MnSO_4 + 2\ H_2SO_4 + 3\ H_2O$$

The structure of sulfur dioxide is a good example of a *resonance* structure. Its molecule is depicted as shown in Figure 39.

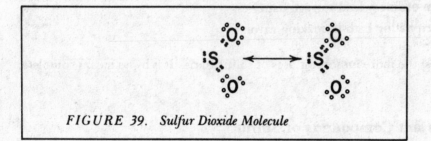

FIGURE 39. Sulfur Dioxide Molecule

You will notice that the covalent bonds between sulfur and oxygen are shown in one drawing as single bonds and in the other as double bonds. This signifies that the bonds between the sulfur and oxygens have been shown by experimentation to be neither single nor double bonds, but "hybrids" of the two. Sulfur trioxide also is a resonance structure.

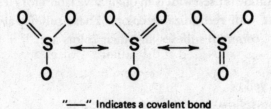

"——" Indicates a covalent bond

HALOGEN FAMILY

The members of the halogen family are shown in Table 13 with some important facts concerning them.

Item	Fluorine	Chlorine	Bromine	Iodine
Molecular formula	F_2	Cl_2	Br_2	I_2
Atomic number	9	17	35	53
Activity	Most active	← to →		Least active
Outer shell structure	7 electrons	7 electrons	7 electrons	7 electrons
State and color at room temperature	Gas — pale yellow	Gas — green	Liquid — dark red	Solid — purplish black crystals

TABLE 13. Halogen Family

Because all the halogens lack one electron in their outer shell, they usually are acceptors of electrons (oxidizing agents). Fluorine is the most active nonmetal in the periodic chart. The halogens can be prepared as follows:

1. Fluorine can be prepared by electrolysis of a mixture of KHF_2 and liquid anhydrous hydrogen fluoride. Electrolysis is necessary because the extreme electronegativity of fluorine makes it impossible to prepare by simple displacement.

2. Chlorine, bromine, and iodine can be prepared by reacting a halide salt of each, like NaCl, NaBr, or NaI, with $H_2SO_4 + MnO_2$. The ionic equations would be:

$$2\ Cl^- + 2\ H_2SO_4 + MnO_2 \rightarrow SO_4^{2-} + Mn^{2+}SO_4^{2-} + 2\ H_2O + Cl_2 \uparrow$$
$$2\ Br^- + 2\ H_2SO_4 + MnO_2 \rightarrow SO_4^{2-} + Mn^{2+}SO_4^{2-} + 2\ H_2O + Br_2 \uparrow$$
$$2\ I^- + 2\ H_2SO_4 + MnO_2 \rightarrow SO_4^{2-} + Mn^{2+}SO_4^{2-} + 2\ H_2O + I_2 \uparrow$$

The common laboratory preparation of Cl_2 is

$$4\ HCl + MnO_2 \rightarrow MnCl_2 + 2\ H_2O + Cl_2 \uparrow$$

The resulting chlorine is collected by the upward displacement of air since Cl_2 is heavier than air.

Figures 40–42 show the typical laboratory setups for the preparation of chlorine, bromine, and iodine.

The commercial preparation of chlorine involves the electrolysis of a salt solution. When a direct current is passed through the solution, chlorine gas is released at the anode and hydrogen gas is released at the cathode. One of these commercial electrolysis apparatus is called the Hooker cell. The overall reaction is:

$$2\ NaCl + 2\ H_2O \rightarrow H_2 \uparrow + Cl_2 \uparrow + 2\ NaOH$$

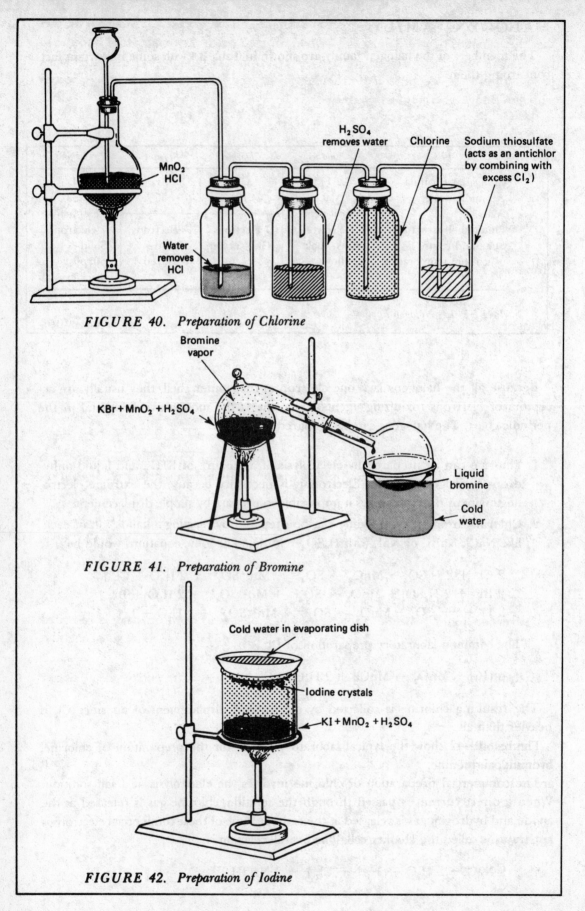

FIGURE 40. *Preparation of Chlorine*

FIGURE 41. *Preparation of Bromine*

FIGURE 42. *Preparation of Iodine*

The commercial methods of preparing bromine and iodine take advantage of the fact that the halogens decrease in activity as you proceed down the family from fluorine to iodine. This means that Cl_2 can displace Br^- and I^- ions and that Br_2 can displace I^- ions in solution. In the preparation of bromine, sea water containing bromide ions is treated with chlorine to liberate the bromine.

$$2\ Br^- + Cl_2 \rightarrow Br_2 + 2\ Cl^-$$

Likewise, iodine is liberated from oil well brines containing iodide salts by treatment with chlorine.

$$2\ I^- + Cl_2 \rightarrow 2\ Cl^- + I_2$$

Testing for Halides

This ability of chlorine to replace I^- and Br^- ions is also used in testing for iodides and bromides. The test involves putting the salt into a water solution, then adding fresh chlorine water and some carbon tetrachloride. Shake the resulting mixture vigorously and look for a purple color or red color in the carbon tetrachloride which will sink to the bottom of the container. Purple indicates a positive test for iodine and deep red indicates the presence of bromine. The reactions for each that release these halogens are:

$$2\ I^- + Cl_2 \rightarrow 2\ Cl^- + I_2 \text{ (purple in } CCl_4)$$
$$2\ Br^- + Cl_2 \rightarrow 2\ Cl^- + Br_2 \text{ (deep red in } CCl_4)$$

Test for fluoride: Add sulfuric acid and test if the gas released, which will be hydrogen fluoride (HF), will etch glass. (Use **caution** not to get on hands.)

Test for chloride: Add a solution of silver salt (usually silver nitrate) and look for a white precipitate which is soluble in ammonium hydroxide, but insoluble in nitric acid.

Some Important Halides and Their Uses

Hydrochloric acid —common acid prepared in the laboratory by reacting sodium chloride with concentrated sulfuric acid. It is used in many important industrial processes.

Silver bromide and
  silver iodide   —halides used on photographic film. Light intensity is recorded by developing as black metallic silver those portions of the film upon which the light fell during exposure.

Hydrofluoric acid —acid used to etch glass by reacting with the SiO_2 to release silicon fluoride gas. Also used to frost light bulbs.

Uses of the Halogens

The halogen chlorine is used to purify water supplies, to bleach, and to prepare pure hydrochloric acid along with many other industrial applications. Its reactions as a bleach are as follows:

$$Cl_2 + H_2O \rightarrow HCl + HClO \text{ (hypochlorous acid)}$$

It is the hypochlorous acid that does the bleaching by oxidation.

$$\text{Colored material} + HClO \rightarrow \text{decolorized material} + HCl$$

Chlorine also has the ability to withdraw hydrogen from hydrocarbons. An example:

$$C_{10}H_{16} + 8\ Cl_2 \rightarrow 10\ C + 16\ HCl$$

          ↑   ↑

warm turpentine   carbon released
as dense black
smoke

Bromine is used as ethylene dibromide in antiknock gasolines to remove the lead deposits that form from the other additive, tetraethyl lead. It is also used as a bromide in many nerve sedatives.

Iodine is most widely known for its use as an antiseptic in an alcohol solution of iodine called tincture of iodine. Iodide is also used in small amounts to treat goiter conditions.

NITROGEN FAMILY

The most common member of this family is nitrogen itself. It is a colorless, odorless, tasteless, and rather inactive gas that makes up about four fifths of the air in our atmosphere. The inactivity of N_2 gas can be explained by noticing that the two atoms of nitrogen are bonded by three covalent bonds ($:N ::N:$) which require a great deal of energy to break. Since nitrogen must be "pushed" into combining with other elements, many of its compounds tend to decompose violently with a release of the energy that went into forming them.

Nitrogen can be prepared in the laboratory by this equation:

$$NaNO_2 + NH_4Cl \rightarrow NaCl + 2\ H_2O + N_2\uparrow$$

It is most often prepared industrially by the fractional distillation of liquid air. This, of course, contains some of the inert gases but is pure enough for most uses.

Nature "fixes" nitrogen, or makes nitrogen combine, by a nitrogen fixing bacteria found in the roots of beans, peas, clover, and other leguminous plants. Discharges of lightning also cause some nitrogen fixation with oxygen to form nitrogen oxides.

Nitric Acid

An important compound of nitrogen is nitric acid. This acid is useful in making dyes, celluloid film, and many of the lacquers on cars.

Its physical properties are: it is a colorless liquid (when pure), it is one and one-half times as dense as water, it has a boiling point of 86°C, the commercial form is about 68% pure, and it is miscible with water in all proportions.

Its outstanding chemical properties are: the dilute acid shows the usual properties of an acid except it rarely produces hydrogen when it reacts with metals, and it is quite unstable and decomposes as follows:

$$4\ HNO_3 \rightarrow 2\ H_2O + 4\ NO_2\uparrow + O_2\uparrow$$

Because of this ease of decomposition, nitric acid is a good oxidizing agent. When it reacts with metals, the nitrogen product formed will depend on the conditions of the reaction, especially the concentration of the acid, the activity of the metal, and the temperature. If the nitric acid is concentrated and the metal is copper, the principal reduction product will be nitrogen dioxide (NO_2), a heavy, red-brown gas with a pungent odor.

$$Cu + 4\ HNO_3 \rightarrow Cu(NO_3)_2 + 2\ NO_2 \uparrow + 2\ H_2O$$

With dilute nitric acid, this reaction would be:

$$3\ Cu + 8\ HNO_3 \rightarrow 3\ Cu(NO_3)_2 + 4\ H_2O + 2\ NO \uparrow$$

The product is called nitric oxide (NO). This NO, which is colorless, is immediately oxidized in air to NO_2 gas.

With still more dilute nitric acid, considerable quantities of nitrous oxide (N_2O) are formed; with an active metal like zinc, the product may be the ammonium ion (NH_4^+).

When nitric acid is mixed with hydrochloric acid, the mixture is called *aqua rigia* because of its ability to dissolve gold. The great activity of aqua regia is due to the formation of the nitrosyl chloride (NOCl) and chlorine (Cl_2). The Cl_2 reacts with gold to form the soluble compound of gold.

Nitric acid reacts with protein to give a yellow color, and also reacts with other organic compounds to form unstable compounds, such as nitrobenzene, trinitrotoluene (TNT), and picric acid. These last two are powerful explosives. Another explosive is made by reacting glycerine with nitric acid to make glyceryl trinitrate ("nitroglycerine").

The preparation of nitric acid in the laboratory involves reacting a nitrate salt with sulfuric acid in a retort.

The industrial preparation, called the Ostwald process, is the greatest source of nitric acid. Ammonia and air are passed through a heated platinum gauze, which acts as a catalyst, to form nitric oxide (NO).

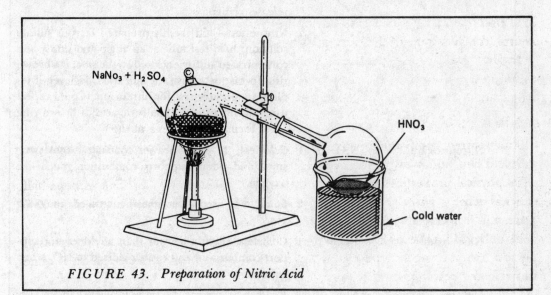

FIGURE 43. Preparation of Nitric Acid

$$4\ NH_3 + 5\ O_2 \rightarrow 4\ NO \uparrow + 6\ H_2O$$

The nitric oxide is further oxidized to NO_2.

$$2\ NO + O_2 \rightarrow 2\ NO_2 \uparrow$$

Then $3\ NO_2 + H_2O \rightarrow 2\ HNO_3 + NO \uparrow$ forms the nitric acid.

Other Important Compounds of Nitrogen

Ammonia is one of the oldest known compounds of nitrogen. In times past, it was prepared by distilling leather scraps, hoofs, and horns. It can be prepared in the laboratory by heating an ammonium salt with a strong base. An example is:

$$Ca(OH)_2 + (NH_4)_2SO_4 \rightarrow CaSO_4 + 2\ H_2O + 2\ NH_3 \uparrow$$

Industrial methods include the Haber process, which combines nitrogen and hydrogen (discussed further on page 122):

$$3\ H_2 + N_2 \rightarrow 2\ NH_3$$

and the destructive distillation of coal which gives off NH_3 as a by-product.

The physical properties of ammonia are: it is a colorless, pungent gas which is extremely soluble in water; it is lighter than air and has a high critical temperature (the critical temperature of a gas is the temperature above which it cannot be liquefied by pressure alone).

Because it can be liquefied easily and it has a high heat of vaporization, one of its uses is as a refrigerant. A water solution of ammonia is used as a household cleaner to cut grease. Another use of ammonia is as a fertilizer.

Some other nitrogen compounds are summarized in Table 14.

Name	Formula	Notes of Importance
Nitrate ion	NO_3^-	1. Nitrates of metals can be decomposed with heat to release oxygen.
		2. Nitrate test — add freshly prepared ferrous sulfate solution; hold test tube at an angle and slowly add concentrated sulfuric acid down the side; if a brown ring forms just above the sulfuric acid level, it indicates a positive test for nitrate ion. (Could repeat using acetic instead of sulfuric acid; if brown ring again forms, it indicates a nitrite.)
Nitrous oxide	N_2O	1. Colorless; $1\frac{1}{2}$ times heavier than air; slight sweet smell and odor; supports combustion much like oxygen.
		2. Sometimes used as an anesthetic called "laughing gas."
Nitric oxide	NO	1. Colorless, slightly heavier than air, does not support combustion, and easily oxidized to NO_2 when exposed to air.
Nitrogen dioxide	NO_2	1. Reddish brown gas, 1.6 times heavier than air, very soluble in water, unpleasant odor, and extremely poisonous.
Nitrogen tetroxide	N_2O_4	1. When NO_2 is cooled, its brown color fades to a pale yellow (N_2O_4). Two molecules of NO_2 combine to form this new compound.

TABLE 14. Some Nitrogen Compounds

Other Members of the Nitrogen Family

As you proceed down this family, the elements change from gases to volatile solids, and then to less volatile solids. At the same time acid-forming properties are decreasing while base-forming properties are increasing. These changes can be attributed to the increased atomic radii and the decreased ability to attract and hold the outer shell electrons.

Name	Symbol	Atomic Number	Properties and Uses
Nitrogen	N	7	(Discussed)
Phosphorus	P	15	Two allotropic forms: white (kindling temperature near room temperature, thus kept under water) and red. When burned in oxygen, forms P_2O_5, the acid anhydride of phosphoric acid. When P_2O_5 reacts with water, it first forms metaphosphoric acid, HPO_3, then more water reacts to form the orthophosphoric acid. Phosphorus is used in making matches.
Arsenic	As	33	Slightly volatile gray solid. Most abundant compound is As_2O_3, which is used as a preservative, in medicines, and glass making (highly poisonous).
Antimony	Sb	51	Lustrous gray solid. Forms very weak acid, mostly basic solutions. Used in type set.
Bismuth	Bi	83	Almost wholly metallic properties. Shiny, dense metal with slight reddish tinge. Low melting point; therefore used in alloys for electric fuses and fire-sprinkler systems. "Woods metal" is the most outstanding of these (melting point 70°C).

TABLE 15. The Nitrogen Family

METALS

Some physical properties of metals are: they have metallic luster, they can conduct heat and electricity, they can be pounded into sheets (malleable), they can be drawn into wires (ductile), most have a silvery color, they have densities usually between 7 and 14 grams per cubic centimeter, and none is soluble in any ordinary solvent without a chemical change.

The general chemical properties of metals are: they are electropositive, and the more active metallic oxides form bases, although some metals form amphoteric hydroxides which can react as both acid and base.

Metals can be removed from their ores by various means depending on their activity. Table 16 summarizes this statement.

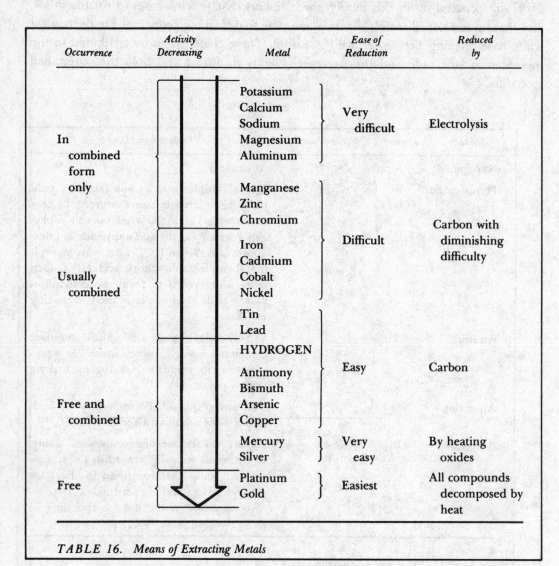

Occurrence	Activity Decreasing	Metal	Ease of Reduction	Reduced by
In combined form only		Potassium Calcium Sodium Magnesium Aluminum	Very difficult	Electrolysis
Usually combined		Manganese Zinc Chromium	Difficult	Carbon with diminishing difficulty
		Iron Cadmium Cobalt Nickel		
Free and combined		Tin Lead HYDROGEN Antimony Bismuth Arsenic Copper	Easy	Carbon
		Mercury Silver	Very easy	By heating oxides
Free		Platinum Gold	Easiest	All compounds decomposed by heat

TABLE 16. Means of Extracting Metals

In some cases the ore is concentrated before it is reduced. This might be a merely physical means, such as flotation, in which the ore is crushed, mixed with a water solution containing a suitable oil like pine oil, and then vigorously agitated so that a frothy mass of oil rises to the top with the adhering metal-bearing portion of the ore. The froth is removed from the top of the machine and filtered to recover the metal ore.

A chemical process is used to convert sulfide or carbonate ores to oxides. It is called *roasting* and consists of heating the compounds in the air. For example:

$$2 \, ZnS + 3 \, O_2 \rightarrow 2 \, ZnO + 2 \, SO_2 \uparrow$$

and

$$PbCO_3 \rightarrow PbO + CO_2 \uparrow$$

These oxides can then be treated with appropriate reducing agents to free the metal.

The following table summarizes the properties of the first two families of metals found in group IA and IIA of the periodic chart. It shows the distinct similarities in their properties and bonding. These are directly related to the similarities that exist in the outer shell electron configuration.

	ALKALI METALS		
Element	Outer Shell	Bonding and Charge	Physical Properties
Lithium Li	$2s^1$	Ionic, 1+	Silvery white
Sodium Na	$3s^1$	Ionic, 1+	Silvery white
Potassium K	$4s^1$	Ionic, 1+	Silvery white
Rubidium Rb	$5s^1$	Ionic, 1+	Silvery white
Cesium Cs	$6s^1$	Ionic, 1+	Silvery white
Francium Fr	$7s^1$	Ionic, 1+	Silvery white
	ALKALINE-EARTH METALS		
Beryllium Be	$2s^2$	Ionic, 2+	Gray-white
Magnesium Mg	$3s^2$	Ionic, 2+	Silvery white
Calcium Ca	$4s^2$	Ionic, 2+	Silvery white
Strontium Sr	$5s^2$	Ionic, 2+	Silvery white
Barium Ba	$6s^2$	Ionic, 2+	Yellow white; lumpy
Radium Ra	$7s^2$	Ionic, 2+	White; radioactive

Some Important Reduction Methods

The Hall process is the electrolytic method of preparing *aluminum*. It uses molten cryolite (Na_3AlF_6) as the solvent for bauxite ore ($Al_2O_3 \cdot 2 H_2O$). This mixture is electrolyzed in a rectangular iron tank lined with carbon which acts as the cathode. The anodes are heavy carbon rods suspended in the tank. The aluminum released at the cathode is tapped periodically and allowed to flow out of the tank so that the process can continue.

The aluminum metal's outstanding properties are its light weight with high strength, its ability to coat itself with a protective aluminum oxide coating that prevents further oxidation, and its ability to conduct an electric current.

The Dow process is an electrolytic method of preparing *magnesium* from sea water. The magnesium salts are precipitated with $Ca(OH)_2$ solution to form $Mg(OH)_2$. The hydroxide is then converted to a chloride by adding hydrochloric acid to the hydroxide. This chloride is mixed with potassium chloride, melted, and electrolyzed to release the magnesium at the cathode and chlorine at the anode.

Magnesium has the important property of being a light, rigid, and inexpensive metal. It, too, can form a protective adherent oxide coating.

Another major metal prepared by electrolysis is *copper*. After the ore is concentrated by flotation and roasted to change sulfides to oxides, it is heated in a converter to get impure copper called "blister" copper. This copper is then used as anodes in the electrolysis apparatus, in which pure copper is used to make the cathode plates. As the process progresses, the copper in the anode becomes ions in the solution which are reduced at the cathode as pure copper.

Copper has wide usage in electrical appliances and wires because of its high conductivity. It is widely used in alloys of bronze (copper and tin) and brass (copper and zinc).

Iron ore is refined by reduction in the blast furnace. The blast furnace is a large, cylinder-shaped furnace which is charged with iron ore (usually hematite, Fe_2O_3), limestone, and coke, A hot air blast, often enriched with oxygen, is blown into the lower part of the furnace through a series of pipes called tuyeres. The chemical reactions that occur can be summarized as follows:

Burning coke:
$$2\ C + O_2 \rightarrow 2\ CO \uparrow$$
$$C + O_2 \rightarrow CO_2 \uparrow$$

Reduction of CO_2: $\quad CO_2 + C \rightarrow 2\ CO \uparrow$

Reduction of ore: $\quad Fe_2O_3 + 3\ CO \rightarrow 2\ Fe + 3\ CO_2 \uparrow$
$$Fe_2O_3 + 3\ C \rightarrow 2\ Fe + 3\ CO \uparrow$$

Formation of slag: $\quad CaCO_3 \rightarrow CaO + CO_2 \uparrow$
$$CaO + SiO_2 \rightarrow CaSiO_3$$

The molten iron from the blast furnace is called "pig" iron.

From pig iron, the molten metal might be taken to one of three steel-making processes which burn out impurities and set the content of carbon, manganese, sulfur, phosphorus, and silicon. Often nickel and chromium are alloyed in steel to give particular properties of hardness needed for tool parts. The three most prominent methods of making steel are the basic oxygen furnace, the open-hearth, and the electric furnace.

The basic oxygen uses a lined "pot" into which the molten pig iron is poured. Then a high speed jet of oxygen is blown from a water cooled lance into the top of the pot. This "burns out" impurities to make a batch of steel rapidly and cheaply.

The open-hearth furnace is a large oven containing a dish-shaped area to hold the molten iron, scrap steel, and other additives with which it is charged. Alternating blasts of flame are directed across the surface of the melted metal until the proper proportions of additives are established for that "heat" so that the steel will have particular properties needed by the customer. The tapping of one of these furnaces holding 50 to 400 tons of steel is a truly beautiful sight.

The final method of making steel is the electric furnace. This method uses enormous amounts of electricity through graphite cathodes which are lowered into the molten iron to purify it and make a high grade of steel.

Alloys are mixtures of two or more metals. In a mixture certain properties of the metals involved are affected. Some of these are:

1. Melting point   The melting point of an alloy is lower than that of its components.

2. Hardness   An alloy is usually harder than the metals which compose it.

3. Crystal structure   The size of the crystalline particles in the alloy determine many of the physical properties. The size of these particles can be controlled by heat treatment. If the alloy cools slowly, the crystalline particles tend to be larger. Thus, by heating and cooling an alloy, its properties can be altered considerably.

Chapter
13
Review

1. The most active nonmetallic element is (1) chlorine (2) fluorine (3) oxygen (4) sulfur.

2. A burning paraffin taper is inserted into a bottle of chlorine and will result in (1) extinguishing the flame (2) production of carbon (3) production of hydrogen (4) production of carbon tetrachloride.

3. The bromide ion is (1) colorless (2) red-orange (3) violet (4) yellow.

4. When chlorine water and carbon tetrachloride are mixed with an unknown, a violet layer is produced. The unknown was a(n) (1) fluoride (2) chloride (3) bromide (4) iodide.

5. The order of decreasing activity of the halogens is (1) Fl, Cl, I, Br (2) F, Cl, Br, I (3) Cl, F, Br, I (4) Cl, Br, I, F.

6. A light-sensitive substance used on photographic films has the formula (1) AgBr (2) CaF_2 (3) CuCl (4) $MgBr_2$.

7. Aqua regia is (1) HCl and H_2SO_4 (2) HCl and HNO_3 (3) HCl and HBr (4) HCl and HF.

8. The metal which has an orange sulfide is (1) Cd (2) As (3) Pb (4) Sb.

9. Sulfur dioxide is the anhydride of (1) hydrosulfuric acid (2) sulfurous acid (3) sulfuric acid (4) hyposulfurous acid.

10. The charring action of sulfuric acid is due to its being (1) a strong acid   (2) an oxidizing agent   (3) a reducing agent   (4) a dehydrating agent.

11. Sulfuric acid is used in the manufacture of many other acids because of its (1) high specific gravity   (2) high boiling point   (3) oxidizing ability   (4) solubility.

12. Nitrogen for commercial use is generally obtained from (1) ammonia   (2) liquid air   (3) sodium nitrate   (4) nitrogen-fixing bacteria.

13. Ammonia is prepared commercially by the (1) Ostwald process   (2) arc process   (3) Haber process   (4) contact process.

14. A nitrogen compound which has a color is (1) nitric oxide   (2) nitrous oxide   (3) nitrogen dioxide   (4) ammonia.

15. If a student heats a mixture of ammonium chloride and calcium hydroxide in a test tube, he will detect (1) no reaction   (2) the odor of ammonia   (3) the odor of rotten eggs   (4) nitric acid fumes.

16. The difference between ammonia and ammonium is (1) an electron   (2) a neutron   (3) a proton   (4) radioactivity.

17. An important ore of iron is (1) bauxite   (2) galena   (3) hematite   (4) smithsonite.

18. The material added to the charge of a blast furnace to react with the sand is (1) calcium carbonate   (2) coke   (3) silicon dioxide   (4) slag.

19. A reducing agent used in the blast furnace is (1) $CaCO_3$ (2) CO   (3) O_2 (4) SiO_2.

20. The most abundant of all metals in the earth's crust is (1) iron   (2) aluminum   (3) copper   (4) silver.

21. The solvent used in obtaining aluminum from its ore is (1) bauxite   (2) cryolite   (3) dolomite   (4) carnalite.

22. The process of extracting aluminum is called (1) the Dow process   (2) the Haber process   (3) the Hall process   (4) electroplating.

23. Aluminum is extracted from its oxide by (1) roasting   (2) reduction with carbon   (3) smelting   (4) electrolysis.

24. The Dow process extracts magnesium from (1) sea shells   (2) dolomite ($CaCO_3$-$MgCO_3$)   (3) sea water   (4) oyster shells.

25. An alloy of bismuth is used in automatic fire sprinklers because (1) it is soft   (2) it has a low melting point   (3) it has a high melting point   (4) it resists corrosion.

• ANSWERS:

1. (2)	10. (4)	18. (1)
2. (2)	11. (2)	19. (2)
3. (1)	12. (2)	20. (2)
4. (4)	13. (3)	21. (2)
5. (2)	14. (3)	22. (3)
6. (1)	15. (2)	23. (4)
7. (2)	16. (3)	24. (3)
8. (4)	17. (3)	25. (2)
9. (2)		

Carbon and Organic Chemistry

CARBON AND FUELS

The element carbon occurs in three allotropic forms. These are diamond, graphite, and amorphous (although some evidence shows the amorphous forms do have some crystalline structure).

The diamond structure has a close-packed crystal structure that gives it its property of extreme hardness. In it each carbon is bonded to four other carbons in a tetrahedron arrangement like this:

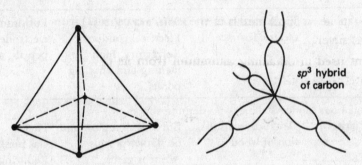

sp³ hybrid of carbon

It has been possible to make man-made diamonds in machines which subject carbon to extremely high pressures and temperatures. Most of these diamonds are used for industrial purposes, such as dies.

The graphite structure is made up of planes of hexagonal structures which are weakly bonded to the planes above and below. This explains graphite's slippery feeling and makes it useful as a dry lubricant. Its structure can be seen below. Graphite also has the property of being an electrical conductor.

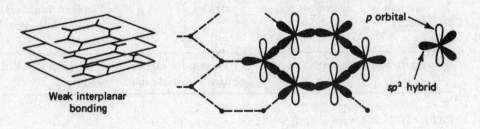

Weak interplanar bonding

p orbital

sp² hybrid

Some of the amorphous forms of carbon are shown in Table 17, along with their properties and uses.

The process of destructive distillation is used in the preparation of both charcoal and coke and should be described. This process involves the heating of a substance in an enclosed container so that volatile materials can be forced out. These volatile materials are the substances which would cause wood and coal to burn with a flame. After their removal by this process, the charcoal and coke merely glow when burned. Many of these volatile substances have proven to be useful by-products of the destructive distillation. From the destructive distillation of coal we get coal gas, coal tar, and ammonia. The coal gas can be used as a fuel; the coal tars are the primary source of most of our synthetic dyes and many other useful organic compounds, and the ammonia is also a useful chemical.

Forms	Occurrence or Preparation	Properties	Uses
1. Crystalline			
a. Diamond	South Africa; machines using very high pressures and temperatures	Hardest substance Brilliant: reflects and refracts light	Abrasive: drill, saw, polish bearings, gem, die
b. Graphite	From hard coal in electric furnace	Soft, gray, greasy Electrical conductor Refractory: high melting and kindling points	Lubricant, crucibles, electrodes, lead pencils, atomic pile
2. Amorphous (noncrystalline)—no definite shape			
a. Charcoal	Destructive distillation of wood	Burns with a glow and no flame; porous Adsorbs gases Reducing agent	Fuel Gas masks (activated) Reducing agent
b. Coke	Destructive distillation of soft coal	Burns with little smoke or flame; porous gray Reducing agent	Fuel and reducing agent in metallurgy Manufacture of water gas, SiC, CaC_2, CS_2
c. Boneblack	Destructive distillation of bones	Adsorbs coloring matter	Decolorizing sugar
d. Lampblack (carbon black)	Incomplete combustion of natural gas (or oil)	Soft, black, minute particles	Rubber tires Pigment: India and printing ink, shoe polish
e. Anthracite coal	Pennsylvania	Almost pure carbon, burns with little smoke	Fuel, production of water gas

TABLE 17. Allotropes of Carbon

Carbon Dixoide

Carbon dioxide is a widely distributed gas which makes up .04 percent of the air. There is a cycle that keeps this figure relatively stable. It is shown in Figure 44.

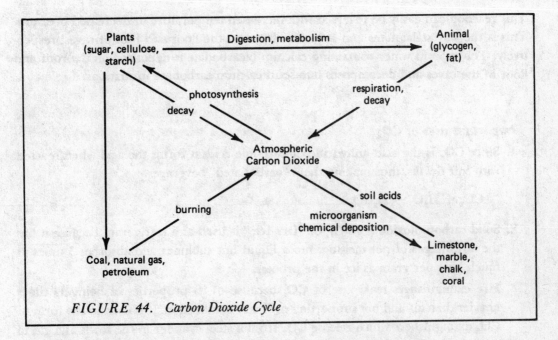

FIGURE 44. Carbon Dioxide Cycle

Laboratory preparation of CO_2. The usual laboratory preparation consists of reacting calcium carbonate (marble chips) with hydrochloric acid, although any carbonate or bicarbonate and any common acid could be used. The gas is collected by water displacement or air displacement. (See Figure 45.)

The test for carbon dioxide consists of passing it through limewater ($Ca(OH)_2$). If CO_2 is present the limewater turns cloudy because of the formation of a white precipitate of finely divided $CaCO_3$:

$$Ca(OH)_2 + CO_2 \rightarrow CaCO_3 \downarrow + H_2O$$

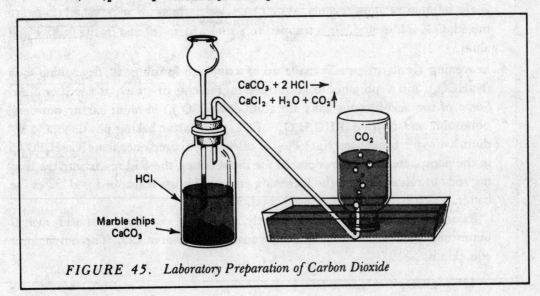

FIGURE 45. Laboratory Preparation of Carbon Dioxide

Continued passing of CO_2 into the solution will clear the cloudy condition because the insoluble $CaCO_3$ becomes soluble calcium bicarbonate $(Ca(HCO_3)_2)$.

$$CaCO_3 \downarrow + H_2O + CO_2 \rightarrow Ca^{2+}(HCO_3)_2{}^-$$

This reaction can easily be reversed with increased temperature or decreased pressure. This is the way stalagmites and stalactites form on the floor and roof of caves, respectively. The ground water containing calcium bicarbonate is deposited on the roof and floor of the caves and decomposes into solid calcium carbonate formations.

Important uses of CO_2

1. Since CO_2 is the acid anhydride of carbonic acid, it forms the acid when reacted with soft drinks, thus making them "carbonated" beverages.

$$CO_2 + H_2O \rightarrow H_2CO_3$$

2. Solid carbon dioxide ($-78°C$), or "Dry Ice," is used as a refrigerant because it has the advantages of not melting into a liquid but sublimes and absorbs 3 times as much heat per gram as ice in the process.

3. Fire extinguishers make use of CO_2 because of its properties of being $1\frac{1}{2}$ times heavier than air and not supporting ordinary combustion. It is used in the form of CO_2 extinguishers which release CO_2 from a steel cylinder in the form of a gas to smother the fire. The soda type extinguisher uses a solution of sodium bicarbonate $(NaHCO_3)$ and a small bottle of sulfuric acid in a cylinder which must be inverted to operate it. When the cylinder is inverted, the acid mixes with the solution to cause the release of CO_2. This gas exerts pressure on the fluid and forces it out of the extinguisher. The equation in ionic form is:

$$2\,Na^+ + 2\,HCO_3{}^- + 2\,H^+ + SO_4{}^{2-} \rightarrow 2\,Na^+ + SO_4{}^{2-} + 2\,H_2O + 2\,CO_2 \uparrow$$

The foam-type extinguisher is similar to the above. It contains a charge of a water solution of sodium bicarbonate and licorice extract and a small bottle containing a water solution of alum (mainly $Al_2(SO_4)_3$), which acts as an acid. When these are mixed, CO_2 is liberated but is trapped in a sticky foam formed by the licorice and alum.

4. Leavening agents are usually made up of a dry acid-forming salt, dry baking soda $(NaHCO_3)$, and a substance to keep these ingredients dry, such as starch or flour. Some of the acid-forming salts are alum $(NaAl(SO_4)_2)$ in alum baking powders, potassium acid tartrate $(KHC_4H_4O_6)$ in cream of tartar baking powders, and sodium hydrogen phosphate (NaH_2PO_4) or calcium hydrogen phosphate $(Ca(H_2PO_4)_2)$ in the phosphate baking powders. In the moist batter, the acid reacts with the baking soda to release CO_2 as the leavening agent. This gas is trapped and raises the batter, which hardens in this raised condition.

Baking soda, when used independently, must have some source of acid, such as buttermilk, molasses, or sour milk, to cause the release of CO_2. The simple ionic equation is:

$$H^+ + HCO_3{}^- \rightarrow H_2O + CO_2 \uparrow$$

5. Plants use up CO_2 in the photosynthesis process. In this process, chlorophyll (the catalyst) and sunlight (the energy source) must be present. The reactants and products of this reaction are:

$$6\ CO_2 + 6\ H_2O \rightarrow C_6H_{12}O_6 + 6\ O_2 \uparrow$$
simple
sugar

OR

$$6\ CO_2 + 5\ H_2O \rightarrow C_6H_{10}O_5 + 6\ O_2 \uparrow$$
cellulose

6. The Solvay process for making $NaHCO_3$ and Na_2CO_3 uses CO_2. The CO_2 is combined with NH_3 and water, and the product is reacted with NaCl. Sodium bicarbonate ($NaHCO_3$) precipitates out of solution since its solubility in the NaCl solution is exceeded. The overall equation can be shown as:

$$NaCl + H_2O + NH_3 + CO_2 \rightarrow NaHCO_3 \downarrow + NH_4Cl$$

Sodium carbonate is made by heating the $NaHCO_3$:

$$2\ Na^+HCO_3^- \rightarrow Na_2^+CO_3^{2-} + H_2O + CO_2 \uparrow$$

ORGANIC CHEMISTRY

Organic chemistry may be defined simply as the chemistry of the compounds of carbon. Since Friedrich Woehler (Wöhler) synthesized urea in 1828, chemists have synthesized thousands of carbon compounds in areas of dyes, plastic, textile fibers, medicines, and drugs. The number of organic compounds have been estimated in the neighborhood of a million and constantly increasing.

The carbon atom (atomic number 6) has four electrons in its outermost shell which show a tendency to be shared (electronegativity of 2.5) in covalent bonds. By this means, carbon bonds to other carbons, hydrogens, halogens, oxygen and other elements to form the many compounds of organic chemistry.

Hydrocarbons

Hydrocarbons, as the name implies, are compounds containing only carbon and hydrogen in their structure. The simplest hydrocarbon is methane, CH_4. This type of formula, which shows the kinds of atoms and their respective numbers, is called an *empirical* formula. In organic chemistry this is not sufficient to identify the compound it is used to represent. A good example of this is C_2H_6O. This formula could denote a structure called either an ether or an ethyl alcohol. For this reason, a *structural* formula is used to indicate how the atoms are arranged in the molecule. The ether of C_2H_6O looks like this:

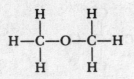

whereas the ethyl alcohol is represented by this structural formula:

$$\begin{array}{c} \ \ \ \ H \ \ H \\ \ \ \ \ | \ \ \ | \\ H-C-C-OH \\ \ \ \ \ | \ \ \ | \\ \ \ \ \ H \ \ H \end{array}$$

For this reason, then, structural formulas are more often used in organic chemistry.

Methane would be
$$\begin{array}{c} \ \ \ H \\ \ \ \ | \\ H-C-H \\ \ \ \ | \\ \ \ \ H \end{array}$$

Alkanes

Methane is the first member of a hydrocarbon series called the alkanes (or paraffin series). The general formula for this series is C_nH_{2n+2}, where n is the number of carbons in the molecule. Table 18 provides some essential information about this series. Since many of the other organic structures use the stem of these alkane names, you should learn these names and structures well. Notice, that as the number of carbons in the chain increases, the boiling point also increases. The first four are gases at room temperature; the subsequent compounds are liquid, then become more viscous with increasing length of the chain.

Since the chain is increased by a carbon and two hydrogens in each subsequent molecule, this is referred to as an *homologous* series.

The alkanes are found in petroleum and natural gas. They are usually extracted by fractional distillation, which separates the compounds by varying the temperature so that each vaporizes at its respective boiling point. Methane, which forms 90% of natural gas, can also be prepared in the laboratory by heating soda lime (containing NaOH) with sodium acetate:

$$NaC_2H_3O_2 + NaOH \rightarrow CH_4\uparrow + Na_2CO_3$$

When the alkanes are burned with sufficient air, the compounds formed are CO_2 and H_2O. An example would be:

$$2\ C_2H_6 + 7\ O_2 \rightarrow 4\ CO_2\uparrow + 6\ H_2O\uparrow$$

The alkanes can be reacted with halogens so that hydrogens are replaced by an halogen atom:

$$\begin{array}{c} \ \ \ H \\ \ \ \ | \\ H-C-H + Br_2 \rightarrow H-C-BR + HBr \\ \ \ \ | \\ \ \ \ H \end{array}$$

Some common substitution compounds of methane are:

methyl chloride
monochloromethane

chloroform
trichloromethane

carbon tetrachloride
tetrachloromethane

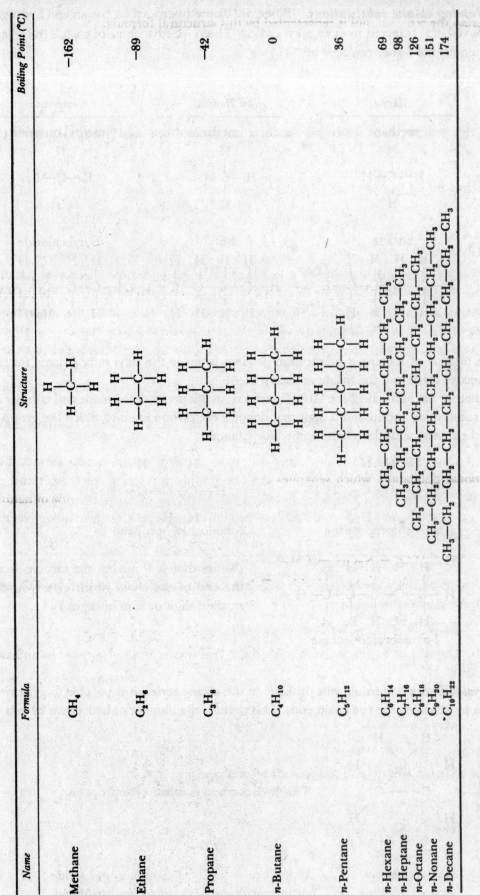

Name	Formula	Structure	Boiling Point (°C)
Methane	CH_4		−162
Ethane	C_2H_6		−89
Propane	C_3H_8		−42
n-Butane	C_4H_{10}		0
n-Pentane	C_5H_{12}		36
n-Hexane	C_6H_{14}	$CH_3{-}CH_2{-}CH_2{-}CH_2{-}CH_2{-}CH_3$	69
n-Heptane	C_7H_{16}	$CH_3{-}CH_2{-}CH_2{-}CH_2{-}CH_2{-}CH_2{-}CH_3$	98
n-Octane	C_8H_{18}	$CH_3{-}CH_2{-}CH_2{-}CH_2{-}CH_2{-}CH_2{-}CH_2{-}CH_3$	126
n-Nonane	C_9H_{20}	$CH_3{-}CH_2{-}CH_2{-}CH_2{-}CH_2{-}CH_2{-}CH_2{-}CH_2{-}CH_3$	151
n-Decane	$C_{10}H_{22}$	$CH_3{-}CH_2{-}CH_2{-}CH_2{-}CH_2{-}CH_2{-}CH_2{-}CH_2{-}CH_2{-}CH_3$	174

TABLE 18. The Alkanes

Naming alkane substitutions. When an alkane hydrocarbon has an end hydrogen removed, it is referred to as an alkyl radical. The respective name of each is the alkane name with the "-ane" replaced by "-yl."

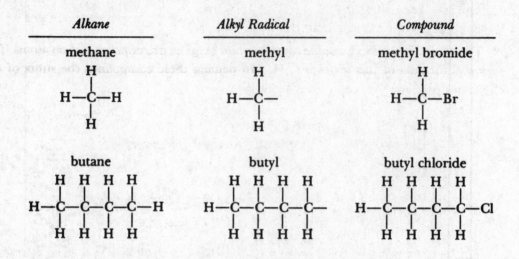

Alkane	*Alkyl Radical*	*Compound*
methane	methyl	methyl bromide
butane	butyl	butyl chloride

One method of naming a substitution product is to use the alkyl radical name for the respective chain and the halide as shown above.

Another method, called the IUPAC method, uses a prefix of the name and number of the substitution group and a digit that denotes the carbon atom to which the group is attached along with the alkane name. For example:

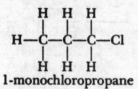

1-monochloropropane

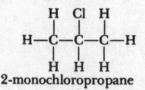

2-monochloropropane

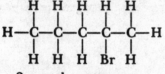

2-monobromopentane

(Notice that you number the carbons from the end of the chain which gives you the smallest digit or sum of digits.)

Cycloalkanes. Starting with propane in the alkane series, it is possible to get a ring form by attaching the two chain ends. This reduces the number of hydrogens by two.

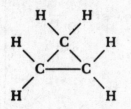

This hydrocarbon is called cyclopropane.

Since all the alkane series are made up of single covalent bonds, this series and all such structures are said to be *saturated*. If the hydrocarbon molecule contains double or triple covalent bonds, it is referred to as *unsaturated*.

Alkene Series (unsaturated)

The alkene series has a double covalent bond between two adjacent carbon atoms. The general formula of this series is C_nH_{2n}. In naming these compounds the suffix of the alkane is replaced by "-ene." Two examples:

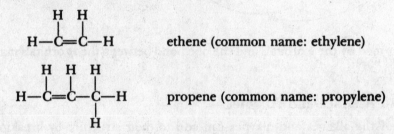

ethene (common name: ethylene)

propene (common name: propylene)

The bonding is more complex in the double covalent bond than in the single bonds in the molecule. Using the orbital pictures of the atom, we can show this as follows:

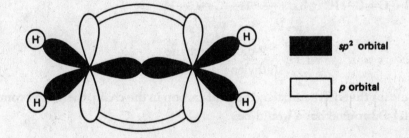

sp^2 orbital

p orbital

The two p lobes attached above and below constitute *one* bond called a pi (π) bond.
The sp^2 orbital bonds between the carbons and with each hydrogen are referred to as sigma (σ) bonds.

Alkyne Series (unsaturated)

The alkyne series has a triple covalent bond between two adjacent carbons. The general formula of this series is C_nH_{2n-2}. In naming these compounds, the alkane suffix is replaced by "-yne." A few examples:

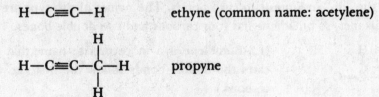

ethyne (common name: acetylene)

propyne

The orbital structure can be shown as follows:

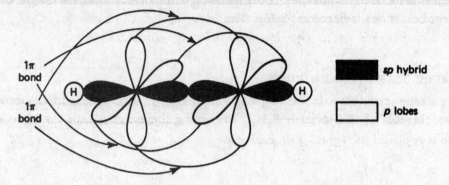

The bonds formed by the *p* orbitals and the one bond between the *sp* orbitals make up the triple bond.

Additions to Alkenes and Alkynes

Unsaturates of the alkenes and alkynes can add to their structure by breaking the double or triple bond present. For example:

1,2-dibromoethane
(ethylene dibromide)

The 1,2- means that on the first and second carbon in the chain, there is a bromine atom bonded. 1,1-Dibromoethane would be:

$$\begin{array}{cc} Br & H \\ | & | \\ H-C-C-H \\ | & | \\ Br & H \end{array}$$

Addition to ethyne could be:

$$H-C\equiv C-H + 2\ Br_2 \rightarrow \begin{array}{cc} Br & Br \\ | & | \\ H-C-C-H \\ | & | \\ Br & Br \end{array}$$

Alkadienes have two double covalent bonds in each molecule. The "-ene" indicates the double bond, and the "di-" indicates two such bonds. The names of this type are also derived from the alkanes, so butadiene has four carbons and two double bonds.

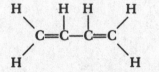

(1,3-Butadiene is a more precise name that indicates the double bonds follow the first and third carbons.)

This compound is used in making synthetic rubber.

Aromatics

The aromatic compounds are unsaturated ring structures. The simplest compound of the aromatic series is benzene (C_6H_6). The basic formula of this series is C_nH_{2n-6}. The benzene structure is a resonance structure which is represented like this:

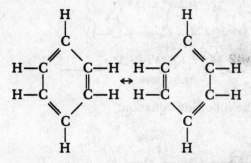

Note: The carbon-to-carbon bonds are neither single nor double bonds but hybrid bonds. This is called *resonance*.

The orbital structure can be represented like this:

- sp^2 orbital
- p orbital
- s orbital

Most of the aromatics have an aroma, thus the name "aromatic."

The C_6H_5 group is a radical called phenyl. This is the benzene structure with one hydrogen missing. If the phenyl radical adds a methyl group, the compound is called toluene or methyl benzene.

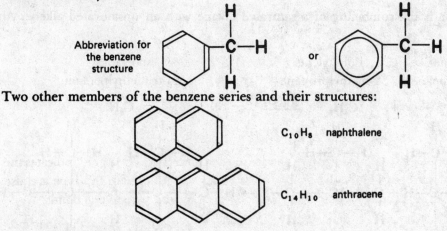

Abbreviation for the benzene structure

or

Two other members of the benzene series and their structures:

$C_{10}H_8$ naphthalene

$C_{14}H_{10}$ anthracene

Changing Hydrocarbons

Isomers

Many of the chain hydrocarbons can have the same formula, but their structures may differ. For example, butane is the first compound that can have two different structures or *isomers* for the same formula.

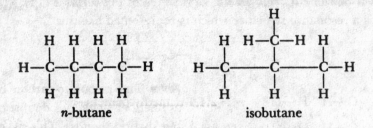

n-butane   isobutane

This isomerization can be shown by the following equation:

$$CH_3—CH_2—CH_2—CH_3 \xrightarrow[70-100°C]{AlCl_3} CH_3—\overset{\displaystyle CH_3}{\overset{|}{CH}}—CH_3$$

butane   isobutane

The isomers have different properties, both physical and chemical, from the normal structure.

Cracking

Under proper conditions of temperature and pressure and often in the presence of a catalyst, long chains of hydrocarbons may be made into more useful smaller molecules. This process is called cracking. An example:

$$C_{16}H_{34} \rightarrow C_8H_{18} + C_8H_{16}$$

hexadecane   octane   octene

Notice that the products formed by cracking a saturated hydrocarbon of the alkanes are a saturated hydrocarbon and an unsaturated alkene.

Alkylation

Alkylation is the combining of a saturated alkane with an unsaturated alkene. An example:

isobutane   isobutylene   isooctane
2-methylpropane   2-methylpropene   2,2,4-trimethylpentane

$$C_4H_{10} \quad + \quad C_4H_8 \xrightarrow{H_2SO_4} \quad\quad C_8H_{18}$$

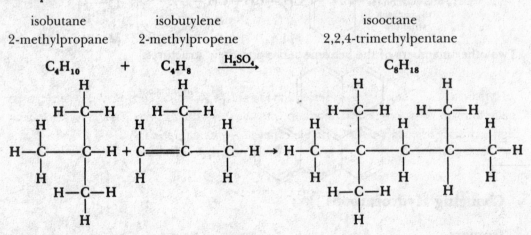

Notice that the product formed by alkylation is a useful chain hydrocarbon without the previous double bond.

Polymerization

The combination of two or more unsaturated molecules to form a larger chain molecule is called polymerization. An example is:

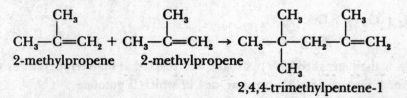

$$CH_3-\underset{\underset{}{}}{\overset{\overset{CH_3}{|}}{C}}=CH_2 + CH_3-\overset{\overset{CH_3}{|}}{C}=CH_2 \rightarrow CH_3-\overset{\overset{CH_3}{|}}{\underset{\underset{CH_3}{|}}{C}}-CH_2-\overset{\overset{CH_3}{|}}{C}=CH_2$$

2-methylpropene   2-methylpropene

2,4,4-trimethylpentene-1

In the product name, 2,4,4-trimethylpentene-1, the first three numbers indicate the carbons which have the trimethyl (three methyl radicals) attached—one each. The -1 at the end indicates that the first carbon has the double bond attaching it to the next carbon. In numbering compounds of the alkenes and alkynes, you always number the carbon atoms from the end which has the double or triple bond.

Hydrogenation

Hydrogenation is the process of adding hydrogens to an unsaturated hydrocarbon in the presence of a suitable catalyst. It is often used to make liquid unsaturated fats into solid saturated fats. An example for hydrocarbons is:

$$CH_3-\overset{\overset{CH_3}{|}}{C}=CH_2 + H_2 \rightarrow CH_3-\overset{\overset{CH_3}{|}}{\underset{\underset{H}{|}}{C}}-CH_3$$

2-monomethylpropene-1   2-monomethylpropane

Dehydrogenation

Dehydrogenation is the removal of hydrogens from a chain molecule in the presence of a catalyst to form an unsaturated molecule. An example is:

$$CH_3-CH_2-CH_2-CH_3 \rightarrow CH_2=CH-CH_2-CH_3 + H_2$$
butane   butene-1

Aromatization

Hydrocarbons of the alkane series having six or more carbons in the chain can be made to form an aromatic hydrocarbon with the loss of hydrogen. This must be done at high temperatures and with a suitable catalyst. An example is:

$$CH_3-CH_2-CH_2-CH_2-CH_2-CH_2-CH_3 \rightarrow$$
heptane

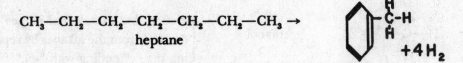

$+4H_2$

toluene

Fischer-Tropsch Process

This process was developed to make gasoline from natural gas, coal, or lignite by converting the original carbon compound to water gas. For example:

$$2 CH_4 + O_2 \rightarrow 2 CO + 4 H_2$$
water gas

This water gas is then enriched with hydrogen and reacted at high temperatures with a catalyst to form a mixture of hydrocarbons, one of which is gasoline.

Alcohols

Alcohols are alkanes which have one or more hydrogen atom replaced by the hydroxyl group, —OH. This is called its functional group.

Methanol is the simplest alcohol. Its structure is

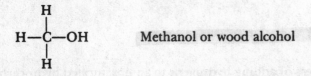

Methanol or wood alcohol

• LABORATORY PREPARATION

$$CH_3Cl + NaOH \rightarrow NaCl + CH_3OH$$

chloromethane   methanol
(methylchloride)   (methyl alcohol)

• INDUSTRIAL PREPARATION

For many years methanol, wood alcohol, was obtained by the destructive distillation of wood. Today it is made by the hydrogenation of oxides of carbon in the presence of a suitable catalyst.

$$CO + 2 H_2 \xrightarrow[\substack{300-400°C \\ 200 \text{ atm.}}]{\text{catalyst}} CH_3OH$$

• PROPERTIES AND USES

Methanol is a colorless, flammable liquid with a boiling point of 65°C. It is miscible with water, is exceedingly poisonous, and can cause blindness if taken internally. It can be used as a fuel, as a solvent, and as a denaturant to make denatured ethyl alcohol unsuitable for drinking.

Ethanol is the best known and most used alcohol. Its structure is

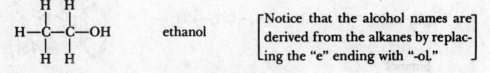

ethanol

[Notice that the alcohol names are derived from the alkanes by replacing the "e" ending with "-ol."]

Its common names are ethyl alcohol or grain alcohol.

• LABORATORY PREPARATION

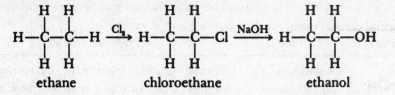

ethane   chloroethane   ethanol

Simple sugars can be converted to ethanol by the action of an enzyme (zymase) which is found in yeast.

$$C_6H_{12}O_6 \rightarrow 2\ C_2H_5OH + 2\ CO_2 \uparrow$$

sugar   ethanol

After fermentation, the alcohol is distilled off. Grains, potatoes, and other starch plants, which can be treated with an acid-water solution to form sugars, can be converted into ethyl alcohol or "whiskey."

• INDUSTRIAL PREPARATION

The usual process for producing industrial alcohol consists of treating ethyene (or ethylene) with concentrated H_2SO_4 and then hydrolyzing the resulting ethyl hydrogen sulfate to ethanol.

$$C_2H_4 \xrightarrow{H_2SO_4} C_2H_5HSO_4 \xrightarrow{H_2O} C_2H_5OH + H_2SO_4$$

• PROPERTIES AND USES

Ethanol is a colorless, flammable liquid with a boiling point of 78°C. It is miscible with water and is a good solvent for a wide variety of substances (these solutions are often referred to as "tinctures"). It is used as an antifreeze because of its low freezing point, −115°C, and for making acetaldehyde and ether.

Other alcohols. Isomeric alcohols have similar formulas, but different properties because of their differences in structure. If the —OH is attached to an end carbon, the alcohol is called a primary alcohol. If attached to a "middle" carbon, it is called a secondary alcohol. Some examples:

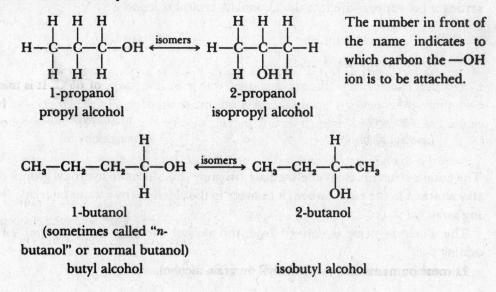

The number in front of the name indicates to which carbon the —OH ion is to be attached.

1-propanol   2-propanol
propyl alcohol   isopropyl alcohol

1-butanol   2-butanol
(sometimes called "*n*-butanol" or normal butanol)
butyl alcohol   isobutyl alcohol

Phenols are alcohols derived from the aromatic hydrocarbons. Some important examples:

Structure	Name	Properties or Uses
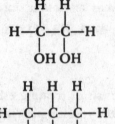	Phenol Carbolic acid Hydroxybenzene	Slightly acidic, extremely corrosive, poisonous. Used to make synthetic resins, plastics, drugs, dyes, and photographic developers. Good disinfectant.

Other alcohols with more than one —OH group:

Structure	Name	Properties or Uses
H H H—C—C—H OH OH	ethylene glycol 1,2-ethanediol	A colorless liquid, high boiling point, low freezing point. Used as permanent antifreeze in automobiles.
H H H H—C—C—C—H OH OH OH	glycerine glycerol 1,2,3-propanetriol	Colorless liquid, odorless, viscous, sweet taste. Used to make nitroglycerine, resins for paint, and cellophane.

Aldehydes

The functional group of an aldehyde is the $-C\begin{smallmatrix}O\\H\end{smallmatrix}$, formyl group. The general formula is RCHO, where R represents an hydrocarbon radical.

• PREPARATION FROM AN ALCOHOL

Aldehydes can be prepared by the oxidation of an alcohol. This can be done by inserting a hot copper wire into the alcohol. A typical reaction is:

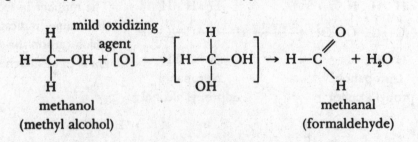

methanol (methyl alcohol)   methanal (formaldehyde)

The middle structure is an intermediate structure, but since two hydroxyl groups do not stay attached to the same carbon, it changes to the aldehyde by a water molecule "breaking away."

The aldehyde name is derived from the alcohol name by dropping the "-ol" and adding "-al."

Ethanol forms ethanal (acetaldehyde) in the same manner.

Organic Acids or Carboxylic Acids

The functional group of an organic acid is the 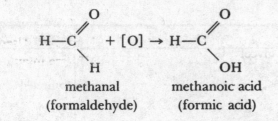 , carboxyl group. The general formula is R—COOH.

• PREPARATION FROM AN ALDEHYDE

Organic acids can be prepared by the mild oxidation of an aldehyde. The simplest acid is methanoic acid, which is present in ants, bees, and other insects. A typical reaction is:

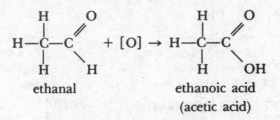

methanal methanoic acid
(formaldehyde) (formic acid)

Notice that the name is derived from the alkane stem by adding "-oic."

Ethanal can be oxidized to ethanoic acid:

$$H-\underset{\underset{H}{|}}{\overset{\overset{H}{|}}{C}}-C\underset{OH}{\overset{O}{\diagup}} + [O] \rightarrow H-\underset{\underset{H}{|}}{\overset{\overset{H}{|}}{C}}-C\underset{OH}{\overset{O}{\diagup}}$$

ethanal   ethanoic acid
(acetic acid)

Acetic acid, as ethanoic acid is commonly called, is a mild acid which, in the concentrated form, is called glacial acetic acid. Glacial acetic is used in many industrial processes, such as making cellulose acetate. Vinegar is a 4% to 8% solution of acetic acid which can be made by fermenting alcohol.

$$C_2H_5OH + O_2 \rightarrow CH_3COOH + H_2O$$
 ethanol   ethanoic acid
     (acetic acid)

One of the aromatic acids is benzoic acid with a carboxyl group replacing one of the hydrogens:

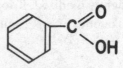

Summary of Oxygen Derivatives

$$\overset{\text{Functional group}}{R-H \rightarrow R-Cl \rightarrow R-OH \rightarrow R^1CHO \rightarrow R^1-COOH}$$
hydrocarbon chlorine alcohol aldehyde acid
    substitution
    product (ending "-ol") (ending "-al") (ending "-oic")

Note: R^1 indicates a radical different from R by one less carbon in the chain.

An actual example using ethane:

$$C_2H_6 \xrightarrow{Cl_2} C_2H_5Cl \xrightarrow{NaOH} C_2H_5OH \xrightarrow{[O]} CH_3CHO \xrightarrow{[O]} CH_3COOH$$

ethane   monochloro-   ethanol   ethanal   ethanoic
       ethane   acid

Complex Organic Acids

There are some more complex organic acids that are important to mention. These are listed in Table 19.

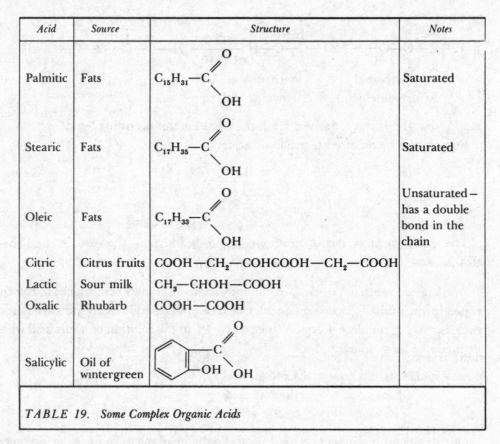

Acid	Source	Structure	Notes
Palmitic	Fats	$C_{15}H_{31}-COOH$	Saturated
Stearic	Fats	$C_{17}H_{35}-COOH$	Saturated
Oleic	Fats	$C_{17}H_{33}-COOH$	Unsaturated — has a double bond in the chain
Citric	Citrus fruits	$COOH-CH_2-COHCOOH-CH_2-COOH$	
Lactic	Sour milk	$CH_3-CHOH-COOH$	
Oxalic	Rhubarb	$COOH-COOH$	
Salicylic	Oil of wintergreen	(benzene ring)—COOH with OH	

TABLE 19. Some Complex Organic Acids

Ketones

When a secondary alcohol is slightly oxidized, it forms a compound having the functional group $R-\overset{\overset{O}{\|}}{C}-R^1$, called a ketone. The R^1 indicates that this radical need not be the same as R. An example is:

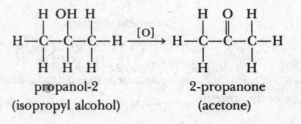

      propanol-2   2-propanone

 (isopropyl alcohol)   (acetone)

The name of the ketone in the IUPAC method has an ending "-one" with a digit indicating the carbon that has the double-bonded oxygen. Another method of naming a ketone is to name the radicals on either side of the ketone structure and use the word "ketone." This would be dimethyl ketone.

Ethers

When a primary alcohol, such as ethanol, is dehydrated with sulfuric acid, an ether forms. The functional group is R—O—R^1, in which R^1 may be the same hydrocarbon group as shown below in the first example, or a different hydrocarbon group as shown in example 2.

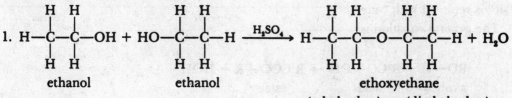

2. Another ether with unlike radicals, R—O—R^1:

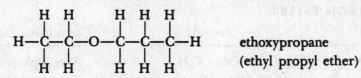

The ether name is made up of the first radical's stem, then "-oxy," and then the alkane name for the second radical.

Amines and Amino Acids

The radical NH$_2^-$ is called the amide ion or the amino group. Under the proper conditions, this ion can replace a hydrogen in a hydrocarbon compound. This resulting compound is called an *amine*. An example:

H—C—NH$_2$ with H above and H below   methyl amine

or

aniline (benzene ring with —NH$_2$)

Amino acids are organic acids which contain one or more amino groups. The simplest uncombined amino acid is glycine, or amino acetic acid, NH$_2$CH$_2$—COOH. More than 20 amino acids are known, about half of which are necessary in the human diet because they are needed to make up the body proteins.

Esters

Esters are often compared to inorganic salts because their preparations are similar. To make a salt, you react the appropriate acid and base. To make an ester, you react the appropriate organic acid and alcohol. For example:

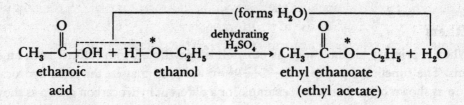

$$CH_3-\overset{\overset{O}{\|}}{C}-OH + H-O-C_2H_5 \xrightarrow[H_2SO_4]{dehydrating} CH_3-\overset{\overset{O}{\|}}{C}-\overset{*}{O}-C_2H_5 + H_2O$$

ethanoic acid   ethanol   ethyl ethanoate (ethyl acetate)

The name is made up of the alkyl radical of the alcohol and the acid name in which "-ic" is replaced with "-ate."

The general equation is:

$$R\overset{*}{O}-H + R^1CO-OH \rightarrow R^1CO\overset{*}{O}-R + HOH$$
alcohol   acid   ester

Esters usually have a sweet smell and are used in perfumes and flavor extracts.

• SOME COMMON ESTERS:

Name	Formula	Characteristic Odor
Ethyl butyrate	$C_3H_7COO \cdot C_2H_5$	Pineapple
Amyl acetate	$CH_3COO \cdot C_5H_{11}$	Banana, pear
Methyl salicylate	$C_6H_4(OH)COO \cdot CH_3$	Wintergreen
Amyl valerate	$C_4H_9COO \cdot C_5H_{11}$	Apple
Octyl acetate	$CH_3COO \cdot C_8H_{17}$	Orange
Methyl anthranilate	$C_3H_4(NH_2)COO \cdot CH_3$	Grape

Some esters found in the seeds of plants and the bodies of animals are fats. Stearin is such an ester.

$$
\begin{array}{lll}
C_{17}H_{35}-COOH & HO-CH_2 & C_{17}H_{35}COO-CH_2 \\
C_{17}H_{35}-COOH + HO-CH & \rightarrow 3\ H_2O + C_{17}H_{35}COO-CH \\
C_{17}H_{35}-COOH & HO-CH_2 & C_{17}H_{35}COO-CH_2 \\
\text{stearic acid} & \text{glycerol} & \text{glycerol stearate} \\
& & \text{(stearin)}
\end{array}
$$

• SOME OTHER EXAMPLES:

Olein or glyceryl oleate $= (C_{17}H_{35}COO)_3C_3H_5$
Butyrin or glyceryl butyrate $= (C_3H_7COO)_3\ C_3H_5$

Soap was made in the olden days by housewives using a fat (like stearin) and lye (sodium hydroxide). This saponification reaction is:

$$(C_{17}H_{35}COO)_3C_3H_5 + 3\ NaOH \rightarrow 3\ C_{17}H_{35}COONa + C_3H_5(OH)_3$$
stearin   lye   soap   glycerine
(sodium stearate)

Carbohydrates

Carbohydrates are made up of carbon, hydrogen, and oxygen. Usually the hydrogen to oxygen ratio is 2:1. The simple carbohydrates are more or less sweet, and are referred to as saccharides (meaning "sweet").

The monosaccharides are simple compounds having six carbons in the formula, but their properties differ because of structural differences. This is shown in dextrose and levulose, which have the same formula (isomers) but different properties.

Dextrose (glucose)

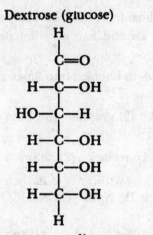

Levulose (fructose)

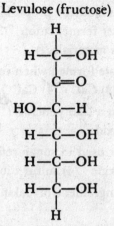

Not as sweet as ordinary sugar (sucrose)

Does not need to be digested.

Twice as sweet as dextrose

Must be changed to dextrose in the body before available for body use.

The sugars that contain an aldehyde structure (—C=O), such as dextrose, or a group that readily changes to this structure act as mild reducing agents and are called *reducing sugars*. They can be identified by their ability to reduce copper (II) hydroxide, in Fehling's solution or Benedict's solution, to copper (I) oxide which is recognized by its brick-red color. This test is often used to detect the disease known as diabetes.

The "double" sugars, those having 12 carbons in their structures, are called disaccharides. Some of these are sucrose, lactose (found in milk), and maltose.

Sucrose can be converted to simple sugars by treating a solution of it with a little acid and boiling. This is called *inversion*, and the products, dextrose and levulose, are called *invert* sugars. The general name for this type of reaction with dissaccharides is hydrolysis.

$$C_{12}H_{22}O_{11} + H_2O \rightarrow C_6H_{12}O_6 + C_6H_{12}O_6$$

sucrose   dextrose   levulose

Polysaccharides

These complex carbohydrates are made up of some multiple of $C_6H_{10}O_5$. This is shown in the general formula $(C_6H_{10}O_5)_n$, in which n represents the variable multiple. This group includes starch, dextrin, and cellulose.

Chapter
14
Review

1. Carbon atoms usually (1) lose 4 electrons   (2) gain 4 electrons   (3) form 4 covalent bonds   (4) share the 2 electrons in their *K* shell.

2. Coke is produced from bituminous coal by (1) cracking   (2) synthesis   (3) substitution   (4) destructive distillation.

3. The usual method for preparing carbon dioxide in the laboratory is (1) heating a carbonate   (2) fermentation   (3) reacting an acid and a carbonate   (4) burning carbonaceous materials.

4. The precipitate formed when carbon dioxide is bubbled into limewater is (1) $CaCl_2$ (2) H_2CO_3 (3) CaO   (4) $CaCO_3$.

5. The "lead" in a lead pencil is (1) boneblack   (2) graphite and clay   (3) lead oxide   (4) lead peroxide.

6. A decolorizer used in sugar refining is (1) boneblack   (2) chlorine water   (3) hydrogen peroxide   (4) sulfur dioxide.

7. A common ingredient of baking powders is (1) NaCl   (2) $NaHCO_3$ (3) Na_2CO_3 (4) NaOH.

8. The first and simplest alkane is (1) ethane   (2) methane   (3) C_2H_2 (4) methene   (5) CCl_4.

9. Slight oxidation of a primary alcohol gives (1) a ketone   (2) an organic acid   (3) an ether   (4) an aldehyde   (5) an ester.

10. The characteristic group of the organic ester is (1) —CO—   (2) —COOH   (3) —CHO   (4) —O—   (5) —COO—.

11. Fermentation of glucose gives (1) CO_2 and H_2O (2) CO and alcohol   (3) CO_2 and CH_3OH (4) CO and C_2H_5OH (5) CO_2 and C_2H_5OH.

12. The organic acid that can be made from ethanol is (1) acetic acid   (2) formic acid   (3) C_3H_7OH (4) found in bees and ants   (5) butanoic acid.

13. An ester can be prepared by the reaction of (1) two alcohols   (2) an alcohol and an aldehyde   (3) an alcohol and an organic acid   (4) an organic acid and an aldehyde   (5) an acid and a ketone.

14. Phenol is a derivative of an (1) alkane   (2) alkene   (3) aliphatic (chain) hydrocarbon   (4) aromatic hydrocarbon   (5) alkyne.

15. Sucrose ia a (1) reducing sugar   (2) monosaccharide   (3) sugar substitute with a ketone group   (4) disaccharide   (5) sugar with an aldehyde group.

16. Compounds that have the same composition but differ in structural formulas (1) are used for substitution products   (2) are called isomers   (3) are called polymers   (4) have the same properties   (5) are usually alkanes.

17. Ethene is the first member of the (1) alkane series   (2) saturated hydrocarbons   (3) alkyne series   (4) unsaturated hydrocarbons   (5) aromatic hydrocarbons.

18. Fehling's solution gives a positive test with (1) glucose (2) sucrose (3) copper (I) oxide (4) starch.

• ANSWERS

1. (3)	7. (2)	13. (3)
2. (4)	8. (2)	14. (4)
3. (3)	9. (4)	15. (4)
4. (4)	10. (5)	16. (2)
5. (2)	11. (5)	17. (4)
6. (1)	12. (1)	18. (1)

Terms You Should Know

alcohol $-OH$	aromatization $-Ar$	hydrocarbon RH
aldehyde $-CHO$	carbohydrate $-CHO; -C=O$	hydrogenation $+H_2$
alkane C_nH_{2n+2}	cracking	isomer
alkene C_nH_{2n}	dehydrogenation $-H_2$	ketone $>C=O$
alkylation $-R$	destruction distillation	photosynthesis
alkyne C_nH_{2n-2}	ester $-COO-$	polymerization
amine $-NH_2$	ether $-O-$	polysaccharide
amino acid $-NH_2; COOH$	Fischer-Tropsch process	Solvay process
aromatics $Ar-$		

15
Nucleonics

RADIOACTIVITY

The discovery of radioactivity came in 1896 (less than two months after Röntgen had published an announcement of the discovery of X-rays). Röntgen had pointed out that the X-rays came from a spot on a glass tube where a beam of electrons, in his experiments, was hitting, and that this spot simultaneously showed strong fluorescence. It occurred to Becquerel and others that X-rays might in some way be related to fluorescence (emission of light when exposed to some exciting agency) and to phosphorescence (emission of light *after* the exciting agency is removed).

Becquerel accordingly tested a number of phosphorescent substances to determine whether they emitted X-rays while phosphorescing. He had no success until he tried a compound of uranium; he found that the uranium compound, whether or not it was allowed to phosphoresce by exposure to light, continuously emitted something that could penetrate lightproof paper and even thicker materials.

He found that the compounds of uranium and the element itself produced *ionization* in the surrounding air. Thus either the ionizing effect, as indicated by the rate of discharge of a charged electroscope, or the degree of darkening of a photographic plate, could be used to measure the intensity of the invisible emission. Moreover, the emission from the uranium was continuous and perhaps even permanent, and required no energy from any external source. Yet, probably because of the current interest and excitement over X-rays, Becquerel's work received little attention for some two years following his discovery, until early in 1898 when the Curies entered the picture (Pierre Curie was one of Becquerel's colleagues in Paris). Their painstaking efforts in isolating several other radioactive compounds resulted in the isolation of polonium and radium.

The Nature of Radioactive Emissions

While the early separation experiments were in progress, an understanding was slowly being gained of the nature of the spontaneous emission from the various radioactive elements. Becquerel thought at first that it was simply X-rays, but it was soon found that there are THREE different kinds of radioactive emission, now called *alpha particles*, *beta particles*, and *gamma rays*. We now know that the alpha particles are positively charged particles of helium nuclei, the beta particles are streams of high-speed electrons, and the gamma rays are high-energy radiations similar to X-rays.

The important characteristics of each type of radiation can be summarized as follows:

Alpha Particle   (helium nucleus $^4_2$He) Positively charged, 2+

1. Ejection reduces the atomic number by 2, the atomic weight by 4 amu.
2. High energy, relatively low velocity.
3. Range: about 5 cm in air.
4. Shielding needed: stopped by the thickness of a sheet of paper, skin.
5. Interactions: produces about 100,000 ionizations per centimeter; repelled by the positively charged nucleus; attracts electrons, but does not capture them until its speed is much reduced.
6. An example: Thorium-230 has an ustable nucleus and undergoes radioactive decay through alpha emission. The nuclear equation that describes this reaction is:

$$^{230}_{90}\text{Th} \rightarrow ^4_2\text{He} + ^{226}_{88}\text{Ra}$$

In a decay reaction like this, the initial element (thorium-230) is called the parent nuclide and the resulting element (radium-226) is called the daughter nuclide.

Beta Particle   (fast electron) Negatively charged, 1−

1. Ejected when a *neutron* decays into a proton and an electron.
2. High velocity, low energy.
3. Range: 30−40 ft.
4. Shielding needed: stopped by 1 cm of aluminum or thickness of average book.
5. Interactions: weak due to high velocity, but produces about 100 ionizations per centimeter.
6. An example: Protactinium-234 is a radioactive nuclide that undergoes beta emission. The nuclear equation is:

$$^{234}_{91}\text{Pa} \rightarrow ^{234}_{92}\text{U} + ^{\ \ 0}_{-1}\text{e}$$

Gamma Radiation   (electromagnetic radiation identical with light; high energy) No charge

1. Beta particles and gamma rays are usually emitted together; after a beta is emitted, a gamma ray follows.
2. Arrangement in nucleus is unknown. Same velocity as visible light.
3. Range: no specific range.
4. Shielding needed: about 5 in. of lead.
5. Interactions: weak of itself . . . gives energy to electrons, which then perform the ionization.

Methods of Detection of Alpha, Beta, and Gamma Rays

All methods of detection of these radiations rely on their ability to ionize. There are a number of methods in common use.

1. Photographic plate. The fogging of a photographic emulsion led to the discovery of radioactivity. If this emulsion is viewed under a high-power microscope, it is seen that beta and gamma rays cause the silver bromide grains to develop in a scattered fashion. However, alpha particles, owing to the dense ionization they produce, leave a definite track of exposed grains in the emulsion. Hence not only is the alpha particle detected, but also its range (in the emulsion) can be measured. Special emulsions are capable of showing the beta-particle tracks.

2. Scintillations. A fluorescent screen (e.g., ZnS) will show the presence of electrons and X-rays, as we have already seen. If the screen is viewed with a magnifying eyepiece, small flashes of light, called scintillations, will be observed. By observing the scintillations, one not only can detect the presence of alpha particles, but also can actually count them.

3. The cloud chamber. One of the most useful instruments for detecting and measuring radiation is the Wilson cloud chamber. Its operation depends on the well-known fact that moisture tends to condense around ions (the probable explanation for some formation of clouds in the sky). If an enclosed region of air is saturated with water vapor (this is always the case if water is present) and the air is cooled suddenly, it becomes supersaturated, that is, it contains for the instant more water vapor than it can hold permanently, and a fog of water droplets develops around the ions in the chamber.

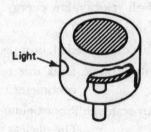

For example, an alpha particle traveling through such a supersaturated atmosphere supplies a trail of ions on which water droplets will condense. This trail is thus made visible. These trails are usually photographed by a camera that operates whenever a piston moves downward causing the air to expand and cool.

4. The bubble chamber. This device utilizes a liquid such as ether, ethyl alcohol, pentane, or propane, which is superheated to well above the boiling point. When the pressure is released quickly, the liquid is in a highly unstable condition, ready to boil violently. An ionizing particle passing through the liquid at this instant leaves a trail of tiny bubbles which may be photographed.

5. The electroscope. One of the simplest and one of the first used in work on radioactivity consists of two thin gold leaves which are charged by a battery with the same type of charge. Radioactive materials produce ions in proportion to their radioactivity. A negatively charged electroscope becomes discharged when ions in the air take electrons from the electroscope, causing the leaves to gradually collapse.

6. The Geiger counter. This instrument is perhaps the most widely used at the present time for counting individual radiation. It consists of a fine wire of tungsten mounted along the axis of the tube, which contains a gas at reduced pressure. A difference of potential of about 1000 volts is applied in such a way as to make the metal tube negative with respect to the wire. The voltage is high enough so that the electrons produced are accelerated by the electric field. Near the wire, where the field is strongest, the accelerated particles produce more ions, positive ones going to the walls, negative ones being collected by the wire. Any particle that will produce an ion gives rise to the same avalanche of ions, so the type of particle cannot be detected. However, you can detect each individual particle.

Source of Radioactivity

The nuclei of uranium, radium and other radioactive elements are continually disintegrating. It should be emphasized that spontaneous disintegration produces RADON. The time it takes for half of the material to disintegrate is called its *half-life*.

For example, for radium, we know that, on the average, half of all the radium nuclei present will have disintegrated to radon in 1590 years. In another 1590 years, half of this remainder will decay, and so on. When the radium atom disintegrates, it loses an alpha particle, which eventually becomes a neutral helium atom upon gaining two electrons. The remainder of the atom is the gas known as radon.

Such a conversion of an element to a new element (because of a change in the number of protons) is called a *transmutation*. This transmutation can be produced artificially by bombarding the nuclei of a substance with various particles from a particle accelerator, such as the cyclotron.

The following uranium-radium disintegration series shows how a radioactive atom may change when it loses each kind of particle. Note how an atomic number is shown by a subscript ($_{92}$U), and the isotopic weight is shown by a superscript (^{238}U). The alpha particle is shown by the Greek symbol "α" and the beta particle by "β".

$$^{238}_{92}\text{U} \xrightarrow{-\alpha} {}^{234}_{90}\text{Th} \xrightarrow{-\beta} {}^{234}_{91}\text{Pa} \xrightarrow{-\beta} {}^{234}_{92}\text{U} \xrightarrow{-\alpha} {}^{230}_{90}\text{Th} \xrightarrow{-\alpha}$$

$$^{226}_{88}\text{Ra} \xrightarrow{-\alpha} {}^{222}_{86}\text{Rn} \xrightarrow{-\alpha} {}^{218}_{84}\text{Po} \xrightarrow{-\alpha} {}^{214}_{82}\text{Pb} \xrightarrow{-\beta} {}^{214}_{83}\text{Bi} \xrightarrow{-\beta}$$

$$^{214}_{84}\text{Po} \xrightarrow{-\alpha} {}^{210}_{82}\text{Pb} \xrightarrow{-\beta} {}^{210}_{83}\text{Bi} \xrightarrow{-\beta} {}^{210}_{84}\text{Po} \xrightarrow{-\alpha} {}^{206}_{82}\text{Pb} \text{ (stable)}$$

Particle Lost	Weight Change	Atomic Number Change
Alpha (α)	Loses 4 amu	Decreases by 2
Beta (β)	None	Increases by 1

The stability of an atom seems to be related to binding energy. The binding energy is the amount of energy released when a nucleus is formed from its component particles. If you add the mass of the components and compare this sum to the actual mass of the nucleus formed, there will be a small difference in these figures. This difference in mass can be converted to its energy equivalent using Einstein's equation, $E = mc^2$, where E is the energy, m is the mass, and c is the velocity of light. It is this energy that is called the binding energy.

It has been found that the lightest and heaviest elements have the smallest binding energy per nuclear particle and thus are less stable than the elements of intermediate atomic weights, which have the greatest binding energy.

The relationship of even-odd number of protons to the number of neutrons affects the stability of a nucleus. Many of the stable nuclei have even numbers of protons and neutrons, while stability is less frequent in nuclei which have an even number of protons and an odd number of neutrons, or vice versa. Only a few stable nuclei are known that have odd numbers of protons and neutrons.

NUCLEAR ENERGY

Since Einstein predicted that matter could be converted to energy over a half century ago, men have tried to unlock this energy source. This prediction was verified in 1932 by Cockcroft and Walton, who produced small quantities of helium and a tremendous amount of energy from the reaction

$$_3^7Li + {}_1^1H \rightarrow 2\ _2^4He + energy$$

In fact, the energy released was almost exactly the amount calculated from Einstein's equation.

In 1942, Fermi and his co-workers discovered a sustained chain reaction of fission could be controlled to produce large quantities of energy. This led to the development of the nuclear fission bomb, which brought World War II to an end.

Conditions for Fission

When fissionable material, like ^{235}U, is bombarded with a "slow" neutron, fission occurs, giving off different fission products. An example of such a reaction is shown in Figure 46.

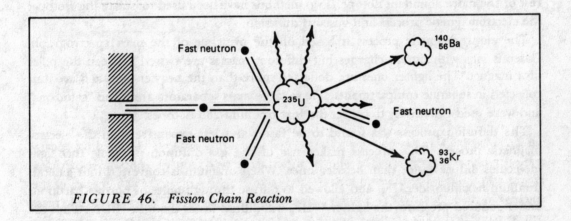

FIGURE 46. Fission Chain Reaction

As long as one of the released neutrons produces another reaction, the chain reaction will continue. If each fission starts more than one, the reaction becomes tremendously powerful in a very short time. This occurs in the atomic bomb.

To obtain the neutron "trigger" for this reaction, neutrons must be slowed down so that they do not pass through the nucleus without effect. This "slowing down" or moderating is best done by letting fast neutrons collide with many relatively light atoms, such as hydrogen, deuterium, and carbon. Graphite, paraffin, ordinary water, and "heavy water" (containing deuterium instead of ordinary hydrogen) are all suitable moderators.

Nuclear fission can be made to occur in an uncontrolled explosion or in a controlled nuclear reactor. In both cases enough fissionable material must be present so that once the reaction starts, it can at least sustain itself. This amount of fissionable material is called the critical mass. In the bomb, the number of reactions increases tremendously, whereas in a reactor the rate of fissions is controlled.

The typical nuclear reactor or "pile" is made up of the following kinds of material:

1. Fissionable material—sustains the chain reaction.
2. Moderator—slows down fission neutrons.
3. Control rods (cadmium or boron steel rods)—absorb excess neutrons and control rate of reaction.
4. Concrete encasement—provides shielding from radiation.

There are many variations of this basic reactor as reactors become more efficient. Many United States cities now receive some of their electrical power from a nuclear power station.

Methods of Obtaining Fissionable Material

The natural abundance of ^{235}U is extremely small: about .0005% of the earth's crust, and .7% of natural uranium. Therefore the problem is to separate this isotope from the rest of the more abundant isotope. Two methods have been used to isolate this isotope: the electromagnetic process and gaseous diffusion.

The electromagnetic process is based on the principle of the mass spectrograph. Gaseous ions with similar charges but different masses are passed between the poles of a magnet. The lighter ones are deflected more than the heavier ones and are thus collected in separate compartments. These large mass separators are called "calutrons" and were used extensively to collect the first uranium-235 isotopes.

The diffusion process was found to be faster and less expensive than the electromagnetic process. This process makes use of the gas diffusion principle that light molecules diffuse faster than heavier ones. When uranium is converted into gaseous uranium hexafluoride, UF_6, and allowed to diffuse through miles of porous partitions under reduced pressure, the lighter $^{235}UF_6$ will diffuse faster than the $^{238}UF_6$. Thus the gas coming through first will be richer in ^{235}U than when it started. The ^{235}U is then retrieved from the fluoride compound for use in atomic devices or reactors.

Fusion is the opposite of fission in that, instead of splitting a nucleus, two nuclei join to form a new nucleus with a great amount of released energy. The sun's energy results from such a fusion reaction, in which four hydrogen protons are eventually made into one helium nucleus with the liberation of a large amount of energy. The reactions are:

$$^{12}C + {}^{1}H \rightarrow {}^{13}N + Energy$$
$$^{13}N \rightarrow {}^{13}C + e^+$$
$$^{13}C + {}^{1}H \rightarrow {}^{14}N + Energy$$
$$^{14}N + {}^{1}H \rightarrow {}^{15}O + Energy$$
$$^{15}O \rightarrow {}^{15}N + e^+$$
$$^{15}N + {}^{1}H \rightarrow {}^{12}C + {}^{4}He$$

Man has only been able to produce a nuclear fusion in an uncontrolled explosion, the hydrogen bomb. Scientists in every major country of the world are attempting to find a means of controlling a sustained fusion reaction.

In the meantime, nuclear fission is being adapted to produce energy for electrical power stations, merchant ships, aircraft carriers, and submarines. Experimentation is proceeding on the nuclear aircraft engine and a nuclear rocket engine.

NEW SUBATOMIC PARTICLES

As scientists continue the search for more information about the atom and its makeup, they have been especially interested in the subatomic particles. In the 1960s, with more powerful accelerators to investigate the atom, they found more than 100 different subatomic particles. More recently, research has suggested a new order that brings a simplicity to the particles that compose matter. By studying the tracks made on photographic plates by the particles that result from nuclear collisions, Gelf-Mann at the California Institute of Technology and Zweig at CERN in Switzerland arrived at a new way of accounting for these subatomic particles. Their theory is that all small particles are combinations of still smaller particles called *quarks*. A quark is a particle with a fractional charge of either 1/3 or 2/3 the charge of an electron. This theory simplifies the multitude of particles.

At first three quarks were postulated, but the number has grown to six. The usual subatomic particles presented in a first-year chemistry book account for only two quarks. Each quark is described or named by its particular characteristic, which is referred to as its "flavor." The following table lists the six quarks and their fractional charges.

SIX KINDS OF QUARKS	
Name	*Electric Charge*
Up	2/3
Down	−1/3
Charmed	2/3
Strangeness	−1/3
Top (truth)	2/3
Bottom (beauty)	−1/3

The two quarks described as "up" and "down" are thought to make up most of the ordinary matter in the universe. Although scientists have never isolated individual quarks, they believe their failure to do so has occurred because the binding force that holds them together is so great.

The electron is now believed to be composed of smaller particles, which are classified as elementary particles called *leptons*. This is also true of the neutrino, which is similar in mass to the electron but has a neutral charge.

1. Radioactive changes differ from ordinary chemical changes because radioactive changes (1) involve changes in the nucleus   (2) are explosive   (3) absorb energy   (4) release energy.

2. Isotopes of uranium have different (1) atomic numbers   (2) atomic weights   (3) numbers of planetary electrons   (4) numbers of protons.

3. Pierre and Marie Curie discovered (1) oxygen   (2) hydrogen   (3) chlorine   (4) radium.

4. The number of protons in the nucleus of an atom of atomic number 32 is (1) 4   (2) 32   (3) 4   (4) 73.

5. Atoms of ^{235}U and ^{238}U differ in structure by three (1) electrons   (2) isotopes   (3) neutrons   (4) protons.

6. In a hydrogen bomb, hydrogen is converted into (1) uranium   (2) helium   (3) barium   (4) plutonium.

7. The use of radioactive isotopes has already produced promising results in the treatment of certain types of (1) cancer   (2) heart disease   (3) pneumonia   (4) diabetes.

8. The emission of a beta particle results in a new element with the atomic number (1) increased by 1   (2) increased by 2   (3) decreased by 1   (4) decreased by 2.

9. A substance used as a moderator in a nuclear reactor is (1) marble   (2) hydrogen   (3) tritium   (4) graphite.

10. A self-sustaining nuclear fission chain reaction depends upon the release of (1) protons   (2) neutrons   (3) electrons   (4) alpha particles.

11. Einstein's formula for the Law of Conservation of Mass and Energy states that $E =$ _mc²_.

12. The energy of the sun is thought to be produced by the fusion of hydrogen atoms into _helium_.

13. Could a mass spectrograph (or calutron) be used to separate ^{54}Mn from ^{54}Cr?
   _No_.

14. The release of energy from the combination of two nuclei into one is called _fusion_.

15. Write the equation for the first observed artificial transmutation by Rutherford:
   _____.

• MATCHING:

A. Crooke
B. alpha particle
C. beta particle
D. Becquerel
E. Einstein
F. chain reaction
G. deuteron

H. gamma ray
I. gaseous diffusion
J. neptunium
K. plutonium
L. ^{235}U
M. ^{238}U

16. Helium nucelus: *B*

17. Stable isotope of uranium: *^{238}U M*

18. One method of separating isotopes: *I*

19. Most penetrating ray from radioactive decay: *H*

20. Mass-energy conversion: *E*

• ANSWERS:

1. (1)
2. (2)
3. (4)
4. (2)
5. (3)
6. (2)
7. (1)

8. (1)
9. (4)
10. (2)
11. Mc^2
12. He
13. No
14. fusion

15. $^{14}_{7}N + ^{4}_{2}He \rightarrow ^{17}_{8}O + ^{1}_{1}H$
16. B
17. M
18. I
19. H
20. E

Terms You Should Know

alpha particle
beta particle
bubble chamber
cloud chamber
critical mass

electroscope
fluorescence
gamma ray
Geiger counter

half-life
lepton
phosphorescence
quark
transmutation

16

Representative Laboratory Setups

Figures 47–61 show some of the basic laboratory setups used in beginning chemistry. The purpose of these diagrams is to review the basic techniques of assembling equipment with regard to some knowledge about the reactants and products involved.

In the equations, the letters in parentheses have the following meanings:

(s) = solid   (l) = liquid   (g) = gas

1. *Preparation of a gaseous product, nonsoluble in water, by water displacement from solid reactants.*

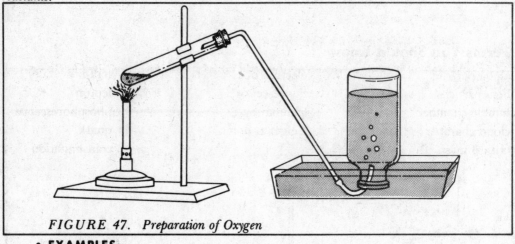

FIGURE 47.   Preparation of Oxygen

• **EXAMPLES:**

Preparation of oxygen (O_2).

$$2\ KClO_3(s) + MnO_2(s) \rightarrow 2\ KCl + 3\ O_2(g)\uparrow\ +\ MnO_2$$

2. *Preparation of a gaseous product, nonsoluble in water, by water displacement from at least one reactant in solution.*

(*Note:* Purpose of thistle tube shown in Figure 48:

1. Introduction of more liquid without "opening" the reacting vessel.
2. Safety valve to indicate blocked delivery tube by the rise of liquid in the thistle tube.)

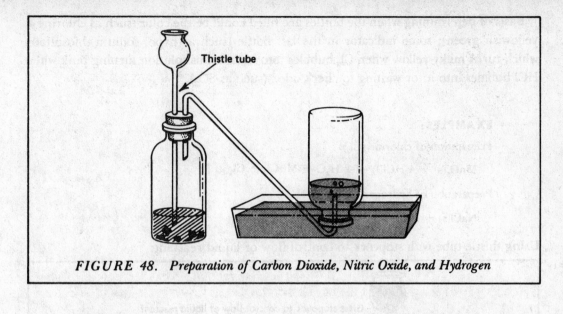

FIGURE 48. *Preparation of Carbon Dioxide, Nitric Oxide, and Hydrogen*

- **EXAMPLES:**

Preparation of carbon dioxide (CO_2).

$$CaCO_3(s) + 2\ HCl(l) \rightarrow CaCl_2 + H_2O + CO_2(g)\uparrow$$

Preparation of nitric oxide (NO).

$$3\ Cu(s) + 8\ dilute\ HNO_3(l) \rightarrow 3\ Cu(NO_3)_2 + 4\ H_2O + 2\ NO(g)\uparrow$$

Preparation of hydrogen (H_2).

$$Zn(s) + 2\ HCl(l) \rightarrow ZnCl_2 + H_2(g)\uparrow$$

3. *Preparation of a gaseous product heavier than air which can best be collected by the upward displacement of air.*

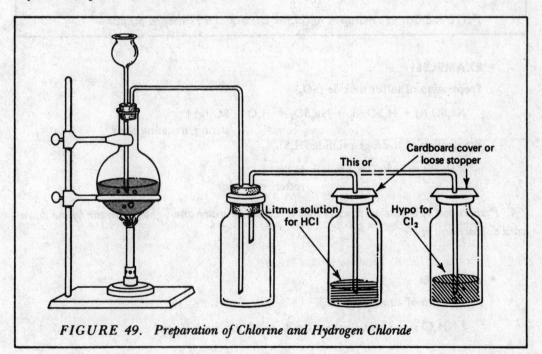

FIGURE 49. *Preparation of Chlorine and Hydrogen Chloride*

Basis of determining when the bottles are filled could be the color (such as chlorine — yellowish green); some indicator in the last bottle (such as hypo, sodium thiosulfate, which turns milky yellow when Cl_2 bubbles into it); a litmus solution turning pink when HCl bubbles into it; or wafting to check odor (such as SO_2).

- **EXAMPLES:**

    Preparation of chlorine (Cl_2).

    $$MnO_2(s) + 4\ HCl(l) \rightarrow 2\ H_2O + MnCl_2 + Cl_2(g) \uparrow$$

    Preparation of hydrogen chloride (HCl).

    $$NaCl(s) + H_2SO_4(l) \rightarrow NaHSO_4 + HCl(g)$$

Using thistle tube with stopcock to control flow of liquid reactant:

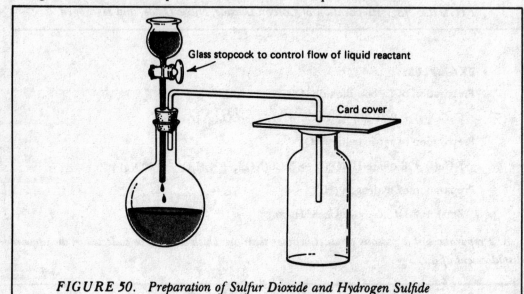

Glass stopcock to control flow of liquid reactant

Card cover

FIGURE 50. Preparation of Sulfur Dioxide and Hydrogen Sulfide

- **EXAMPLES:**

    Preparation of sulfur dioxide (SO_2).

    $$Na_2SO_3(s) + H_2SO_4(l) \rightarrow Na_2SO_4 + H_2O + SO_2(g) \uparrow$$
    $$\text{strong, irritating odor}$$

    Preparation of hydrogen sulfide (H_2S).

    $$2\ HCl(l) + FeS(s) \rightarrow FeCl_2 + H_2S(g) \uparrow$$
    $$\text{rotten egg odor}$$

4. *Preparation of a gaseous product that is soluble in water and lighter than air by the downward displacement of air.*

- **EXAMPLE:**

    Preparation of ammonia (NH_3).

    $$2\ NH_4Cl(s) + Ca(OH)_2(s) \rightarrow CaCl_2 + 2\ H_2O + 2\ NH_3(g) \uparrow$$

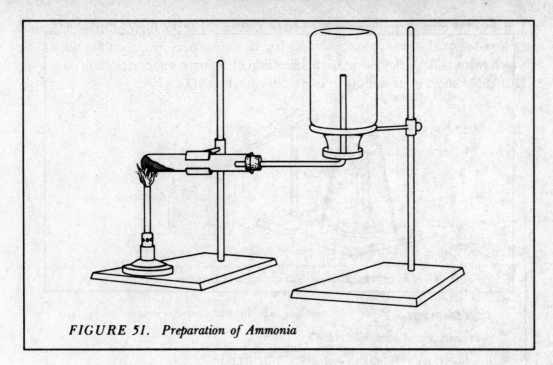

FIGURE 51. Preparation of Ammonia

5. *Distillation of a liquid.*

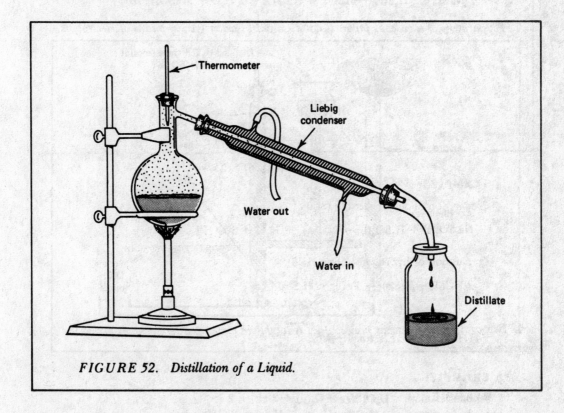

FIGURE 52. Distillation of a Liquid.

Removes dissolved impurities, which remain in the flask. Does not remove volatile materials, which vaporize and pass over into the distillate.

6. *Preparation of a corrosive substance which would attack rubber or cork in distillation.*

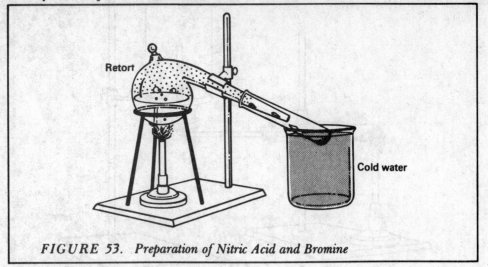

FIGURE 53. *Preparation of Nitric Acid and Bromine*

- **EXAMPLES:**

  Preparation of nitric acid (HNO_3).

  $$NaNO_3(s) + H_2SO_4(l) \rightarrow NaHSO_4 - HNO_3(g) \uparrow$$

  Preparation of bromine (Br_2).

  $$2\ NaBr + 2\ H_2SO_4 + MnO_2 \rightarrow Na_2SO_4 + 2\ H_2O + MnSO_4 + Br_2 \uparrow$$

7. *Preparation of a gaseous product, not dissolved by water gas, by means of electrolysis.*

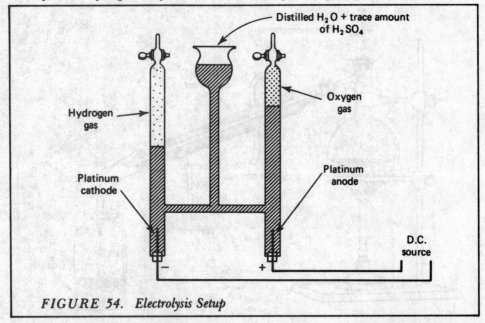

FIGURE 54. *Electrolysis Setup*

- **EXAMPLE:**

  Anode reaction: $H_2O(l) \rightarrow \frac{1}{2} O_2(g) + 2\ H^+ + 2\ e^-$
  Cathode reaction: $2\ H_2O(l) + 2\ e^- \rightarrow H_2(g) + 2\ OH^-$
  Cell reaction: $3\ H_2O(l) \rightarrow \frac{1}{2} O_2(g) + H_2(g) + 2\ H_2O(l)$

  or $H_2O(l) \rightarrow \frac{1}{2} O_2(g) + H_2(g)$

8. *Separation of a mixture by chromatography.*

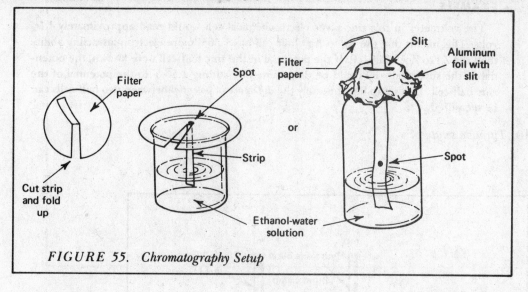

FIGURE 55. *Chromatography Setup*

• EXAMPLE:

Chromatography is a process used to separate parts of a mixture. The component parts separate as the solvent carrier moves past the spot by capillary action. Because of variations in solubility, attraction to the filter paper, and density, each fraction moves at a different rate. Once separation occurs, the fractions are either identified by their color or removed for other tests. A usual example is the use of Shaeffer Skrip Ink No. 32, which separates into yellow, red, and blue streaks of dyes.

9. *Measuring potentials in electrochemical cells.*

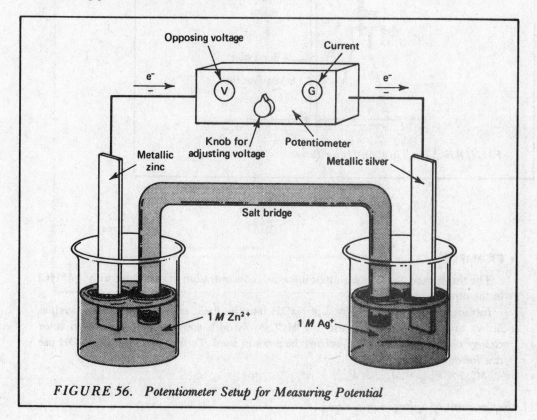

FIGURE 56. *Potentiometer Setup for Measuring Potential*

• **EXAMPLE:**

The voltmeter in this zinc-silver electrochemical cell would read approximately 1.56 volts. This means that the Ag to Ag^+ half-cell has 1.56V more electron-attracting ability than the Zn to Zn^{2+} half-cell. If the potential of the zinc half-cell were known, the potential of the silver half-cell could be determined by adding 1.56 V to the potential of the zinc half-cell. In a setup like this, only the *difference in potential* between two half-cells can be measured.

10. *Titration setup.*

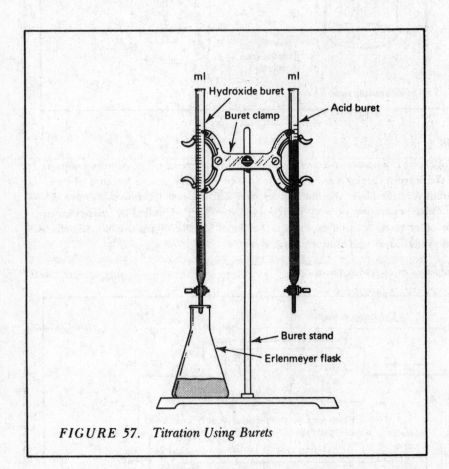

FIGURE 57. Titration Using Burets

• **EXAMPLE:**

The titration of NaOH solution of unknown concentration in one buret with .1 M HCl in the other.

Introduce approximately 15mL of NaOH into the flask, and add an indicator such as litmus or phenolphthalein. Add the HCl slowly with constant swirling. When color change occurs and is retained, record the amount used. To find the M of the NaOH use this formula:

$$M_{acid} \times V_{acid} = M_{base} \times V_{base}$$

11. *Replacement of hydrogen by a metal.*

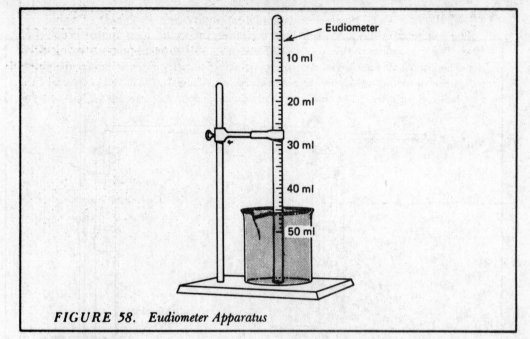

FIGURE 58.   Eudiometer Apparatus

• **EXAMPLE:**

Measure the mass of a strip of magnesium with an analytical balance to the nearest .001 g. Using a strip with a mass of about .040 g produces about 40 ml of H_2. Pour 5 ml of concentrated HCl into the eudiometer, and slowly fill the remainder with water. Try to minimize mixing. Lower the coil of Mg strip into the tube, invert it, and lower it to the bottom of the battery jar. After the reaction is complete, you can measure the volume of the gas released and calculate the mass of hydrogen replaced by the magnesium. (Refer to Chapter 5 for a discussion of gas laws.)

12. *Drying gases to remove water vapor.*

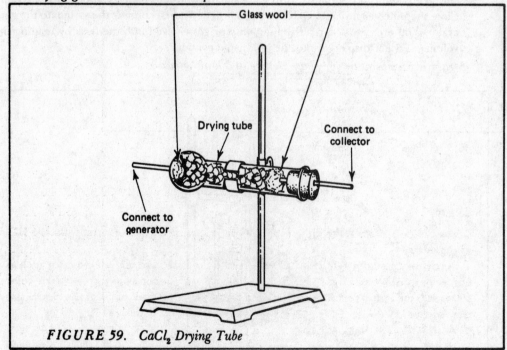

FIGURE 59. $CaCl_2$ Drying Tube

Calcium chloride is hydroscopic and absorb water vapor as the gas passes through this tube.

13. *Using displaced water to measure the volume of a gas that is produced.*

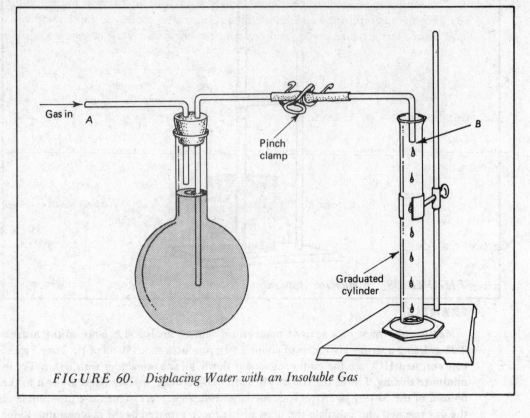

FIGURE 60. *Displacing Water with an Insoluble Gas*

• **EXAMPLE:**

In this setup, before the gas begins to be generated, you should release the pinch clamp and gently blow in the tube at *A* to get water up and into the remainder of the system. Close the pinch clamp once water starts to flow at *B;* then connect the system to the generator. You are ready to read the quantity of gas evolved into the flask by reading the volume of water displaced into the graduated cylinder.

14. *Measuring temperature for phase change and heating curves.*

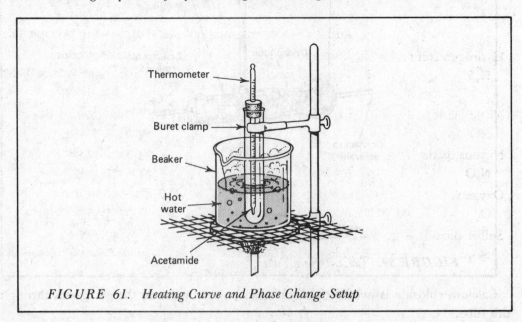

FIGURE 61. *Heating Curve and Phase Change Setup*

• **EXAMPLE:**

This setup can be used with either acetamide or paradichlorobenzene in the test tube. By keeping a careful time and temperature observation chart, you can obtain data to plot a time-temperature graph, and can note the effect of a phase change on this plot.

SUMMARY OF QUALITATIVE TESTS

I. Identification of Gases

Gas	Test	Result
Ammonia NH_3	1. Smell cautiously. 2. Test with litmus. 3. Expose to HCl fumes.	1. Sharp odor. 2. Red litmus turns blue. 3. White fumes form (NH_4Cl).
Carbon dioxide CO_2	1. Pass through limewater, $Ca(OH)_2$.	1. White precipitate forms, $CaCO_3$.
Carbon monoxide CO	1. Burn it and pass product through limewater, $Ca(OH)_2$.	1. White precipitate forms, $CaCO_3$.
Chlorine Cl_2	1. Smell cautiously. 2. Observe color.	1. Irritating odor. 2. Yellowish green.
Hydrogen H_2	1. Allow it to mix with some air, then ignite. 2. Burn it—trap product.	1. Gas explodes. 2. Burns with blue flame— product H_2O turns cobalt chloride paper from blue to pink.
Hydrogen chloride HCl	1. Smell cautiously. 2. Exhale over the gas. 3. Dissolve in water and test with litmus. 4. Add $AgNO_3$ to the solution.	1. Choking odor. 2. Vapor fumes form. 3. Blue litmus turns red. 4. Forms white precipitate.
Hydrogen sulfide H_2S	1. Smell cautiously. 2. Test with moist lead acetate paper.	1. Rotten egg odor. 2. Turns brown-black (PbS).
Nitric oxide NO	1. Expose to the air.	1. Colorless gas turns reddish brown.
Nitrous oxide N_2O	1. Insert glowing splint. 2. Add nitric oxide gas.	1. Bursts into flame. 2. Remains colorless.
Oxygen O_2	1. Insert glowing splint. 2. Add nitric oxide gas.	1. Bursts into flame. 2. Turns reddish brown.
Sulfur dioxide SO_2	1. Smell cautiously. 2. Allow it to bubble into purple potassium permanganate solution.	1. Choking odor. 2. Solution becomes colorless.

II. Identification of Negative Ions

Ion	Test	Result
Acetate $C_2H_3O_2^-$	Add conc. H_2SO_4 and warm gently.	Odor of vinegar released.
Bromide Br^-	Add chlorine water and some CCl_4; shake.	Reddish brown color concentrated in CCl_4 layer.
Carbonate CO_3^-	Add HCl acid; pass released gas through limewater.	White cloudy ppt. forms (ppt. = precipitate).
Chloride Cl^-	1. Add silver nitrate solution. 2. Then add nitric acid, later followed by ammonium hydroxide.	1. White ppt. forms. 2. Precipitate insol. in HNO_3 but dissolves in NH_4OH.
Hydroxide OH^-	Test with red litmus paper.	Turns blue.
Iodide I^-	Add chlorine water and some CCl_4; shake.	Purple color concentrated in CCl_4 layer.
Nitrate NO_3^-	Add freshly made ferrous sulfate sol., then conc. H_2SO_4 carefully down the side of the tilted tube.	Brown ring forms at junction of layers.
Nitrite NO_2^-	Add dilute H_2SO_4.	Brown fumes (NO_2) released.
Sulfate SO_4^-	Add sol. of $BaCl_2$, then HCl.	White ppt. forms; insoluble in HCl.
Sulfide S^{2-}	Add HCl and test gas released with lead acetate paper.	Gas, with rotten egg odor, turns paper brown-black.
Sulfite SO_3^{2-}	Add HCl and pass gas into purple $KMnO_4$ sol.	Solution turns colorless.

III. Identification of Positive Ions

Ion	Test	Result
Ammonium NH_4^+	Add strong base (NaOH); heat gently.	Odor of ammonia.
Ferrous Fe^{2+}	Add sol. of potassium ferricyanide, $K_3Fe(CN)_6$.	Dark blue ppt. forms (Turnball's blue).
Ferric Fe^{3+}	Add sol. of potassium ferrocyanide, $K_4Fe(CN)_6$.	Dark blue ppt. forms (Prussian blue).
Hydrogen H^+	Test with blue litmus paper.	Turns red.

IV. Qualitative Tests for Metals

Flame tests. Carefully clean a platinum wire by dipping it into dil. HNO_3 and heating in the Bunsen flame. Repeat until the flame is colorless. Dip heated wire into the substance being tested (either solid or solution), and then hold it in the hot outer part of the Bunsen flame.

Compound of	Color of Flame
Na	Yellow
K	Violet (use cobalt-blue glass to screen out Na impurities)
Li	Crimson
Ca	Orange-red
Ba	Green
Sr	Bright red

Borax bead tests. Make a borax bead by heating some borax in a platinum wire loop. Dip the bead in the substance being tested, and heat in the outer part of the Bunsen flame. Check color.

Compound of	Color of Bead
Co	Blue
Mn	Amethyst (violet)
Cr	Green
Iron(-ic)	Yellow
Ni	Brown
Cu	Blue

Cobalt nitrate tests. Scoop out a small cavity in a plaster block or piece of charcoal, place in it some of the substance being tested, and heat strongly by means of a blow pipe. Moisten with a few drops of cobalt nitrate solution and reheat. Check color.

Compound of	Color of Substance
Al	Blue
Zn	Green
Mg	Pink

Hydrogen sulfide tests. Bubble hydrogen sulfide gas through the solution of a salt of the metal being tested. Check color of the precipitate formed.

Compound of	Color of Sulfide Precipitate
Lead (Pb)	Brown-black (PbS)
Copper (Cu)	Black (CuS)
Silver (Ag)	Black (Ag_2S)
Mercury (Hg)	Black (HgS)
Nickel (Ni)	Black (NiS)
Iron (Fe)	Black (FeS)
Cadmium (Cd)	Yellow (CdS)
Arsenic (As)	Light yellow (As_2S_3)
Antimony (Sb)	Orange (Sb_2S_3)
Zinc (Zn)	White (ZnS)
Bismuth (Bi)	Brown (Bi_2S_3)

*Practice
College Board
Achievement
Tests in
Chemistry*

GENERAL INFORMATION

THE COLLEGE BOARD ACHIEVEMENT TEST IN CHEMISTRY IS PLANNED to test the principles and concepts drawn from the factual material found largely in inorganic chemistry and, to a much lesser extent, those found in organic chemistry. Only a few questions are asked concerning industrial or analytical chemistry.

According to a description of the achievement test by the College Entrance Examination Board, the tests include: kinetic molecular theory and the three states of matter; atomic theory and structure and the periodic table; nuclear reactions; quantitative relations as applied to chemical formulas and equations; chemical bonding and molecular structure, and their relations to properties; the nature of chemical reactions, including acid-base reactions, oxidation-reduction reactions, ionic reactions, and other chemical changes occurring in solution; energy changes accompanying chemical reactions; interpretation of chemical equilibria and reaction rates; solution phenomena; electrochemistry, nuclear chemistry, and radiochemistry; physical and chemical properties of the more familiar metals, transition elements, and nonmetals and of the more familiar compounds; understanding and interpretation of laboratory procedures and observations.

You will be provided with a periodic table to use during the test. All necessary information regarding atomic numbers and atomic weights is given on the chart. Plan to use the chart provided with the diagnostic test in the front of this book when taking each practice test.

The test has 85 multiple-choice questions that are distributed in the following manner among the topics covered:

Topics Covered	Approximate Percentage of Test	Approximate Number of Questions
I. Atomic Theory and Structure, including periodic relationships	10	8
II. Nuclear Reactions	2	2
III. Chemical Bonding and Molecular Structure	11	9
IV. States of Matter and Kinetic Molecular Theory	9	8
V. Solutions, including concentration units, solubility, and colligative properties	6	5
VI. Acids and Bases	9	8
VII. Oxidation-reduction and Electrochemistry	7	6
VIII. Stoichiometry, including the mole concept, Avogadro's number, empirical and molecular formulas, percentage composition, stoichiometric calculations, and limiting reagents	11	9
IX. Reaction Rates, including rate equations and factors affecting rates	2	2
X. Equilibrium, including mass action expressions, ionic equilibria, and LeChatelier's prinicple	6	5
XI. Thermodynamics, including energy changes in chemical reactions, randomness, and criteria for spontaneity	4	3
XII. Descriptive Chemistry: physical and chemical properties of elements and their more familiar compounds, including simple examples from organic chemistry; periodic properties	16	14
XII. Laboratory: equipment, procedures, observations, safety, calculations, and interpretation of results	7	6

Note: Every edition contains approximately five questions on equation balancing and/or predicting products of chemical reactions. These are distributed among the various content categories.

BASIC ABILITIES TESTED FOR

1. The ability to demonstrate an understanding of basic scientific principles and concepts.

2. The ability to apply scientific concepts and principles.

3. The ability to handle quantitative relationships (to use mathematical relationships and to solve problems involving chemical principles and concepts).

4. The ability to organize and interpret information obtained by observation and experimentation so as to discern cause-and-effect relationships.

5. The ability to interpret so as to draw conclusions or make inferences from experimental data, including data presented in graphic or tabular form.

6. The ability to apply laboratory procedures.

Each of the following tests measures these abilities with various types of questions. Before attempting the tests, read the Introduction in the front of this book, especially the section entitled "Final Preparation—The Day before the Test."

ANSWER SHEET FOR PRACTICE TEST 1

Determine the correct answer for each question. Then, using a No. 2 pencil, blacken completely the oval containing the letter of your choice.

1. (A) (B) (C) (D) (E)	16. (A) (B) (C) (D) (E)	31. (A) (B) (C) (D) (E)
2. (A) (B) (C) (D) (E)	17. (A) (B) (C) (D) (E)	32. (A) (B) (C) (D) (E)
3. (A) (B) (C) (D) (E)	18. (A) (B) (C) (D) (E)	33. (A) (B) (C) (D) (E)
4. (A) (B) (C) (D) (E)	19. (A) (B) (C) (D) (E)	34. (A) (B) (C) (D) (E)
5. (A) (B) (C) (D) (E)	20. (A) (B) (C) (D) (E)	35. (A) (B) (C) (D) (E)
6. (A) (B) (C) (D) (E)	21. (A) (B) (C) (D) (E)	36. (A) (B) (C) (D) (E)
7. (A) (B) (C) (D) (E)	22. (A) (B) (C) (D) (E)	37. (A) (B) (C) (D) (E)
8. (A) (B) (C) (D) (E)	23. (A) (B) (C) (D) (E)	38. (A) (B) (C) (D) (E)
9. (A) (B) (C) (D) (E)	24. (A) (B) (C) (D) (E)	39. (A) (B) (C) (D) (E)
10. (A) (B) (C) (D) (E)	25. (A) (B) (C) (D) (E)	40. (A) (B) (C) (D) (E)
11. (A) (B) (C) (D) (E)	26. (A) (B) (C) (D) (E)	41. (A) (B) (C) (D) (E)
12. (A) (B) (C) (D) (E)	27. (A) (B) (C) (D) (E)	42. (A) (B) (C) (D) (E)
13. (A) (B) (C) (D) (E)	28. (A) (B) (C) (D) (E)	43. (A) (B) (C) (D) (E)
14. (A) (B) (C) (D) (E)	29. (A) (B) (C) (D) (E)	44. (A) (B) (C) (D) (E)
15. (A) (B) (C) (D) (E)	30. (A) (B) (C) (D) (E)	45. (A) (B) (C) (D) (E)

46. (A) (B) (C) (D) (E)	61. (A) (B) (C) (D) (E)	76. (A) (B) (C) (D) (E)
47. (A) (B) (C) (D) (E)	62. (A) (B) (C) (D) (E)	77. (A) (B) (C) (D) (E)
48. (A) (B) (C) (D) (E)	63. (A) (B) (C) (D) (E)	78. (A) (B) (C) (D) (E)
49. (A) (B) (C) (D) (E)	64. (A) (B) (C) (D) (E)	79. (A) (B) (C) (D) (E)
50. (A) (B) (C) (D) (E)	65. (A) (B) (C) (D) (E)	80. (A) (B) (C) (D) (E)
51. (A) (B) (C) (D) (E)	66. (A) (B) (C) (D) (E)	81. (A) (B) (C) (D) (E)
52. (A) (B) (C) (D) (E)	67. (A) (B) (C) (D) (E)	82. (A) (B) (C) (D) (E)
53. (A) (B) (C) (D) (E)	68. (A) (B) (C) (D) (E)	83. (A) (B) (C) (D) (E)
54. (A) (B) (C) (D) (E)	69. (A) (B) (C) (D) (E)	84. (A) (B) (C) (D) (E)
55. (A) (B) (C) (D) (E)	70. (A) (B) (C) (D) (E)	85. (A) (B) (C) (D) (E)
56. (A) (B) (C) (D) (E)	71. (A) (B) (C) (D) (E)	
57. (A) (B) (C) (D) (E)	72. (A) (B) (C) (D) (E)	
58. (A) (B) (C) (D) (E)	73. (A) (B) (C) (D) (E)	
59. (A) (B) (C) (D) (E)	74. (A) (B) (C) (D) (E)	
60. (A) (B) (C) (D) (E)	75. (A) (B) (C) (D) (E)	

Note: For all questions involving solutions and/or chemical equations, assume that the system is in water unless otherwise stated.

Part A

Directions: Each set of lettered choices below refers to the numbered statements or formulas immediately following it. Select the one lettered choice that best fits each statement or formula and then fill in the corresponding oval on the answer sheet. A choice may be used once, more than once, or not at all in each set.

Questions 1–9

$$IA \quad IIA \quad IIIA \quad IVA \quad VA \quad VIA \quad VIIA \quad VIII$$

| ^3Li | | | | | | | ^{10}Ne |

        (A)   (C)   (D)

(B) ^{20}Ca (E)

1. The most electronegative element

2. The element with a possible oxidation number of -2

3. The element that would react in a one-to-one ratio with (D)

4. The element with the smallest ionic radius

5. The element with the smallest first ionization potential

6. The element with a complete p orbital as its outermost energy level

7. A member of the alkali-metals family

8. An inert gas

9. The element that would react most actively when placed in water to form a strong base

Questions 10–12 refer to the following heating curve:

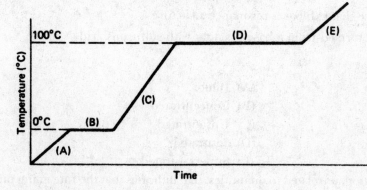

10. In which part of the curve is the state of H_2O only a solid?

11. When is the heat to change the state of H_2O greater?

12. Where is the temperature of H_2O changing at 1°C/cal/g?

Questions 13–15 refer to the following diagram:

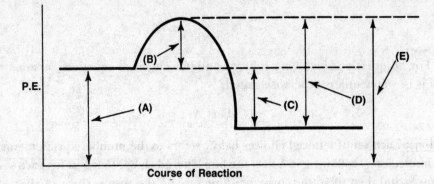

Course of Reaction

13. Indicates the activation energy of this reaction

14. Indicates the portion of the curve that would be directly affected by the addition of a catalyst

15. Indicates the net energy released from the reaction

Questions 16–18

      (A) Iron
      (B) Gold
      (C) Sodium
      (D) Helium
      (E) Uranium

16. An element that resists reaction with acids

17. A nonoatomic element with filled p orbitals

18. A transition element that occurs when the inner $3d$ orbital is partially filled

Questions 19–20

      (A) Rhombic sulfur
      (B) Monoclinic sulfur
      (C) Sulfur trioxide
      (D) Sulfate
      (E) Sulfite

19. A substance that exhibits a resonance structure

20. A product formed from a base reacting with sulfurous acid

Questions 21–23

      (A) Dilute
      (B) Concentrated
      (C) Unsaturated
      (D) Saturated
      (E) Supersaturated

21. The condition, unrelated to quantities, that indicates that the rate going into solution is equal to the rate coming out of solution

22. The condition that exists when a water solution that has been at equilibrium is heated to a higher temperature with a higher solubility, but no additional solute is added

23. The descriptive term that indicates there is a large quantity of solute, compared to the amount of solvent, in a solution

Part B

<u>Directions:</u> Each question below consists of an <u>assertion</u> (statement) in the left-hand column and a <u>reason</u> in the right-hand column. On the appropriate line of the answer sheet fill in oval

    A if both assertion and reason are true statements and the reason is *a correct explanation* of the assertion;

    B if both assertion and reason are true statements but the reason is *NOT a correct explanation* of the assertion;

    C if the assertion is true, but the reason is a false statement;

    D if the assertion is false, but the reason is a true statement;

    E if both assertion and reason are false statements.

Directions Summarized

	Assertion	Reason	
A	True	True	Reason is *a correct explanation*
B	True	True	Reason is *NOT a correct explanation*
C	True	False	
D	False	True	
E	False	False	

Assertion		Reason
24. Nonmetallic oxides are usually acid anhydrides	BECAUSE	they form acids when placed in water.
25. When HCl gas and NH_3 gas come in contact, a white smoke forms	BECAUSE	they combine to form a white solid, ammonium chlorate.
26. The reaction of barium chloride and sodium sulfate does not go to completion	BECAUSE	the compound barium sulfate is formed.
27. When two elements react exothermically to form a compound, the compound should be relatively stable	BECAUSE	the release of energy indicates the compound is at a lower energy level than the reactants and thus relatively stable.
28. The ion of a nonmetallic atom is larger in radius than the atom	BECAUSE	each shell in the atom may have several subshells.
29. Oxidation and reduction occur together	BECAUSE	in these processes electrons must be gained and lost.

Directions Summarized

	Assertion	Reason	
A	True	True	Reason is *a correct explanation*
B	True	True	Reason is *NOT a correct explanation*
C	True	False	
D	False	True	
E	False	False	

	Assertion		*Reason*
30.	Decreasing the atmospheric pressure on a pot of boiling water causes it to stop boiling	BECAUSE	changes in pressure are directly related to the boiling point of water.
31.	Catalysts speed up or slow down a reaction	BECAUSE	they change the temperature of the reaction.
32.	Atoms of different elements can have the same mass number	BECAUSE	the atoms of each element have a characteristic number of protons in the nucleus.
33.	Isotopes have different atomic numbers	BECAUSE	there each has a different number of neutrons.
34.	The most active nonmetal in the halogen family is fluorine	BECAUSE	it has the greatest electronegativity.
35.	The Cu^{2+} ion needs to be oxidized to form Cu metal	BECAUSE	oxidation is a gain of electrons.
36.	The volume of a gas at 100°C and 600 mm Hg pressure will be decreased at STP	BECAUSE	both changes will cause the volume to decrease.
37.	The pH of a 0.01 molar solution of HCl will be 2	BECAUSE	HCl is made up of two essentially ionic particles.
38.	Nuclear fusion on the sun converts hydrogen to helium with a release of energy	BECAUSE	Some mass is converted to energy.
39.	The "bullet" usually used to initiate the fusion of ^{235}U is a neutron	BECAUSE	capture of the neutron by the nucleus causes an unstable condition that leads to its disintegration.

Part C

<u>Directions:</u> Each of the questions or incomplete statements below is followed by five suggested answers or completions. Select the one that is best in each case and then fill in the corresponding oval on the answer sheet.

40. What is the approximate formula weight of $Ca(NO_3)_2$?
    (A) 70
    (B) 82
    (C) 102
    (D) 150
    (E) 164

41. In this reaction: $XClO_3 + A \rightarrow XCl + O_2 \uparrow + A$, which substance is the catalyst?
    (A) X
    (B) $XClO_3$
    (C) A
    (D) XCl
    (E) O_2

42. The normal electron configuration for ethyne (acetylene) is
    (A) H:C::C:H
    (B) H:C̈:C̈:H
    (C) H·C:::C·H
    (D) H:C̈:C:H
    (E) H:Ċ:Ċ:H

43. According to the kinetic molecular theory, molecules increase in kinetic energy when they
    (A) are mixed with other molecules at lower temperature
    (B) are frozen into a solid
    (C) are condensed into a liquid
    (D) are melted from a solid to a liquid state
    (E) collide with each other in a container at low temperature

44. How many atoms are represented in the formula $Ca_3(PO_4)_2$?
    (A) 5
    (B) 8
    (C) 9
    (D) 12
    (E) 13

45. All of the following have covalent bonds EXCEPT
    (A) HCl
    (B) CCl_4
    (C) H_2O
    (D) CsF
    (E) CO_2

46. Which of the following bonds is (are) the weakest attractive force?
    (A) van der Waals forces
    (B) coordinate covalent bonding
    (C) covalent bonding
    (D) polar covalent bonding
    (E) ionic bonding

47. **Which of these resembles the molecular structure of the water molecule?**

48. The two most important considerations in deciding whether a reaction will occur spontaneously are
    (A) the stability and state of the reactants
    (B) the energy gained and the heat evolved
    (C) the exothermic energy and the randomness of the products
    (D) the endothermic energy and the randomness of the products
    (E) the endothermic energy and the structure of the products

49. **The reaction of an acid like HCl and a base like NaOH always**
    (A) **forms a precipitate**
    (B) **forms a volatile product**
    (C) **forms a soluble salt and water**
    (D) **forms a sulfate salt and water**
    (E) **forms a salt and water**

50. The oxidation number of sulfur in H_2SO_4 is
    (A) +2
    (B) +3
    (C) +4
    (D) +6
    (E) +8

51. **Balance this reaction by using the oxidation-reduction method of electron exchange:**

    $$KMnO_4 + H_2SO_3 \rightarrow K_2SO_4 + MnSO_4 + H_2SO_4 + H_2O$$

    Which of the following partial equations is the correct reduction half-reaction for the *balanced* equation?
    (A) $5\ SO_3^{2-} + 5\ H_2O \rightarrow 5\ SO_4^{2-} + 2\ H^+ + 10\ e^-$
    (B) $2\ MnO_4^- + 16\ H^+ + 10\ e^- \rightarrow 2\ Mn^{2+} + 8\ H_2O$
    (C) $SO_3^{2-} \rightarrow SO_4^{2-} + 2\ e^-$
    (D) $SO_3^{2-} + 2\ H^+ \rightarrow SO_4^{2-} + H_2O + 2\ e^-$
    (E) $Mn^{7+} \rightarrow Mn^{2+} + 5\ e^-$

52. Acid solutions in water are exemplified by which of the following?
    I. HCl
    II. Excess H_3O^+
    III. $CuSO_4$
       (A) I only
       (B) III only
       (C) I and II only
       (D) II and III only
       (E) I, II, and III

53. The property of matter that is independent of its surrounding conditions and position is
    (A) volume
    (B) density
    (C) mass
    (D) weight
    (E) velocity

54. Where are the highest ionization energies found in the periodic chart?
    (A) upper left corner
    (B) lower left corner
    (C) upper right corner
    (D) lower right corner
    (E) middle of transition elements

55. Which of the following pairs of compounds can be used to illustrate the Law of Multiple Proportions?
    (A) NO and NO_2
    (B) CH_4 and CO_2
    (C) ZnO_2 and $ZnCl_2$
    (D) NH_4 and NH_4Cl
    (E) H_2O and HCl

56. In this equilibrium reaction: $A + B \rightleftharpoons AB + heat$ (in a closed container), how could the forward reaction rate be increased?
    I. By increasing the concentration of AB
    II. By increasing the concentration of A
    III. By removing some of product AB
       (A) I only
       (B) III only
       (C) I and III only
       (D) II and III only
       (E) I, II, and III

57. In the reaction of sodium with water, the balanced equation has which of the following coefficients?
    I. 1
    II. 2
    III. 3
    (A) I only
    (B) III only
    (C) I and II only
    (D) II and III only
    (E) I, II, and III

58. If 10 liters of CO gas react with sufficient oxygen to completely react, how many liters of CO_2 gas are formed?
    (A) 5 liters
    (B) 10 liters
    (C) 15 liters
    (D) 20 liters
    (E) 40 liters

59. If 49 grams of H_2SO_4 react with 80 grams of NaOH, how much reactant will be left over after the reaction is complete?
    (A) 24.5 g H_2SO_4
    (B) none of either compound
    (C) 20 g NaOH
    (D) 40 g NaOH
    (E) 60 g NaOH

60. The K_{sp} of AgCl is 1.9×10^{-10}. What is the molar concentration of Ag^+ in 1 liter of a saturated water solution of silver chloride?
    (A) 0.44×10^{-5}
    (B) 0.95×10^{-5}
    (C) 1.38×10^{-5}
    (D) 0.95×10^{-10}
    (E) 3.81×10^{-10}

61. If the density of a diatomic gas is 1.43 grams/liter, what is its gram-molecular weight?
    (A) 16 g
    (B) 32 g
    (C) 48 g
    (D) 64 g
    (E) 14.3 g

62. From 2 moles of $KClO_3$ how many liters of O_2 can be produced by decomposition of all the $KClO_3$?
    (A) 11.2
    (B) 22.4
    (C) 33.6
    (D) 44.8
    (E) 67.2

63. The reaction Fe $\rightarrow$ Fe^{2+} + 2e$^-$ (+ 0.44 volt) would occur spontaneously with which of the following?

    I. Pb $\rightarrow$ Pb^{2+} + 2e$^-$ (+ 0.13 volt)

    II. Cu $\rightarrow$ Cu^{2+} + 2e$^-$ ($-$ 0.34 volt)

    III. 2Ag + 2e$^-$ $\rightarrow$ 2Ag$^\circ$ (+ 0.80 volt)

      (A) I only

      (B) III only

      (C) I and III only

      (D) II and III only

      (E) I, II, and III

Questions 64–68 refer to the following experimental setup and data:

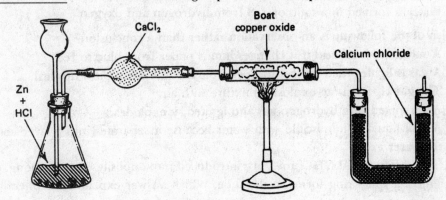

Recorded data:

Before: CuO + porcelain boat = 62.869 g

           CaCl$_2$ + U tube   = 80.483 g

After: Porcelain boat + contents = 54.869 g

           CaCl$_2$ + U tube   = 89.483 g

Reactions: Zn + 2 HCl $\rightarrow$ ZnCl$_2$ + H$_2$ $\uparrow$ =

           H$_2$ + CuO $\rightarrow$ H$_2$O + Cu

64. What type of reaction occurred in the porcelain boat?

    (A) electrolysis

    (B) double displacement

    (C) reduction and oxidation

    (D) decomposition or analysis

    (E) combination or synthesis

65. Why was the CaCl$_2$ tube placed between the generator and the tube containing the porcelain boat?

    (A) to absorb evaporated HCl

    (B) to absorb evaporated H$_2$O

    (C) to slow down the gases released

    (D) to adsorb the evaporated Zn particles

    (E) to remove the initial air that passes through the tube

66. How many grams of hydrogen were used in the formation of the water that was a product?
    (A) 1
    (B) 2
    (C) 4
    (D) 8
    (E) 9

67. What conclusion can you draw from this experiment?
    (A) Hydrogen diffuses faster than oxygen.
    (B) Hydrogen is lighter than oxygen.
    (C) The gram-molecular weight of oxygen is 32 g.
    (D) Water is a triatomic molecule with polar characteristics.
    (E) Water is formed in a ratio of 1:8 from hydrogen and oxygen.

68. Which of the following is an observation rather than a conclusion?
    (A) A substance is an acid if it changes litmus paper from blue to red.
    (B) A gas is lighter than air if it escapes from a bottle left mouth upward.
    (C) The gas H_2 forms an explosive mixture with air.
    (D) Air is mixed with hydrogen gas and ignited; it explodes.
    (E) An oil liquid is immiscible with water because it separates into a layer above the water.

69. When HCl fumes and NH_3 fumes are introduced into opposite ends of a long, dry glass tube, a white ring forms in the tube. Which answer explains this phenomenon?
    (A) NH_4Cl forms.
    (B) HCl diffuses faster.
    (C) NH_3 diffuses faster.
    (D) The ring occurs closer to the end into which HCl was introduced.
    (E) The ring occurs in the very middle of the tube.

70. The correct formula for calcium hydrogen sulfate is
    (A) CaH_2SO_4
    (B) $CaHSO_4$
    (C) $Ca(HSO_4)_2$
    (D) Ca_2HSO_4
    (E) $Ca_2H_2SO_4$

71. Ten grams of sodium hydroxide dissolved in 1 liter of water makes a solution that is
    (A) 0.25 M
    (B) 0.5 M
    (C) 1 M
    (D) 1.5 M
    (E) 4 M

72. For a saturated solution, which statement is true?
    (A) All dissolving has stopped.
    (B) Crystals begin to grow.
    (C) An equilibrium has been established.
    (D) Crytals of the solute will continue to dissolve.
    (E) The solute is exceeding its solubility.

73. In which of the following series is the pi bond found in the bonding structure?
    I. Alkane
    II. Alkene
    III. Alkyne
       (A) I only
       (B) III only
       (C) I and III only
       (D) II and III only
       (E) I, II, and III

Questions 74–76 refer to the following setup:

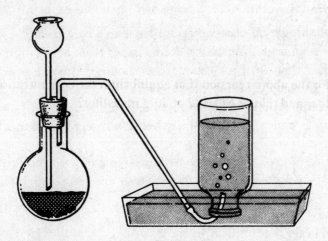

74. Why could you *not* use this setup for preparing H_2 if the generator contained Zn + vinegar?
       (A) Hydrogen would not be produced.
       (B) The setup of the generator is improper.
       (C) The generator must be heated with a burner.
       (D) The delivery tube setup is wrong.
       (E) The gas cannot be collected over water.

75. If the setup were corrected, which of these gases could not be collected over H_2O because of its solubility?
       (A) CO_2
       (B) NO
       (C) O_2
       (D) NH_3
       (E) CH_4

76. If the setup were corrected, what reactants would you use to produce CO_2 in this setup?
       (A) $C + HClO_3$
       (B) $Ca + H_2O$
       (C) $CH_4 + H_2O$
       (D) $C + KClO_3$
       (E) $CaCO_3 + HCl$

77. For the following reaction: $N_2O_4(g) \rightleftarrows 2\ NO_2(g)$, the K_e expression is

(A) $K_e = \dfrac{[N_2O_4]}{[NO_2]}$

(B) $K_e = \dfrac{[N_2O_4]}{[NO_2]^2}$

(C) $K_e = \dfrac{[NO_2]}{[N_2O_4]}$

(D) $K_e = \dfrac{[NO_2]^2}{[N_2O_4]}$

(E) $K_e = \dfrac{[N_2O_4]^2}{[NO_2]}$

78. What is the K_e for the above reaction if at equilibrium the concentration of N_2O_4 is 4×10^{-2} mole/liter and that of NO_2 is 2×10^{-2} mole/liter?

(A) 1×10^{-2}
(B) 2×10^{-2}
(C) 4×10^{-2}
(D) 4×10^{-4}
(E) 8×10^{-2}

79. How much water, in liters, must be added to 0.5 liter of 6 M HCl to make it 2 M?

(A) 0.33
(B) 0.5
(C) 1
(D) 1.5
(E) 2

80. Four grams of hydrogen are ignited with 4 grams of oxygen. How many grams of water can be formed?
(A) 0.5 g
(B) 2.5 g
(C) 4.5 g
(D) 8 g
(E) 36 g

81. Which structure is an ester?

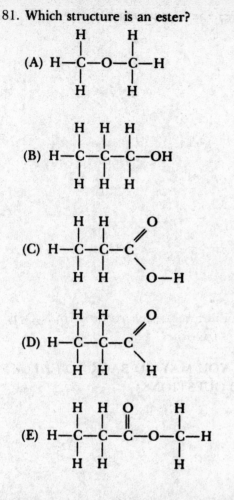

82. What piece of apparatus can be used to introduce more liquid into a reaction and also serve as a pressure valve?
    (A) stopcock
    (B) pinchcock
    (C) thistle tube
    (D) retort
    (E) condenser

83. Which formulas could represent the empirical formula and the molecular formula of a given compound?
    (A) CH_2O and $C_4H_6O_4$
    (B) CHO and $C_6H_{12}O_6$
    (C) CH_4 and C_5H_{12}
    (D) CH_2 and C_3H_6
    (E) CO and CO_2

84. Which pair of atoms are isotopes of element X?

    (A) $^{226}_{90}$X and $^{226}_{91}$X

    (B) $^{226}_{91}$X and $^{227}_{91}$X

    (C) $^{227}_{91}$X and $^{227}_{90}$X

    (D) $^{226}_{90}$X and $^{227}_{91}$X

    (E) $^{226}_{91}$X and $^{227}_{90}$X

85. A binary compound of sodium is
    (A) sodium chlorate
    (B) sodium chlorite
    (C) sodium hypochlorite
    (D) sodium chloride
    (E) sodium perchlorate

STOP

IF YOU FINISH BEFORE ONE HOUR IS UP, YOU MAY GO BACK TO CHECK YOUR WORK OR COMPLETE UNANSWERED QUESTIONS.

Answers and
Explanations to Test

1

1. (D) The most electronegative element (F) would be found in the upper right corner (Group VIIA of the periodic table); the noble gases are exceptions at the far right (VIII).

2. (C) Elements in Group VIA have a possible oxidation number of −2.

3. (B) Elements in Group IA react in a 1:1 ratio with elements in Group VIIA, since one has an electron to lose and the other one needs an electron to complete its outer shell.

4. (A) (A) loses 2 electrons to form an ion whose remaining electrons, being close to the nucleus, are pulled in closer because of the unbalanced +2 charge.

5. (B) Since (B) has only one electron in the outer $4s$ orbital, it can more easily be removed than can an electron from the $3s$ orbital of (A), which is closer to the positive nucleus.

6. (E) All Group VIII elements have a complete p orbital as the outer energy level. This explains why these elements are "inert."

7. (B) The Group IA elements comprise the alkali-metals family.

8. (E) The Group VIII elements are the inert or noble gases.

9. (B) The alkali metals react with water to form a strong base.

10. (A) H_2O is ice in Part A.

11. (D) H_2O changes state at Parts B and D. The heat of vaporization at D (540 cal/g) is greater than the heat of fusion (80 cal/g) at B.

12. (C) Water is heating at 1°C/cal/g in Part C.

13. (B) This is the energy needed to start the reaction.

14. (B) The addition of a catalyst affects only Part B, by either raising or lowering the energy needed to commence the reaction.

15. (C) The net energy released is the exothermic quantity shown by Part C.

16. (B) Gold is known as a noble metal because of its resistance to acids. Aqua regia, a mixture of HNO_3 and HCl, will react with gold.

17. (D) Helium is the only element that is monoatomic in the molecular form.

18. (A) Iron has 5 electrons in the d orbitals that are partially filled.

19. (C) Only sulfur trioxide has a resonance structure (as shown here):

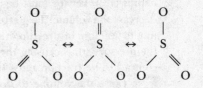

20. (E) Sulfurous acid reacts with a base to form a sulfite salt.

21. (D) The condition described is the equilibrium that exists at saturation.

22. (C) With the increased temperature more solute may go into solution; therefore the solution is now unsaturated.

23. (B) The term "concentrated" means that there is a large amount of solute in the solvent.

24. (A) Nonmetallic oxides are usually acid anhydrides, and they form acids in water.

25. (C) The white smoke is ammonium chloride; no solid is formed.

26. (D) The reaction does go to completion since barium sulfate is a precipitate.

27. (A) The product of an exothermic reaction is relatively stable because it is at a lower energy level than the reactants.

28. (B) The statements are both true but are not related.

29. (A) Both the statements are true, and the reason explains the assertion.

30. (D) The statement is false while the reason is true. Decreasing the pressure on a boiling pot will only cause it to

boil more vigorously.

31. (C) The statement is true, but the reason is false.

32. (B) The statements are true, but the reason doesn't explain the assertion.

33. (D) Isotopes do not have different atomic numbers; they have the same one. The number of neutrons in the nucleus, however, varies.

34. (A) Fluorine is the most active since it has the highest electronegativity.

35. (E) Both the statement and the reason are false.

36. (A) To go from 100°C to 0°C decreases the volume as the gas gets colder; therefore the temperature fraction expressed in kelvins must be $\frac{273}{373}$ to decrease the volume. To go from 600 mm to 760 mm of pressure increases the pressure, thus causing the gas to contract. The fraction must then be $\frac{600}{760}$ to cause the volume to decrease. You could use the formula

$$V_2 = V_1 \times \frac{T_2}{T_1} \times \frac{P_1}{P_2}$$

37. (B) Since HCl is a strong acid and ionizes completely in dilute solutions of water, the $[H^+]$ (H_3O^+ is the same thing) or molar concentration of a .01 molar concentration is 1×10^{-2} mole/liter.

$$pH = -\log [H^+]$$
$$pH = -\log [1 \times 10^{-2}]$$
$$pH = -(-2) = 2$$

The pH is 2, but the reason, although true, does not explain the statement.

38. (A) Both statements are true, and the reason explains the assertion.

39. (D) The statement should refer to fission, not fusion. The reason is true.

40. (E) The total formula weight is:

Ca = 40
2 N = 28
6 O = 96
Total 164

41. (C) The catalyst, by definition, is not consumed in the reaction and ends up in its original form as one of the products.

42. (D) The ethyne molecule is the first member of the acetylene series with a general formula of C_nH_{2n-2}. It contains a triple bond between the two C atoms: $H:C::C:H$.

43. (D) Heating molecules increases their kinetic energy.

44. (E) $3 Ca + 2 P + 8 O = 13$ atoms

45. (D) Cesium and fluorine are from the most electropositive and electronegative portions, respectively, of the periodic chart, and thus form an ionic bond by cesium giving an electron to fluorine to form the respective ions.

46. (A) Van der Waals forces are the weak attraction of the nuclear positive charge of one atom to the negative electron field of an adjacent atom. They are much weaker than the others named.

47. (D) The molecular structure of water is a polar covalent compound with the hydrogens 105° apart.

48. (C) The most important considerations for a spontaneous reaction are (1) that the reaction is exothermic so that once started it tends to continue on its own because of the energy released and (2) that reactions tend to go to the highest state of randomness.

49. (E) Normal H^+ acids and OH^- bases form water and a salt.

50. (D) All compounds have a charge of 0. H usually has +1, and O usually has −2, so

H_2 = +2
O_4 = −8
S = x
Total = 0

$$+2 - 8 + x = 0$$
$$x = -2 + 8 = +6$$

51. (B) In the balanced equation the two half-reactions are:

$5 SO_3^{2-} + 5 H_2O \rightarrow 5 SO_4^{2-} + 10 H^+ + 10e^-$
$2 MnO_4^- + 16 H^+ + 10e^- \rightarrow 2 Mn^{2+} + 8 H_2O$

$2 KMnO_4 + 5 H_2SO_3 \rightarrow$
$K_2SO_4 + 2 MnSO_4 + 2 H_2SO_4 + 3 H_2O$

52. (E) HCl and H_3O^+ give acid solutions, as does $CuSO_4$ when it hydrolyzes in

water. I, II, and III are correct.

53. (C) Mass is a constant and is not dependent on position or surrounding conditions.

54. (C) The complete outer shells of electrons of the smallest inert gases have the highest ionization potential.

55. (A) Only NO and NO_2 fit the definition of the Law of Multiple Proportions, in which one substance stays the same and the other varies in units of whole integers.

56. (D) Increasing the concentration of one or both of the reactants and removing some of the product formed would cause the forward reaction to increase in rate to try to regain the equilibrium condition. II and III are correct.

57. (C) I and II are correct. The equation is $2 NA + 2H_2O \rightarrow 2 NaOH + H_2 \uparrow$. The coefficients include a 1 and a 2.

58. (B) $2 CO + O_2 \rightarrow 2 CO_2$ indicates 2 volumes of CO react with 1 volume of O_2 to form 2 volumes of CO_2. Therefore 10 liters of CO form 10 liters of CO_2.

59. (D) $H_2SO_4 + 2 NaOH \rightarrow 2H_2O + Na_2SO_4$ is the equation for this reaction. 1 mole $H_2SO_4 = 98$ g. 1 mole $NaOH = 40$ g. Then 49 g of $H_2SO_4 = \frac{1}{2}$ mole H_2SO_4. The equation shows that 1 mole of sulfuric reacts with 2 moles of sodium hydroxide or a ratio of 1:2. Therefore, $\frac{1}{2}$ mole of sulfuric reacts with 1 mole of sodium hydroxide in this reaction. 1 mole of NaOH we found to equal 40 g. Since 80 g of NaOH is given, 40 g of it will remain after the reaction has gone to completion.

60. (C) The $K_{sp} = [Ag^+][Cl^-] = 1.9 \times 10^{-10}$. Let $x =$ the molar concentration of Ag^+ and Cl^-.

Then $(x)(x) = x^2 = 1.9 \times 10^{-10}$
$$x = \sqrt{1.9} \times 10^{-10}$$
$$x = 1.38 \times 10^{-5}$$

61. (B) If 1.43 g is the weight of 1 liter, then the weight of 22.4 liters, which is the gram-molecular volume of a gas at STP, will give the gram-molecular weight. So 22.4 liters $\times$ 1.43 g/L = 32 g, the gram-molecular weight.

62. (E) The equation is

$$2 KClO_3 \rightarrow 2 KCl + 3 O_2$$

This shows that 2 moles of $KClO_3$ yields 3 moles of O_2. Three moles of $O_2 = 3 \times 22.4$ L = 67.2 L of O_2.

63. (E) I, II, and III would occur since their reduction reactions would be $-.13$, $+.34$, $+.80$ volt, respectively. These numbers added to $+.44$ volt give a positive E^0 for the reaction with Fe.

64. (C) The CuO was reduced while the H_2 was oxidized, forming $H_2O + Cu$. The reaction is:

$$CuO + H_2 \rightarrow H_2O + Cu$$

65. (B) The purpose of this $CaCl_2$ tube is to absorb any water evaporated from the generator. If it were not present, some water vapor would pass through to the final drying tube and cause the weight of water gained there to be larger than it should be from the reaction alone.

66. (A) Since 8 g of O_2 was lost by the CuO, the weight of water gained by the U-tube came from 8 g of O_2 and 1 g of H_2, to make 9 g of water that it absorbed.

67. (E) The weight ratio of water is 1 g of H_2 to 8 g of O_2 or 1:8. The other statements are true but are not conclusions from this experiment.

68. (D) This is the only observation of the group. All the others are conclusions. Remember that an observation is only what you see, smell, taste, or measure with a piece of equipment.

69. (A) The phenomenon is the formation of the white ring, which is NH_4Cl. Although measuring the distance traveled by each gas could be used to verify Graham's Law of Gaseous Diffusion, this was not asked. The relationship is that the diffusion rate is inversely proportional to the square root of the gas's molecular weight.

70. (C) Since Ca^{2+} and HSO_4^- combine, the formula is $Ca(HSO_4)_2$.

71. (A) NaOH is 40 g/mole. 10 g is 10/40 or .25 mole in 1 liter or .25 M.

72. (C) A saturated solution represents a condition where the solute is going into solution as rapidly as some solute is coming out of the solution.

73. (D) II and III are correct since the double-bonded carbons in the alkene and the triple-bonded carbons in the alkyne series have pi bonds.

74. (B) The thistle tube is not below the level of the liquid in the generator and the gas would escape into the air. (Vinegar is an acid and would produce hydrogen.)

75. (D) NH_3 is very soluble and could not be collected in this manner. All others are not sufficiently soluble to hamper this method of collection.

76. (E) $CaCO_3$ reacts with an acid (HCl here) to form a salt and release CO_2.

77. (D) The K_e expression is made up of the concentration(s) of the products over those of the reactants, with the coefficients becoming exponents. So $K_e = \dfrac{[NO_2]^2}{[N_2O_4]}$.

78. (A) $K_e = \dfrac{[2 \times 10^{-2}]^2}{[4 \times 10^{-2}]} = \dfrac{4 \times 10^{-4}}{4 \times 10^{-2}}$
$= 1 \times 10^{-2}$.

79. (C) In dilution expressions $M_1 \times V_1 = M_2 \times V_2$ can be used. Substituting $(6M)(.5\ L) = (2M)(x\ L)$ gives $x = 1.5$ L total volume. Since there was .5 L to begin with, an additional 1 L must be added.

80. (C) The reaction equation and information given can be set up like this:

Given   Given
$$\begin{array}{cccc} 4\text{ g} & 4\text{ g} & & x\text{g} \\ 2\,H_2 & +\ \ O_2 & \rightarrow & 2\,H_2O \\ 4\text{ g} & 32\text{ g} & & 36\text{ g} \end{array}$$

Studying this shows that the limiting element will be the 4 g of oxygen since 4 g of H_2 would require 32 g of O_2. The solution setup is

$$\frac{4\text{ g }O_2}{32\text{ g }O_2} = \frac{x\text{ g }H_2O}{36\text{ g }H_2O}$$
$$\therefore\ = 4.5\text{ g }H_2O.$$

4 mL   4 mL x mL

81. (E) The functional group is $R\!-\!\overset{\displaystyle O}{\overset{\|}{C}}\!-\!R_1$. This appears only in (E).

82. (C) The thistle tube serves both these purposes.

83. (D) The empirical formula is a representation of the elements in their simplest ratio. Therefore, CH_2 is the simplest ratio of the molecular formula C_3H_6.

84. (B) Isotopes vary in the number of neutrons. This does not involve the atomic number which is the number of protons.

85. (D) Only sodium chloride is made up of two elements and therefore is a binary compound. All the others consist of three elements—sodium, chlorine, and oxygen.

DIAGNOSING YOUR NEEDS

After taking Practice Test 1, check your answers against the correct ones. Then fill in the chart below.

In the space under each question number, place a check if you answered that question correctly.

> • **EXAMPLE:**

If your answer to question 5 was correct, place a check in the appropriate box.

Next, total the check marks for each section and insert the number in the designated block. Now do the arithmetic indicated, and insert your percent for each area.

SUBJECT AREA   (✔) QUESTIONS ANSWERED CORRECTLY

I. Atomic Theory and Structure, including periodic relationships	6	8	28	32	33	34	47	54

☐ No. of checks ÷ 8 × 100 = _____%

II. Nuclear Reactions	38	39

☐ No. of checks ÷ 2 × 100 = _____%

III. Chemical Bonding and Molecular Structure	3	17	18	19	45	46	55	70	73

☐ No. of checks ÷ 9 × 100 = _____%

IV. States of Matter and Kinetic Molecular Theory	10	11	12	30	31	36	43

☐ No. of checks ÷ 7 × 100 = _____%

V. Solutions, including concentration units, solubility, and colligative properties	21	26	60	71	79

☐ No. of checks ÷ 5 × 100 = _____%

VI. Acids and Bases	20	24	25	37	52	49	69	76

☐ No. of checks ÷ 8 × 100 = _____%

VII. Oxidation-Reduction and Electrochemistry	29	35	50	51	63	64

☐ No. of checks ÷ 6 × 100 = _____%

SUBJECT AREA

(✔) QUESTIONS ANSWERED
CORRECTLY

VIII. Stoichiometry	40	44	57	58	59	61	62	66	80

☐ No. of checks ÷ 9 × 100 = _____%

IX. Reaction Rates		41	48

☐ No. of checks ÷ 2 × 100 = _____%

X. Equilibrium		22	56	72	77	78

☐ No. of checks ÷ 5 × 100 = _____%

XI. Thermodynamics: energy changes in chemical reactions, randomness, and criteria for spontaneity		13	14	15	27

☐ No. of checks ÷ 4 × 100 = _____%

XII. Descriptive Chemistry: physical and chemical properties of elements and their familiar compounds; organic chemistry; periodic properties	1	2	4	5	7	9	16
	23	42	53	81	83	84	85

☐ No. of checks ÷ 14 × 100 = _____%

XIII. Laboratory: equipment, procedures, observations, safety, calculations, and interpretation of results	65	67	68	74	75	82

☐ No. of checks ÷ 6 × 100 = _____%

PLANNING YOUR STUDY

The percentages give you an idea of how you have done on the various major areas of the test. Because of the limited number of questions on some parts, these percentages may not be as reliable as the percentages for parts with larger numbers of questions. However, you should now have at least a rough idea of the areas in which you have done well and those in which you need more study.

Subject Area	Chapters to Review
I. Atomic Theory and Structure, including periodic relationships	2
II. Nuclear Reactions	15
III. Chemical Bonding and Molecular Structure	3, 4
IV. States of Matter and Kinetic Molecular Theory	1
V. Solutions, including concentration units, solubility, and colligative properties	7
VI. Acids and Bases	11
VII. Oxidation-reduction and Electrochemistry	12
VIII. Stoichiometry	5, 6
IX. Reaction Rates	9
X. Equilibrium	10
XI. Thermodynamics, including energy changes in chemical reactions, randomness, and criteria for spontaneity	8
XII. Descriptive Chemistry: physical and chemical properties of elements and their familiar compounds; organic chemistry; periodic properties	1, 2, 13, 14
XIII. Laboratory: equipment, procedures, observations, safety, calculations, and interpretation of results	All lab diagrams, 16

ANSWER SHEET FOR PRACTICE TEST 2

Determine the correct answer for each question. Then, using a No. 2 pencil, blacken completely the oval containing the letter of your choice.

1. Ⓐ Ⓑ Ⓒ Ⓓ Ⓔ	16. Ⓐ Ⓑ Ⓒ Ⓓ Ⓔ	31. Ⓐ Ⓑ Ⓒ Ⓓ Ⓔ
2. Ⓐ Ⓑ Ⓒ Ⓓ Ⓔ	17. Ⓐ Ⓑ Ⓒ Ⓓ Ⓔ	32. Ⓐ Ⓑ Ⓒ Ⓓ Ⓔ
3. Ⓐ Ⓑ Ⓒ Ⓓ Ⓔ	18. Ⓐ Ⓑ Ⓒ Ⓓ Ⓔ	33. Ⓐ Ⓑ Ⓒ Ⓓ Ⓔ
4. Ⓐ Ⓑ Ⓒ Ⓓ Ⓔ	19. Ⓐ Ⓑ Ⓒ Ⓓ Ⓔ	34. Ⓐ Ⓑ Ⓒ Ⓓ Ⓔ
5. Ⓐ Ⓑ Ⓒ Ⓓ Ⓔ	20. Ⓐ Ⓑ Ⓒ Ⓓ Ⓔ	35. Ⓐ Ⓑ Ⓒ Ⓓ Ⓔ
6. Ⓐ Ⓑ Ⓒ Ⓓ Ⓔ	21. Ⓐ Ⓑ Ⓒ Ⓓ Ⓔ	36. Ⓐ Ⓑ Ⓒ Ⓓ Ⓔ
7. Ⓐ Ⓑ Ⓒ Ⓓ Ⓔ	22. Ⓐ Ⓑ Ⓒ Ⓓ Ⓔ	37. Ⓐ Ⓑ Ⓒ Ⓓ Ⓔ
8. Ⓐ Ⓑ Ⓒ Ⓓ Ⓔ	23. Ⓐ Ⓑ Ⓒ Ⓓ Ⓔ	38. Ⓐ Ⓑ Ⓒ Ⓓ Ⓔ
9. Ⓐ Ⓑ Ⓒ Ⓓ Ⓔ	24. Ⓐ Ⓑ Ⓒ Ⓓ Ⓔ	39. Ⓐ Ⓑ Ⓒ Ⓓ Ⓔ
10. Ⓐ Ⓑ Ⓒ Ⓓ Ⓔ	25. Ⓐ Ⓑ Ⓒ Ⓓ Ⓔ	40. Ⓐ Ⓑ Ⓒ Ⓓ Ⓔ
11. Ⓐ Ⓑ Ⓒ Ⓓ Ⓔ	26. Ⓐ Ⓑ Ⓒ Ⓓ Ⓔ	41. Ⓐ Ⓑ Ⓒ Ⓓ Ⓔ
12. Ⓐ Ⓑ Ⓒ Ⓓ Ⓔ	27. Ⓐ Ⓑ Ⓒ Ⓓ Ⓔ	42. Ⓐ Ⓑ Ⓒ Ⓓ Ⓔ
13. Ⓐ Ⓑ Ⓒ Ⓓ Ⓔ	28. Ⓐ Ⓑ Ⓒ Ⓓ Ⓔ	43. Ⓐ Ⓑ Ⓒ Ⓓ Ⓔ
14. Ⓐ Ⓑ Ⓒ Ⓓ Ⓔ	29. Ⓐ Ⓑ Ⓒ Ⓓ Ⓔ	44. Ⓐ Ⓑ Ⓒ Ⓓ Ⓔ
15. Ⓐ Ⓑ Ⓒ Ⓓ Ⓔ	30. Ⓐ Ⓑ Ⓒ Ⓓ Ⓔ	45. Ⓐ Ⓑ Ⓒ Ⓓ Ⓔ

46. Ⓐ Ⓑ Ⓒ Ⓓ Ⓔ	61. Ⓐ Ⓑ Ⓒ Ⓓ Ⓔ	76. Ⓐ Ⓑ Ⓒ Ⓓ Ⓔ
47. Ⓐ Ⓑ Ⓒ Ⓓ Ⓔ	62. Ⓐ Ⓑ Ⓒ Ⓓ Ⓔ	77. Ⓐ Ⓑ Ⓒ Ⓓ Ⓔ
48. Ⓐ Ⓑ Ⓒ Ⓓ Ⓔ	63. Ⓐ Ⓑ Ⓒ Ⓓ Ⓔ	78. Ⓐ Ⓑ Ⓒ Ⓓ Ⓔ
49. Ⓐ Ⓑ Ⓒ Ⓓ Ⓔ	64. Ⓐ Ⓑ Ⓒ Ⓓ Ⓔ	79. Ⓐ Ⓑ Ⓒ Ⓓ Ⓔ
50. Ⓐ Ⓑ Ⓒ Ⓓ Ⓔ	65. Ⓐ Ⓑ Ⓒ Ⓓ Ⓔ	80. Ⓐ Ⓑ Ⓒ Ⓓ Ⓔ
51. Ⓐ Ⓑ Ⓒ Ⓓ Ⓔ	66. Ⓐ Ⓑ Ⓒ Ⓓ Ⓔ	81. Ⓐ Ⓑ Ⓒ Ⓓ Ⓔ
52. Ⓐ Ⓑ Ⓒ Ⓓ Ⓔ	67. Ⓐ Ⓑ Ⓒ Ⓓ Ⓔ	82. Ⓐ Ⓑ Ⓒ Ⓓ Ⓔ
53. Ⓐ Ⓑ Ⓒ Ⓓ Ⓔ	68. Ⓐ Ⓑ Ⓒ Ⓓ Ⓔ	83. Ⓐ Ⓑ Ⓒ Ⓓ Ⓔ
54. Ⓐ Ⓑ Ⓒ Ⓓ Ⓔ	69. Ⓐ Ⓑ Ⓒ Ⓓ Ⓔ	84. Ⓐ Ⓑ Ⓒ Ⓓ Ⓔ
55. Ⓐ Ⓑ Ⓒ Ⓓ Ⓔ	70. Ⓐ Ⓑ Ⓒ Ⓓ Ⓔ	85. Ⓐ Ⓑ Ⓒ Ⓓ Ⓔ
56. Ⓐ Ⓑ Ⓒ Ⓓ Ⓔ	71. Ⓐ Ⓑ Ⓒ Ⓓ Ⓔ	
57. Ⓐ Ⓑ Ⓒ Ⓓ Ⓔ	72. Ⓐ Ⓑ Ⓒ Ⓓ Ⓔ	
58. Ⓐ Ⓑ Ⓒ Ⓓ Ⓔ	73. Ⓐ Ⓑ Ⓒ Ⓓ Ⓔ	
59. Ⓐ Ⓑ Ⓒ Ⓓ Ⓔ	74. Ⓐ Ⓑ Ⓒ Ⓓ Ⓔ	
60. Ⓐ Ⓑ Ⓒ Ⓓ Ⓔ	75. Ⓐ Ⓑ Ⓒ Ⓓ Ⓔ	

<u>Note</u>: For all questions involving solutions and/or chemical equations, assume that the system is in water unless otherwise stated.

Part A

<u>Directions</u>: Each set of lettered choices below refers to the numbered statements or formulas immediately following it. Select the one lettered choice that best fits each statement or formula and then fill in the corresponding oval on the answer sheet. A choice may be used once, more than once, or not at all in each set.

Questions 1–4

     (A) Law of Definite Composition
     (B) High dielectric constant
     (C) van der Waals forces
     (D) Graham's Law of Diffusion
     (E) Triple point

1. At a particular temperature and pressure, three states of a compound may coexist

2. Water is a good solvent

3. The rate of movement of hydrogen gas compared to oxygen gas is 4:1

4. The molecules of nitrous oxide and nitrogen dioxide differ by a multiple of the weight and oxygen

Questions 5–7 refer to the following diagram:

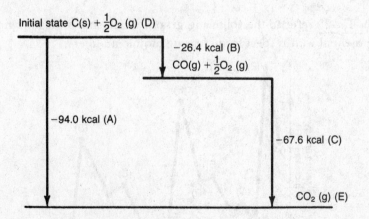

Initial state $C(s) + \frac{1}{2}O_2$ (g) (D)

−26.4 kcal (B)
$CO(g) + \frac{1}{2}O_2$ (g)

−94.0 kcal (A)

−67.6 kcal (C)

CO_2 (g) (E)

5. The ΔH_r of the reaction to form CO from C + O$_2$

6. The ΔH_r of the reaction to form CO$_2$ from CO + O$_2$

7. The ΔH_r of the reaction to form CO$_2$ from C + O$_2$

Questions 8–11

(A) Hydrogen bond
(B) Ionic bond
(C) Polar covalent bond
(D) Pure covalent bond
(E) Metallic bond

8. The type of bond between atoms of potassium and chloride in a crystal of potassium chloride

9. The type of bond between the atoms in a nitrogen molecule

10. The type of bond between the atoms in a molecule of CO_2 (electronegativity difference = 1)

11. The type of bond between the atoms of calcium in a crystal of calcium

Questions 12–14 refer to the following diagram, which represents data obtained when a solid was heated at a constant rate below its melting point:

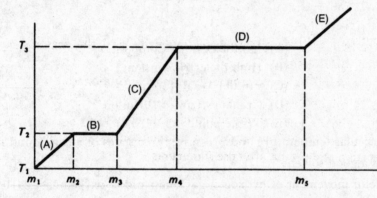

12. Represents the temperature at which the solid melts

13. Represents the section of the curve where the molecules are at the lowest kinetic energy

14. Represents the temperature at which boiling begins

Questions 15–23 refer to the following graph, which shows the variation of the first ionization potential with respect to increasing atomic numbers:

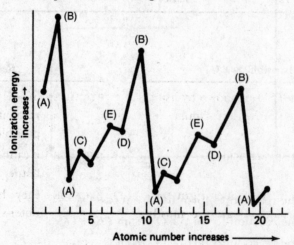

15. The atoms likely to react with water to release hydrogen
16. Nonmetals that are all found in the gaseous state at STP
17. The inert or noble gases
18. The alkali metals
19. The half-filled condition of the p orbitals
20. The filled s orbitals with the exception of He
21. The beginning of pairing in the p orbitals
22. The most active metals
23. The filled p orbitals

Part B

Directions: Each question below consists of an <u>assertion</u> (statement) in the left-hand column and a <u>reason</u> in the right-hand column. On the appropriate line of the answer sheet fill in oval

A if both assertion and reason are true statements and the reason is *a correct explanation* of the assertion;

B if both assertion and reason are true statements but the reason is *NOT a correct explanation* of the assertion;

C if the assertion is true but the reason is a false statement;

D if the assertion is false but the reason is a true statement;

E if both assertion and reason are false statements.

Directions Summarized			
	Assertion	*Reason*	
A	True	True	Reason is *a correct explanation*
B	True	True	Reason is *NOT a correct explanation*
C	True	False	
D	False	True	
E	False	False	

Assertion		*Reason*
24. The structure of SO_3 is shown by using more than one structural formula	BECAUSE	SO_3 is very unstable and resonates between these possible structures.
25. Boric acid is a strong acid	BECAUSE	it cannot be kept in a glass bottle.
26. The elements in the upper right portion of the periodic chart are mostly nonmetallic	BECAUSE	they have fairly similar outer shell configurations.

Directions Summarized			
	Assertion	*Reason*	
A	True	True	Reason is *a correct explanation*
B	True	True	Reason is *NOT a correct explanation*
C	True	False	
D	False	True	
E	False	False	

Assertion		*Reason*
27. Hydrosulfuric acid is often used in qualitative tests	BECAUSE	it is weak acid.
28. Molecules of sodium chloride go into solution in water as ions	BECAUSE	the sodium has a +1 charge and the chloride has a −1 charge and they are hydrated by the water molecules.
29. In an equilibrium reaction, if the concentration of the reactants is increased, the reaction will increase its forward rate	BECAUSE	when a stress is applied to a reaction in equilibrium, the equilibrium shifts in the direction which opposes the stress.
30. The heat ($\triangle H_r$) of a particular reaction can be arrived at by the summation of the $\triangle H_r$ values of two or more reactions that, added together, give the $\triangle H_r$ of the particular reaction	BECAUSE	the First Law of Thermodynamics states that the total energy of the universe is a constant.
31. In a reaction that has both a forward and a reverse reaction, $A + B \rightleftarrows AB$, when only A and B are introduced into a reacting vessel, the forward reaction rate is the highest at the beginning and begins to decrease from that point until equilibrium is reached	BECAUSE	the reverse reaction does not begin until equilibrium is reached.
32. At equilibrium, the forward reaction and reverse reaction stop	BECAUSE	the reactants and products have reached the equilibrium concentrations.
33. The hybrid form of carbon in acetylene is believed to be the *sp* form	BECAUSE	it is a linear compound with a triple bond between the carbons.
34. The weakest of the bonds between molecules is coordinate covalent bonds	BECAUSE	they represent the weak attractive force of the electrons of one molecule for

Assertion		Reason
		the positively charged nucleus of another.
35. A saturated solution is not necessarily concentrated	BECAUSE	dilute and concentrated are terms that relate only to the relative amount of solute dissolved in the solvent.
36. Lithium is the most active metal in the first group of the periodic chart	BECAUSE	it has only one electron in the outer shell.
37. The anions migrate to the cathode in an electrochemical reaction	BECAUSE	positively charged ions are attracted to the negatively charged cathode.
38. The atomic number of a neutral atom that has a mass of 39 and has 19 electrons is 19	BECAUSE	the number of protons in a neutral atom is equal to the number of electrons.
39. For an element with an atomic number of 17, the most probable oxidation number is +1	BECAUSE	the outer shell has a tendency to add one electron to itself.

Part C

Directions: Each of the questions or incomplete statements below is followed by five suggested answers or completions. Select the one that is best in each case and then fill in the corresponding oval on the answer sheet.

40. All of the following involve a chemical change EXCEPT
    (A) the formation of HCl from H_2 and Cl_2
    (B) the color change when NO is exposed to air
    (C) the formation of steam from burning H_2 and O_2
    (D) the solidification of "Crisco" at low temperatures
    (E) the oder of NH_3 when NH_4Cl is rubbed together with $Ca(OH)_2$ powder

41. **When most fuels burn, the products include carbon dioxide and**
    (A) **hydrocarbons**
    (B) **hydrogen**
    (C) **water**
    (D) **hydroxide**
    (E) **hydrogen peroxide**

42. **In the metric system, the prefix "kilo-" means**
    (A) 10^0
    (B) 10^{-1}
    (C) 10^{-2}
    (D) 10^2
    (E) 10^3

43. How many atoms are in 1 mole of water?
    (A) 3
    (B) 54
    (C) 6.02×10^{23}
    (D) $2(6.02 \times 10^{23})$
    (E) $3(6.02 \times 10^{23})$

44. **Which of the following atoms normally forms monoatomic molecules?**
    (A) Cl
    (B) H
    (C) O
    (D) N
    (E) He

45. **Which method is often employed in the separation of the hydrocarbons found in petroleum?**
    (A) alkylation
    (B) hydrogenation
    (C) catalytic cracking
    (D) fractional distillation
    (E) polymerization

46. The complete loss of an electron of one atom to another atom with the consequent formation of electrostatic charges is said to be
    (A) a covalent bond
    (B) a polar covalent bond
    (C) an ionic bond
    (D) a coordinate covalent bond
    (E) a pi bond between p orbitals

47. In the elctrolysis of water, the cathode reaction is
    (A) $2\ H_2O(1) + 2\ e^- \rightarrow H_2(g) \uparrow + 2\ OH^-$

    (B) $2\ H_2O(1) \rightarrow \frac{1}{2}O_2(g) \uparrow + 2\ H^+ + 2\ e^-$

    (C) $2\ OH^- + 2\ e^- \rightarrow O_2(g) \uparrow + H_2(g)$
    (D) $2\ H^+ + 2e^- \rightarrow H_2(g) \uparrow$
    (E) $2\ H_2O(1) + 4e^- \rightarrow O_2(g) \uparrow + 2\ H_2(g) \uparrow$

Questions 48–49 refer to a solution of 0.100 M acetic acid.

48. What is the H_3O^+ concentration? ($K_a = 1.8 \times 10^{-5}$)
    (A) 1.8×10^{-5}
    (B) 1.8×10^{-4}
    (C) 1.3×10^{-2}
    (D) 1.3×10^{-3}
    (E) 0.9×10^{-3}

49. What is the percent dissociation of the acetic acid if the $[H_3O^+]$ is 1×10^{-3} mole/liter?
    (A) 0.01%
    (B) 0.1%
    (C) 1.0%
    (D) 1.5%
    (E) 2.0%

50. What is the pH of an acetic acid solution if the $[H_3O^+] = 1 \times 10^{-4}$ mole/liter?
    (A) 1
    (B) 2
    (C) 3
    (D) 4
    (E) 5

51. The polarity of water is useful in explaining which of the following?
    I. The solution process
    II. The ionization process
    III. The high conductivity of distilled water
    (A) I only
    (B) II only
    (C) I and II only
    (D) II and III only
    (E) I, II, and III

52. When sulfur dioxide is bubbled through water, the solution will contain
    (A) sulfurous acid
    (B) sulfuric acid
    (C) hyposulfuric acid
    (D) persulfuric acid
    (E) anhydrous sulfuric acid

53. Four grams of hydrogen gas at STP contain
    (A) 6.02×10^{23} atoms
    (B) 12.04×10^{23} atoms
    (C) 12.04×10^{46} atoms
    (D) 1.2×10^{22} molecules
    (E) 12.04×10^{23} molecules

54. Analysis of a gas gave: $C = 85.7\%$ and $H = 14.3\%$. If the formula weight of this gas is 42 amu, what are the empirical formula and the true formula?
    (A) CH ; C_4H_4
    (B) CH_2 ; C_3H_6
    (C) CH_3 ; C_3H_9
    (D) C_2H_2 ; C_3H_6
    (E) C_2H_4 ; C_3H_6

55. Which fraction would be used to correct a given volume of gas at 30°C to its new volume when it is heated to 60°C and the pressure is kept constant?

   (A) $\dfrac{30}{60}$

   (B) $\dfrac{60}{30}$

   (C) $\dfrac{273}{333}$

   (D) $\dfrac{303}{333}$

   (E) $\dfrac{333}{303}$

56. What would be the predicted freezing point of a solution which has 684 grams of sugar (1 mole = 342 grams) dissolved in 1500 grams of water?

   (A) 1.86°C
   (B) −0.93°C
   (C) −1.39°C
   (D) −2.48°C
   (E) −2.79°C

57. What is the approximate pH of a 0.003 M solution of H_2SO_4?

   (A) 1
   (B) 2
   (C) 5
   (D) 9
   (E) 13

58. How many grams of NaOH are needed to make 100 grams of a 5% solution?

   (A) 2
   (B) 5
   (C) 20
   (D) 40
   (E) 95

59. For the Haber process: $N_2 + 3\ H_2 \rightleftharpoons 2\ NH_3 + heat$ (at equilibrium), which statement concerning the reaction rate is true?

   (A) The reaction to the right will decrease when more N_2 is added.
   (B) The reaction to the right will decrease when the pressure is increased.
   (C) The reaction to the right will decrease when the temperature is increased.
   (D) The reaction to the right will decrease when more H_2 is added.
   (E) The reaction to the right will decrease when NH_3 is removed.

60. If you titrate 1 M H_2SO_4 solution against 50 milliliters of 1 M NaOH solution, what volume of H_2SO_4, in milliliters, will be needed for neutralization?

   (A) 10
   (B) 25
   (C) 40
   (D) 50
   (E) 100

61. How many grams of CO_2 can be prepared from 150 grams of calcium carbonate reacting with an excess of hydrochloric acid solution?
    (A) 11
    (B) 22
    (C) 33
    (D) 44
    (E) 66

Questions 62–63 refer to the following:

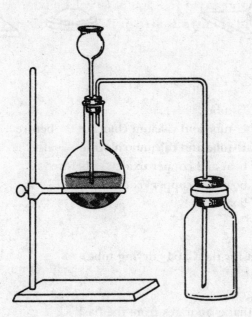

62. **The above diagram represents a setup that may be used to prepare and collect**
    (A) NH_3
    (B) NO
    (C) H_2
    (D) SO_3
    (E) CO_2

63. **The above apparatus would have to be heated with a Bunsen burner to be used for the preparation and collection of**
    (A) HCl
    (B) H_2S
    (C) Br_2
    (D) H_2
    (E) NH_3

Questions 64–66 refer to the following experimental setup and data:

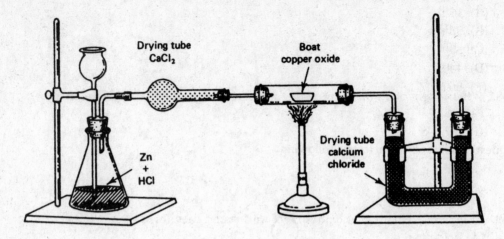

Recorded data:

Weight of U-tube . . . . 20.36 g

Weight of U-tube and calcium chloride   before . . . . 39.32 g

Weight of U-tube and calcium chloride   after . . . . 57.32 g

Weight of boat and copper oxide   before . . . . 30.23 g

Weight of boat and copper oxide   after . . . . 14.23 g

Weight of boat . . . . 5.00 g

64. What is the reason for the $CaCl_2$ drying tube?
    (A) generate water
    (B) absorb hydrogen
    (C) absorb water that evaporates from the flask
    (D) decompose the water from the flask
    (E) act as a catalyst for the combination of hydrogen and oxygen

65. What conclusion can be derived from the data collected?
    (A) Oxygen was lost from the $CaCl_2$.
    (B) Oxygen was generated in the U-tube.
    (C) Water was formed from the reaction.
    (D) Hydrogen was absorbed by the $CaCl_2$.
    (E) CuO was formed in the decomposition.

66. What is the ratio of the weight of water formed to the weight of hydrogen used in the formation of water?
    (A) 1:8
    (B) 1:9
    (C) 8:1
    (D) 9:1
    (E) 8:9

67. What is the weight, in grams, of 1 mole of $KAl(SO_4)_2 \cdot 12\ H_2O$?
    (A) 132
    (B) 180
    (C) 394
    (D) 474
    (E) 516

68. What weight of aluminum will be completely oxidized by 44.8 liters of oxygen?
    (A) 18 g
    (B) 37.8 g
    (C) 50.4 g
    (D) 72.0 g
    (E) 100.8 g

69. In general, when metal oxides react with water, they form solutions that are
    (A) acidic
    (B) basic
    (C) neutral
    (D) unstable
    (E) colored

Questions 70–72 refer to the following diagram:

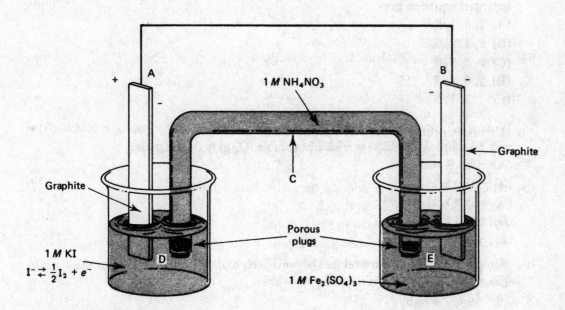

70. The oxidation reaction will occur at
    (A) A
    (B) B
    (C) C
    (D) D
    (E) E

71. The apparatus at *C* is called the
    (A) anode
    (B) cathode
    (C) salt bridge
    (D) ion bridge
    (E) osmotic bridge

72. Starch solution added to beaker *D* after the reaction has occurred for some time will
    (A) remain clear
    (B) turn red
    (C) turn blue-black
    (D) cause the current to reverse
    (E) cause an increase in activity at the pole

73. How many liters of oxygen (STP) can you prepare from the decomposition of 213 grams of sodium chlorate?
    (A) 11.2
    (B) 22.4
    (C) 44.8
    (D) 67.2
    (E) 78.4

74. In this equation: $Al(OH)_3 + H_2SO_4 \rightarrow Al_2(SO_4)_3 + H_2O$, the coefficients of the balanced equation are
    (A) 1, 3, 1, 2
    (B) 2, 3, 2, 6
    (C) 2, 3, 1, 6
    (D) 2, 6, 1, 3
    (E) 1, 3, 1, 6

75. What is ΔH_r for the decomposition of sodium chlorate? (H_f° values are: $NaClO_3(s) = -85.7$ kcal/mole, $NaCl(s) = -98.2$ kcal/mole, $O_2(g) = 0$ kcal/mole)
    (A) −183.9 kcal
    (B) −91.9 kcal
    (C) +45.3 kcal
    (D) +22.5 kcal
    (E) −12.5 kcal

76. Isotopes of an element are related because which of the following is (are) the same in these isotopes?
    I. Atomic weight
    II. Atomic number
    III. Arrangement of orbital electrons
    (A) I only
    (B) II only
    (C) I and III only
    (D) II and III only
    (E) I, II, and III

77. In the reaction of Zn with dilute HCl to form H_2, which of the following will increase the reaction rate?
    I. Increasing the temperature
    II. Increasing the exposed surface of Zn
    III. Using a more active metal instead of Zn
       (A) I only
       (B) II only
       (C) I and III only
       (D) II and III only
       (E) I, II, and III

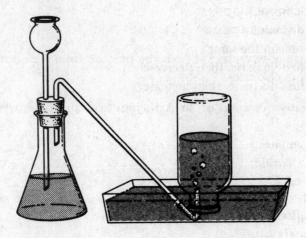

78. The laboratory setup shown above can be used to prepare
    (A) gas lighter than air
    (B) gas heavier than air
    (C) gas soluble in water
    (D) gas insoluble in water
    (E) gas that reacts with water

79. In this reaction: $CaCO_3 + 2\ HCl \rightarrow CaCl_2 + H_2O + CO_2$, if 3 moles of HCl are available to the reaction with an unlimited supply of $CaCO_3$, how many moles of CO_2 can be produced at STP?
    (A) 1
    (B) 1.5
    (C) 2
    (D) 2.5
    (E) 3

80. A saturated solution of $BaSO_4$ at 25°C contains 3.9×10^{-5} mole/liter of B^{2+} ions. What is the K_{sp} of this salt?
    (A) 3.9×10^{-5}
    (B) 3.9×10^{-6}
    (C) 2.1×10^{-7}
    (D) 1.5×10^{-8}
    (E) 1.5×10^{-9}

81. Which of the following will cause the volume of a gas to increase?
    I. Decrease the pressure with the temperature held constant.
    II. Increase the pressure with a temperature decrease.
    III. Increase the temperature with a pressure increase.
    (A) I only
    (B) II only
    (C) I and III only
    (D) II and III only
    (E) I, II and III

82. If 0.1 mole of K_2SO_4 was added to the solution in question 80, what would happen to the Ba^{2+} concentration?
    (A) It would increase.
    (B) It would decrease.
    (C) It would remain the same.
    (D) It would first increase, then decrease.
    (E) It would first decrease, then increase.

83. In a neutralization reaction of an Arrhenius acid and base, what are the products?
    I. H_2O + a sulfate
    II. H_2O + a chloride
    III. H_2O + a salt
    (A) I only
    (B) II only
    (C) I and III only
    (D) II and III only
    (E) III only

84. Which of the following particles has the least mass?
    (A) alpha particle
    (B) beta particle
    (C) proton
    (D) neutron
    (E) gamma ray

85. Which emission from a radioactive source is an electromagnetic wave and is not affected by an electric field?
    (A) alpha particle
    (B) beta particle
    (C) proton
    (D) neutron
    (E) gamma ray

STOP

IF YOU HAVE FINISHED BEFORE ONE HOUR IS UP, YOU MAY GO BACK TO CHECK YOUR WORK OR COMPLETE UNANSWERED QUESTIONS.

Answers and
Explanations to Test

2

1. (E) The triple point of a phase diagram shows that all three states can exist at this point.

2. (B) The high dielectric constant of water is due to the polar nature of the water molecules. It is this property that is largely responsible for the ability of water to dissolve so many solutes.

3. (D) According to Graham's Law of Gaseous Diffusion, the rate of diffusion is inversely proportional to the square root of the molecular weight. So.

$$\frac{rate_H}{rate_O} = \frac{\sqrt{mol.\ wt.\ O_2}}{\sqrt{mol.\ wt.\ H_2}} = \frac{\sqrt{32}}{\sqrt{2}} = \sqrt{\frac{16}{1}} = \frac{4}{1}$$

4. (A) The Law of Definite Composition states that, when compounds form, they always form in the same ratio by weight. Water, for instance, always forms in a ratio of 1:8 of hydrogen to oxygen by *weight*. In nitrous oxide (NO) and nitrogen dioxide (NO_2) the difference in molecular weight is one atomic weight of oxygen.

5. (B) The first step is the ΔH_r for C + $\frac{1}{2}O_2 \rightarrow$ CO. It releases 26.4 kcal of heat. This is written as -26.46.

6. (C) This is the second step on the chart.

7. (A) To arrive at the ΔH_r of this, you can take the total drop (-94.0 kcal) or algebraically add these reactions:

$$\begin{array}{ll} C + \frac{1}{2}O_2 \rightarrow CO & -26.4\ kcal \\ CO + \frac{1}{2}O_2 \rightarrow CO_2 & -67.6\ kcal \\ \hline C + O_2 \rightarrow CO_2 & -94.0\ kcal \end{array}$$

8. (B) Potassium and chlorine have a large enough difference in their electronegativities to form ionic bonds. The respective positions of these two elements in the periodic chart also are indicative of the large difference in their electronegativity values.

9. (D) Two atoms of an element which forms a diatomic molecule always have a pure covalent bond between them since the electron attraction or electronegativity of each atom is the same.

10. (C) Electronegativity differences between .5 and 1.7 are usually indicative of polar covalent bonding. CO_2 is an interesting example of a *nonpolar molecule* with polar covalent bonds since the bonds are symmetrical in the molecule.

11. (E) Calcium is a metal and forms a metallic bond between atoms.

12. (B) In typical heating curves, the first of two plateaus indicates the change of state from a solid to a liquid.

13. (A) Since Part A is at lower temperature, it represents an average lower kinetic energy.

14. (D) Boiling occurs at the very beginning of the second plateau.

15. (A) These are the alkali metals.

16. (B) Because of complete outer shells these elements have very low boiling points.

17. (B) The complete outer shell makes it difficult to remove an electron; thus the high peaks would be the noble gases.

18. (A) With only one electron in the outer shell, the alkali metals lose an electron relatively easily.

19. (E) There is a slight increase in the stability of the *p* orbitals when each has one electron—thus the little peaks at E.

20. (C) The filled *s* orbital shows a slight increase in stability—thus the little peaks at C.

21. (D) The dip at the D areas on the chart shows the first pairing of electrons in the *p* orbitals and indicates this first paired electron has a lower ionization potential than in the atom, which has only one electron in each *p* orbital.

22. (A) The most active metals, the alkalis, have the lowest ionization potentials.

23. (B) The filled *p* orbitals occur in the noble gases.

24. (C) Sulfur trioxide is shown by three structural formulas because each bond is a "hybrid" of a single and a double bond. Resonance in chemistry does not mean that the bonds resonate between the structures shown in the structural drawing.

25. (E) Boric acid is a weak acid because it ionizes only slightly. Hydrofluoric acid is the acid that reacts with glass.

26. (A) Statement and reason are true, and the reason explains the statement.

27. (B) Hydrosulfuric acid is a weak acid but is used in qualitative tests because of the distinctly colored precipitates of sulfides that it forms with many metallic ions.

28. (D) Sodium chloride is an ionic crystal, not a molecule, and its ions are hydrated by the polar water molecules.

29. (A) This statement and its reason fit the rules of Le Châtelier's Principle.

30. (A) The statement is true, and the reason is also true and explains the statement.

31. (C) The statement is true, but not the reason. In an equilibrium reaction, concentrations can be shown to progress like this:

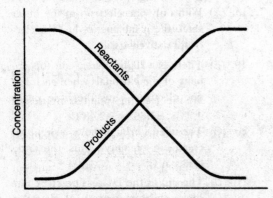

until equilibrium is reached. Then the concentrations stabilize.

32. (D) The forward and reverse reactions are occurring at equal rates when equilibrium is reached. The reactions do not stop. The concentrations remain the same at this point.

33. (A) Since acetylene is known to be a linear molecule with a triple bond between the two carbons, the *sp* orbitals along the central axis with the hydrogens bonded on either end fit the experimental evidence.

34. (E) The weakest bonds between molecules are called van der Waals forces.

35. (A) The terms "dilute" and "concentrated" merely indicate a relatively large amount of solvent or small amount of solvent, respectively. You can have a dilute saturated solution if the solute is only slightly soluble.

36. (D) Cs is the most active Group I metal because it has (a) the largest atomic radius, thus making it easier to lose its outer shell electron, and (b) the intermediate electrons help screen the positive attraction of the nucleus, also increasing the ease with which the outer electron is lost.

37. (D) The cations are positively charged ions and migrate to the cathode, while the anions are negatively charged and migrate to the anode.

38. (A) There are as many electrons as there are protons in a neutral atom, and the atomic number represents the number of each.

39. (D) The first two energy shells fill up at 2 and 8 electrons, respectively. That leaves 7 electrons to fill the $3s$ and $3p$ orbitals like this: $3s^2, 3p^5$. With only one electron missing in the $3p$ orbitals, the most likely oxidation number is -1.

40. (D) The solidification of "Crisco" is merely a physical change just like the formation of ice from liquid water at lower temperatures. All the others involve actual recombinations of the atoms and thus are chemical changes.

41. (C) Water is formed because most common fuels contain hydrogen in their structure.

42. (E) The other choices, in order, would be 1, $\frac{1}{10}$ or deci-, $\frac{1}{100}$ or centi-, and 100.

43. (E) One mole of any substance contains 6.02×10^{23} molecules. Since each water molecule is triatomic, there would be $3(6.02 \times 10^{23})$ atoms present.

44. (E) The "noble" gases are all monoatomic because of their complete outer shell. A rule to help you remember diatomic gases is: Gases ending in -gen or -ine usually form diatomic molecules.

45. (D) Fractional distillation separates the hydrocarbons by taking advantage of their difference in boiling points. All other processes mentioned are used to modify the molecular structure of the hydrocarbons.

46. (C) The complete loss and gain of electrons is an ionic bond. All other bonds indicated are "sharing of electrons" type bonds or some form of covalent bonding.

47. (D) The cathode reaction releases only H_2 gas. The half-reaction is as given in (D).

48. (D) $K_a = 1.8 \times 10^{-5} = \dfrac{[H_3O^+][Ac^-]}{[HAc]} = \dfrac{(x)(x)}{.1-x}$. Since acetic acid is a weak acid, the .1 molar concentration is much greater than the amount ionized (x), so $.1 - x$ is essentially equal to .1. Therefore $1.8 \times 10^{-5} = \dfrac{(x)(x)}{.1} = \dfrac{x^2}{.1}$ and $x = 1.3 \times 10^{-3}$.

49. (C) Percent dissociation $= \dfrac{\text{moles/liter of } H_3O^+}{\text{original concentration}} \times 100$. Percent dissociation $= \dfrac{1 \times 10^{-3}}{1 \times 10^{-1}} \times 100 = 1\%$.

50. (D) $pH = -\log[H_3O^+] = -\log[1 \times 10^{-4}] = -(-4) = 4.$

51. (C) Only I and II are true. Distilled water does not significantly conduct an electric current. The polarity of the water molecule is helpful in ionization and in causing substances to go into solution.

52. (A) SO_2 is the acid anhydride of H_2SO_3 or sulfurous acid. $H_2O + SO_2 \rightarrow H_2SO_3$.

53. (E) Four grams of hydrogen gas at STP represent 2 moles of hydrogen since 2 grams is the gram-molecular weight of hydrogen. Each mole of a gas contains 6.02×10^{23} molecules, and 2 moles would contain $2 \times 6.02 \times 10^{23}$ or 12.04×10^{23} molecules.

54. (B) To solve % composition problems, first divide the % given by the atomic mass:

$$12)\overline{85.7} \qquad 1)\overline{14.3}$$
$$7.14 \qquad\qquad 14.3$$

Then divide by the smallest quotient to get small whole numbers:

$$7.14)\overline{7.14} \qquad 7.14)\overline{14.3}$$
$$1.0 \qquad\qquad 2.0$$

So the empirical formula is CH_2. Since the molecular weight is 42 and the empirical formula has a molecular weight of 14, the true formula must be 3 times the empirical formula, or C_3H_6.

55. (E) Since the temperature (in kelvins) increases from 303 K to 333 K, the volume of the gas should increase with pressure held constant. The correct fraction is $\frac{333}{303}$.

56. (D) One mole of dissolved substance (which does not ionize) causes a 1.86 degree drop (Celsius) in the freezing point of a 1 molal solution. Since 1500 grams of water were used, the solution has $\frac{684}{342}$ or 2 moles in 1500 grams of water. Then

$$\frac{2 \text{ moles}}{1500 \text{ g}} = \frac{x \text{ moles}}{1000 \text{ g}}$$

$$x = \frac{2000}{1500} = 1.33 \text{ moles/1000 g}$$

So the freezing point is depressed $1.33 \times -1.86° = -2.48°C$.

57. (B) The pH is $-\log[H^+]$. A .003 molar solution of H_2SO_4 ionizes in a dilute solution to release two H^+ ions per molecule of H_2SO_4. Therefore the molar concentration of H^+ ion is $2 \times .003$ mole/liter or .006 mole/liter. Substituting this in the formula, you have:

$$pH = -\log[.006] = -\log[6 \times 10^{-3}]$$

The log of a number is the exponent

to which the base 10 is raised to express that number in exponential form:

$1 = 10^0$ so the log $1 = 0$ and
$10 = 10^1$ so the log $10 = 1$

The log of 6 must be 10 to a decimal exponent less than 1 and probably close to .7. Then

$-\log [6 \times 10^{-3}] = -\log [10^{.7} \times 10^{-3}] =$
$-[.7 + (-3)] = -[-2.3] = 2.3$ (approx.)

58. (B) If the solution is to be 5% sodium hydroxide, then 5% of 100 g is 5 grams. % is always by weight unless otherwise specified.

59. (C) Since this reaction is exothermic, higher temperatures will decrease the reaction to the right and increase the reaction to the left.

60. (B) One mole of H_2SO_4 contains 2 gram-equivalent wts. of H^+ so the normality is 2 N. The NaOH contains 1 gram-equivalent wt. of OH^- so the normality is 1 N. Using the formula

$N_1 \times V_1 = N_2 \times V_2$

we get

$2 N \times x$ mL = 1 $N \times 50$ mL

x mL $= \dfrac{1 \, N \times 50 \text{ mL}}{2N}$

$= 25$ mL of H_2SO_4

61. (E) The reaction is:
150 g
$CaCO_3 + 2 \text{ HCl} \rightarrow CaCl_2 + H_2O + CO_2 \uparrow$

The gram-molecular weight of calcium carbonate is 100 g. Then 150 g $= \frac{150}{100}$ or 1.5 moles of calcium carbonate. According to the equation, 1 mole of calcium carbonate yields 1 mole of carbon dioxide, so 1.5 moles of calcium carbonate would yield 1.5 moles of carbon dioxide.
The gram-molecular weight of $CO_2 = 44$ g:
1.5 moles of $CO_2 = 1.5 \times 44$ g $=$ 66 g CO_2

62. (E) The other choices are wrong because:
(A) is lighter than air
(B) reacts with air

(C) is lighter than air
(D) needs heat to be evolved

63. (A) The other choices are wrong because:
(B) needs no heat
(C) needs a condenser for collection
(D) is lighter than air
(E) is lighter than air

64. (C) To insure that all the water vapor collected in the U-tube came from the reaction, the first drying tube is placed in the path of the hydrogen to absorb any evaporated water.

65. (C) Calcium chloride is deliquescent, and its weight gain of water indicates that water was formed from the reaction.

66. (D)
$CaCl_2$ and U-tube after $= 57.32$ g
$CaCl_2$ and U-tube before $= 39.32$ g

Wt. of water formed $= 18.00$ g

CuO and boat before $= 30.23$ g
CuO and boat after $= 14.23$ g

Wt. of oxygen reacted $= 16.00$ g

So, wt. of water $= 18$ g
wt. of oxygen $= 16$ g

Wt. of hydrogen $= 2$ g

The ratio of weight of water to weight of hydrogen is 18:2 or 9:1.

67. (D) 1 K $= 39$, 1 Al $= 27$, 2(SO_4) $= 2(32 + 16 \times 4) = 192$, and 12 $H_2O = 12(2 + 16) = 216$. This totals 474 g.
xg   44.8L

68. (D) 4 Al + 3 $O_2 \rightarrow$ 2 Al_2O_3
108 g 67.2 L

$\dfrac{x}{108} = \dfrac{44.8}{67.2}$, so $x = 72$ g.

69. (B) Metal oxides are generally basic anhydrides

70. (A) Since A is marked +, the oxidation (or loss of electrons) will occur on this pole.

71. (C) "Salt bridge" is the correct terminology.

72. (C) Starch solution is used to test for the presence of I_2. Since this half-cell produces I_2, the identifying blue-black color due to the I_2 reacting with the starch will appear.

213 g   x L
73. (D) $2\ NaClO_3 \rightarrow 2\ NaCl + 3\ O_2$

$$\frac{213}{213} = \frac{x}{67.2}, \text{ so } x = 67.2 \text{ L.}$$

74. (C) The balanced equation has the coefficients 2, 3, 1, and 6, respectively.

75. (E) The reaction is $NaClO_3 \rightarrow NaCl + \frac{3}{2}\ O_2$.

$\triangle H_r = \triangle H_{f(\text{products})} - H_{f(\text{reactants})}$
$\triangle H_r = -98.2 + 0 - (-85.7)$
$\triangle H_r = -12.5 \text{ kcal}$

76. (D) II and III are identical; isotopes differ only in the number of neutrons in the nucleus and this affects the atomic weight or mass only.

77. (E) I, II, and III would affect this reaction.

78. (D) This setup depends on water displacement of an insoluble gas.

79. (B) The coefficients give the molar relations, so 2 moles of HCl give off 1 mole of CO_2. Given 3 moles of HCl, you have

$$\frac{3}{2} = \frac{x}{1}; \text{ then } x = \frac{3}{2} = 1.5 \text{ moles.}$$

80. (E) $K_{sp} = [B_2^{2+}]\ [SO_4^{2-}] = (3.9 \times 10^{-5})$
$(3.9 \times 10^{-5}) = 1.5 \times 10^{-9}$
These two will be the same since
$BaSO_4 \rightarrow Ba^{2+} + SO_4^{2-}$.

81. (A) According to the gas laws, only I will cause an increase in the volume of a confined gas.

82. (B) The introduction of the "common ion" SO_4^{2-} at .1 molar forces the equilibrium to shift to the left and reduce the Ba^{2+} concentration.

83. (E) This type of neutralization always involves the formation of H_2O and a salt. I and II would be true only if the acid used was sulfuric and hydrochloric, respectively.

84. (B) The beta particle is a high-speed electron and has the smallest mass of the first four choices. Gamma rays are electromagnetic waves.

85. (E) A gamma ray is not affected by an electric field because it is an electromagnetic wave. The neutron is not an electromagnetic wave; it is a particle with no charge but the approximate mass of a proton.

DIAGNOSING YOUR NEEDS

After taking Practice Test 2, check your answers against the correct ones. Then fill in the chart below.

In the space under each question number, place a check if you answered that question correctly.

• **EXAMPLE:**

If your answer to question 5 was correct, place a check in the appropriate box.

Next, total the check marks for each section and insert the number in the designated block. Now do the arithmetic indicated, and insert your percent for each area.

SUBJECT AREA

(✔) QUESTIONS ANSWERED CORRECTLY

I. Atomic Theory and Structure, including periodic relationships	19	20	21	23	24	38	44	76

☐ No. of checks ÷ 8 × 100 = _____ %

II. Nuclear Reactions	84	85

☐ No. of checks ÷ 2 × 100 = _____ %

III. Chemical Bonding and Molecular Structure	4	8	9	10	11	33	34	46

☐ No. of checks ÷ 8 × 100 = _____ %

IV. States of Matter and Kinetic Molecular Theory	3	12	13	14	43	53	55	81

☐ No. of checks ÷ 8 × 100 = _____ %

V. Solutions, including concentration units, solubility, and colligative properties	2	28	35	56	60

☐ No. of checks ÷ 5 × 100 = _____ %

VI. Acids and Bases	25	48	49	50	52	57	69	83

☐ No. of checks ÷ 8 × 100 = _____ %

VII. Oxidation-Reduction and Electrochemistry	37	39	47	51	70	71

☐ No. of checks ÷ 6 × 100 = _____ %

VIII. Stoichiometry	54	58	61	66	67	68	73	74	79

☐ No. of checks ÷ 9 × 100 = _____ %

SUBJECT AREA	(✔) QUESTIONS ANSWERED CORRECTLY									

IX. Reaction Rates	59	77								
☐ No. of checks ÷ 2 × 100 = _____%										

X. Equilibrium	29	31	32	80	82
☐ No. of checks ÷ 5 × 100 = _____%					

XI. Thermodynamics: energy changes in chemical reactions, randomness, and criteria for spontaneity	5	6	7	30	75
☐ No. of checks ÷ 5 × 100 = _____%					

XII. Descriptive Chemistry: physical and chemical properties of elements and their familiar compounds; organic chemistry; periodic properties	1	15	16	17	18	22	26	27	36	40
						41	42	45	72	
☐ No. of checks ÷ 14 × 100 = _____%										

XIII. Laboratory: equipment, procedures, observations, safety, calculations, and interpretation of results	62	63	64	65	78
☐ No. of checks ÷ 5 × 100 = _____%					

PLANNING YOUR STUDY

The percentages give you an idea of how you have done on the various major areas of the test. Because of the limited number of questions on some parts, these percentages may not be as reliable as the percentages for parts with larger numbers of questions. However, you should now have at least a rough idea of the areas in which you have done well and those in which you need more study.

Start your study with the areas in which you are the weakest. The corresponding chapters are indicated below:

Subject Area	Chapters to Review
I. Atomic Theory and Structure, including periodic relationships	2
II. Nuclear Reactions	15
III. Chemical Bonding and Molecular Structure	3, 4
IV. States of Matter and Kinetic Molecular Theory	1
V. Solutions, including concentration units, solubility, and colligative properties	7

Subject Area	Chapters to Review
VI. Acids and Bases	11
VII. Oxidation-reduction and Electrochemistry	12
VIII. Stoichiometry	5, 6
IX. Reaction Rates	9
X. Equilibrium	10
XI. Thermodynamics, including energy changes in chemical reactions, randomness, and criteria for spontaneity	8
XII. Descriptive Chemistry: physical and chemical properties of elements and their familiar compounds; organic chemistry; periodic properties	1, 2, 13, 14
XIII. Laboratory: equipment, procedures, observations, safety, calculations, and interpretation of results	All lab diagrams, 16

ANSWER SHEET FOR PRACTICE TEST 3

Determine the correct answer for each question. Then, using a No. 2 pencil, blacken completely the oval containing the letter of your choice.

1. Ⓐ Ⓑ Ⓒ Ⓓ Ⓔ   16. Ⓐ Ⓑ Ⓒ Ⓓ Ⓔ   31. Ⓐ Ⓑ Ⓒ Ⓓ Ⓔ
2. Ⓐ Ⓑ Ⓒ Ⓓ Ⓔ   17. Ⓐ Ⓑ Ⓒ Ⓓ Ⓔ   32. Ⓐ Ⓑ Ⓒ Ⓓ Ⓔ
3. Ⓐ Ⓑ Ⓒ Ⓓ Ⓔ   18. Ⓐ Ⓑ Ⓒ Ⓓ Ⓔ   33. Ⓐ Ⓑ Ⓒ Ⓓ Ⓔ
4. Ⓐ Ⓑ Ⓒ Ⓓ Ⓔ   19. Ⓐ Ⓑ Ⓒ Ⓓ Ⓔ   34. Ⓐ Ⓑ Ⓒ Ⓓ Ⓔ
5. Ⓐ Ⓑ Ⓒ Ⓓ Ⓔ   20. Ⓐ Ⓑ Ⓒ Ⓓ Ⓔ   35. Ⓐ Ⓑ Ⓒ Ⓓ Ⓔ
6. Ⓐ Ⓑ Ⓒ Ⓓ Ⓔ   21. Ⓐ Ⓑ Ⓒ Ⓓ Ⓔ   36. Ⓐ Ⓑ Ⓒ Ⓓ Ⓔ
7. Ⓐ Ⓑ Ⓒ Ⓓ Ⓔ   22. Ⓐ Ⓑ Ⓒ Ⓓ Ⓔ   37. Ⓐ Ⓑ Ⓒ Ⓓ Ⓔ
8. Ⓐ Ⓑ Ⓒ Ⓓ Ⓔ   23. Ⓐ Ⓑ Ⓒ Ⓓ Ⓔ   38. Ⓐ Ⓑ Ⓒ Ⓓ Ⓔ
9. Ⓐ Ⓑ Ⓒ Ⓓ Ⓔ   24. Ⓐ Ⓑ Ⓒ Ⓓ Ⓔ   39. Ⓐ Ⓑ Ⓒ Ⓓ Ⓔ
10. Ⓐ Ⓑ Ⓒ Ⓓ Ⓔ   25. Ⓐ Ⓑ Ⓒ Ⓓ Ⓔ   40. Ⓐ Ⓑ Ⓒ Ⓓ Ⓔ
11. Ⓐ Ⓑ Ⓒ Ⓓ Ⓔ   26. Ⓐ Ⓑ Ⓒ Ⓓ Ⓔ   41. Ⓐ Ⓑ Ⓒ Ⓓ Ⓔ
12. Ⓐ Ⓑ Ⓒ Ⓓ Ⓔ   27. Ⓐ Ⓑ Ⓒ Ⓓ Ⓔ   42. Ⓐ Ⓑ Ⓒ Ⓓ Ⓔ
13. Ⓐ Ⓑ Ⓒ Ⓓ Ⓔ   28. Ⓐ Ⓑ Ⓒ Ⓓ Ⓔ   43. Ⓐ Ⓑ Ⓒ Ⓓ Ⓔ
14. Ⓐ Ⓑ Ⓒ Ⓓ Ⓔ   29. Ⓐ Ⓑ Ⓒ Ⓓ Ⓔ   44. Ⓐ Ⓑ Ⓒ Ⓓ Ⓔ
15. Ⓐ Ⓑ Ⓒ Ⓓ Ⓔ   30. Ⓐ Ⓑ Ⓒ Ⓓ Ⓔ   45. Ⓐ Ⓑ Ⓒ Ⓓ Ⓔ

46. Ⓐ Ⓑ Ⓒ Ⓓ Ⓔ   61. Ⓐ Ⓑ Ⓒ Ⓓ Ⓔ   76. Ⓐ Ⓑ Ⓒ Ⓓ Ⓔ
47. Ⓐ Ⓑ Ⓒ Ⓓ Ⓔ   62. Ⓐ Ⓑ Ⓒ Ⓓ Ⓔ   77. Ⓐ Ⓑ Ⓒ Ⓓ Ⓔ
48. Ⓐ Ⓑ Ⓒ Ⓓ Ⓔ   63. Ⓐ Ⓑ Ⓒ Ⓓ Ⓔ   78. Ⓐ Ⓑ Ⓒ Ⓓ Ⓔ
49. Ⓐ Ⓑ Ⓒ Ⓓ Ⓔ   64. Ⓐ Ⓑ Ⓒ Ⓓ Ⓔ   79. Ⓐ Ⓑ Ⓒ Ⓓ Ⓔ
50. Ⓐ Ⓑ Ⓒ Ⓓ Ⓔ   65. Ⓐ Ⓑ Ⓒ Ⓓ Ⓔ   80. Ⓐ Ⓑ Ⓒ Ⓓ Ⓔ
51. Ⓐ Ⓑ Ⓒ Ⓓ Ⓔ   66. Ⓐ Ⓑ Ⓒ Ⓓ Ⓔ   81. Ⓐ Ⓑ Ⓒ Ⓓ Ⓔ
52. Ⓐ Ⓑ Ⓒ Ⓓ Ⓔ   67. Ⓐ Ⓑ Ⓒ Ⓓ Ⓔ   82. Ⓐ Ⓑ Ⓒ Ⓓ Ⓔ
53. Ⓐ Ⓑ Ⓒ Ⓓ Ⓔ   68. Ⓐ Ⓑ Ⓒ Ⓓ Ⓔ   83. Ⓐ Ⓑ Ⓒ Ⓓ Ⓔ
54. Ⓐ Ⓑ Ⓒ Ⓓ Ⓔ   69. Ⓐ Ⓑ Ⓒ Ⓓ Ⓔ   84. Ⓐ Ⓑ Ⓒ Ⓓ Ⓔ
55. Ⓐ Ⓑ Ⓒ Ⓓ Ⓔ   70. Ⓐ Ⓑ Ⓒ Ⓓ Ⓔ   85. Ⓐ Ⓑ Ⓒ Ⓓ Ⓔ
56. Ⓐ Ⓑ Ⓒ Ⓓ Ⓔ   71. Ⓐ Ⓑ Ⓒ Ⓓ Ⓔ
57. Ⓐ Ⓑ Ⓒ Ⓓ Ⓔ   72. Ⓐ Ⓑ Ⓒ Ⓓ Ⓔ
58. Ⓐ Ⓑ Ⓒ Ⓓ Ⓔ   73. Ⓐ Ⓑ Ⓒ Ⓓ Ⓔ
59. Ⓐ Ⓑ Ⓒ Ⓓ Ⓔ   74. Ⓐ Ⓑ Ⓒ Ⓓ Ⓔ
60. Ⓐ Ⓑ Ⓒ Ⓓ Ⓔ   75. Ⓐ Ⓑ Ⓒ Ⓓ Ⓔ

Practice Test

3

Note: For all questions involving solutions and/or chemical equations, assume that the system is in water unless otherwise stated.

Part A

Directions: Each set of lettered choices below refers to the numbered statements or formulas immediately following it. Select the one lettered choice that best fits each statement or formula and then fill in the corresponding oval on the answer sheet. A choice may be used once, more than once, or not at all in each set.

Questions 1–4 refer to the following diagram:

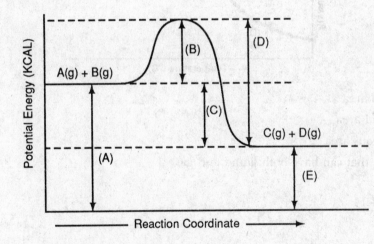

1. The activation energy of the forward reaction
2. The activation energy of the reverse reaction
3. The heat of the reaction for the forward reaction
4. The potential energy of the activated complex
   (A) A + B
   (B) B + C
   (C) C + D
   (D) D + E
   (E) E + A

Questions 5–7 refer to the following diagram:

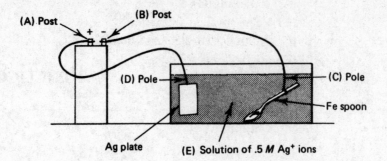

5. To plate silver on the spoon, the position to which the spoon must be connected
6. The position of the anode
7. The position from which silver that is plated out emerges

Questions 8–11 refer to the following graph:

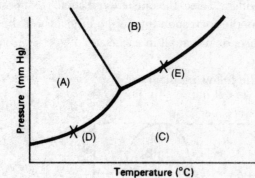

8. The solid area
9. The liquid area
10. The gaseous area
11. The area that can have both liquid and gas

Questions 12–14
   (A) 1
   (B) 2
   (C) 3
   (D) 4
   (E) 5

12. When the following equation: $Cu + HNO_3 \rightarrow Cu(NO_3)_2 + H_2O + NO$ is balanced, the coefficient, in the lowest whole number, of NO

13. When the following equation: $KMnO_4 + H_2SO_3 \rightarrow K_2SO_4 + MnSO_4 + H_2SO_4 + H_2O$ is balanced, the coefficient, in the lowest whole number, of H_2SO_3

14. When the following equation: $Br_2 + SO_2 + H_2O \rightarrow H_2SO_4 + HBr$ is balanced, the coefficient, in the lowest whole number, of HBr

Questions 15–18

        (A) Ionic substance
        (B) Polar covalent substance
        (C) Nonpolar covalent substance
        (D) Amorphous substance
        (E) Metallic network

15. KCl
16. HCl(g)
17. CH_4
18. Lithium

Questions 19–23

        (A) Brownian movement
        (B) Litmus paper reaction
        (C) Phenolphthalein reaction
        (D) Dehydration
        (E) Deliquescent

19. The reason why a blue crystal of $CuSO_4 \cdot 5\ H_2O$ turns white when heated
20. The zigzag path of colloidal particles in light
21. The pink color in a basic solution
22. The pink color in an acid solution
23. The adsorbtion of water to the surface of a crystal

Part B

Directions: Each question below consists of an <u>assertion</u> (statement) in the left-hand column and a <u>reason</u> in the right-hand column. On the appropriate line of the answer sheet fill in oval

A if both assertion and reason are true statements and the reason is *a correct explanation* of the assertion;
B if both assertion and reason are true statements but the reason is *NOT a correct explanation* of the assertion;
C if the assertion is true but the reason is a false statement;
D if the assertion is false but the reason is a true statement;
E if both assertion and reason are false statements.

Directions Summarized

	Assertion	Reason	
A	True	True	Reason is _a correct explanation_
B	True	True	Reason is _NOT a correct explanation_
C	True	False	
D	False	True	
E	False	False	

	Assertion		_Reason_
24.	Elements in the upper right of the periodic table are active nonmetals	BECAUSE	they have larger ionic radii than atomic radii.
25.	A synthesis reaction with a high heat of formation will not occur until some heat is added	BECAUSE	the reaction needs enough activation energy to get it started.
26.	Transition elements in a particular period often have the same oxidation number	BECAUSE	they have a complete outer shell.
27.	When a crystal is added to a supersaturated solution, of itself, the crystal does not change	BECAUSE	the solution cannot hold more solute.
28.	Equilibrium is a static condition	BECAUSE	the forward reaction rate equals the reverse reaction rate.
29.	The ionic bond is the strongest bond	BECAUSE	it has an electrostatic attraction due to the loss and gain of electron(s).
30.	Pressure applied to a system in equilibrium favors the forward reaction	BECAUSE	the product side of an equilibrium would show an increase in gas pressure.
31.	If the forward reaction of an equilibrium is exothermic, adding heat to the system favors the reverse reaction	BECAUSE	additional heat causes a stress on the system, and the system moves in the direction that releases the stress.
32.	The ions of an element of Group VIIA have a larger radius than the atom	BECAUSE	more positive charges in the nucleus than electron charges in the shells will cause the radius of the ion to be different than that of the atom.
33.	The most electronegative elements in	BECAUSE	electronegativity is a mea-

Assertion		Reason
the periodic chart are found among nonmetals		sure of the ability of an atom to draw valence electrons to itself.
34. Basic anhydrides react in water to form bases	BECAUSE	nonmetallic oxides react with water to form a solution that has an excess of hydroxide ions.
35. There are 3 moles of atoms in 18 grams of water	BECAUSE	there are 6×10^{23} atoms in 1 mole.
36. Benzene is a good electrolyte	BECAUSE	a good electrolyte has charged ions that carry the electric current.
37. Normal butyl alcohol and butanol -2 are isomers	BECAUSE	isomers vary in the number of neutrons in the nucleus of the atom.
38. The reaction of the $CaCO_3$ and HCl goes to completion	BECAUSE	reactions that form a precipitate go to completion.
39. A large number of alpha particles were deflected in the Rutherford experiment	BECAUSE	alpha particles that came close to the nucleus of the gold atoms were deflected.

Part C

Directions: Each of the questions or incomplete statements below is followed by five suggested answers or completions. Select the one that is best in each case and then fill in the corresponding oval on the answer sheet.

40. What are the simplest coefficients that balance this equation?

$$\ldots C_4H_{10} + \ldots O_2 \rightarrow \ldots CO_2 + \ldots H_2O$$

   (A) 1, 6, 4, 2
   (B) 2, 13, 8, 10
   (C) 1, 6, 1, 5
   (D) 3, 10, 16, 20
   (E) 4, 26, 16, 20

41. **How many atoms are present in the formula $KAl(SO_4)_2$?**
   (A) 7
   (B) 9
   (C) 11
   (D) 12
   (E) 13

42. All of the following are compounds EXCEPT
    (A) copper sulfate
    (B) carbon dioxide
    (C) washing soda
    (D) air
    (E) lime

43. What volume of gas, in liters, would 1.5 moles of hydrogen occupy at STP?
    (A) 11.2
    (B) 22.4
    (C) 33.6
    (D) 44.8
    (E) 67.2

44. What is the maximum number of electrons held in the d orbitals?
    (A) 2
    (B) 6
    (C) 8
    (D) 10
    (E) 14

45. If an element has an atomic number of 11, it will combine most readily with an element that has an atomic number of
    (A) 12
    (B) 14
    (C) 15
    (D) 16
    (E) 17

46. An example of a physical property is
    (A) rusting
    (B) decay
    (C) souring
    (D) low melting point
    (E) high heat of formation

47. A gas at STP which contains 6.02×10^{23} atoms and forms diatomic molecules will occupy
    (A) 11.2 liters
    (B) 22.4 liters
    (C) 33.6 liters
    (D) 67.2 liters
    (E) 1.06 quarts

48. When excited electrons cascade to lower energy levels in an atom,
    (A) visible light is always emitted
    (B) the potential energy of the atom increases
    (C) the electrons always fall back to the first energy level
    (D) the electrons fall indiscriminately to all levels
    (E) the electrons fall back to the lowest unfilled energy level

49. The spectroscope uses the concept that
    (A) charged particles are evenly deflected in a magnetic field
    (B) charged particles are deflected in a magnetic field inversely to the mass of the particles
    (C) particles of heavier mass are deflected in a magnetic field to a greater degree than lighter particles
    (D) particles are evenly deflected in a magnetic field

50. The bond that includes an upper and a lower sharing of electron orbitals is called
    (A) a pi bond
    (B) a sigma bond
    (C) a hydrogen bond
    (D) a covalent bond
    (E) an ionic bond

51. What is the boiling point of water at the top of Pike's Peak?
    (A) It is 100°C.
    (B) It is >100°C since the pressure is less.
    (C) It is <100°C since the pressure is less.
    (D) It is >100°C since the pressure is greater.
    (E) It is <100°C since the pressure is greater.

52. The atomic structure of the alkane series contains the hybrid orbitals designated as
    (A) sp
    (B) sp^2
    (C) sp^3
    (D) sp^3d^2
    (E) sp^4d^3

53. Which of the following is (are) true for this reaction?

    $$Cu + 4\ HNO_3 \rightarrow Cu(NO_3)_2 + 2\ H_2O + 2\ NO_2 \uparrow$$

    I. It is an oxidation-reduction reaction.
    II. Copper is oxidized.
    III. The oxidation number of nitrogen goes from +5 to +4.
        (A) I, II, and III
        (B) I and II only
        (C) II and III only
        (D) I only
        (E) III only

54. Which of the following properties can be attributed to water?
    I. It is attracted by a magnet.
    II. It is a very good conductor of electricity.
    III. It has its polar covalent bonds with hydrogen on opposite sides of the oxygen atom, so that the molecule is linear.
        (A) I, II, and III
        (B) I and II only
        (C) II and III only
        (D) I only
        (E) III only

55. Which of the following define(s) an acid?
    I. It is a good proton donor.
    II. It is a good electron pair acceptor.
    III. It has an excess of H^+ in solution.
        (A) I, II, and III
        (B) I and II only
        (C) II and III only
        (D) I only
        (E) III only

56. A nuclear reactor must include which of the following parts?
    I. Electric generator
    II. Fissionable fuel elements
    III. Moderator
        (A) I, II, and III
        (B) I and II only
        (C) II and III only
        (D) I only
        (E) III only

57. Which of the following salts will hydrolyze in water to form basic solutions?
    I. NaCl
    II. $CuSO_4$
    III. K_3PO_4
        (A) I, II, and III
        (B) I and II only
        (C) II and III only
        (D) I only
        (E) III only

58. The dissolving of 1 mole of NaCl in 1000 grams of water will change the boiling point of the water to
    (A) 100.51°C
    (B) 101.02°C
    (C) 101.53°C
    (D) 101.86°C
    (E) 103.62°C

59. What is the structure associated with the sp^2 hybrid?
    (A) rectangle
    (B) triangle
    (C) tetrahedron
    (D) octahedron
    (E) square

60. Which of these statements is the best expression for the sp^3 hybridization of carbon electrons?
    (A) The new orbitals are one s orbital and three p orbitals.
    (B) The s electron is promoted to the p orbitals.
    (C) The s orbital is deformed into a p orbital.
    (D) Four new and equivalent orbitals are formed.
    (E) The s orbital electron loses energy to fall back into a partially filled p orbital.

61. The bonding that explains the variation of the boiling point of water from the boiling points of similarly structured molecules is
    (A) hydrogen bonding
    (B) van der Waals forces
    (C) covalent bonding
    (D) ionic bonding
    (E) coordinate covalent bonding

62. If K for the reaction $H_2 + I_2 \rightleftharpoons 2\,HI$ is equal to 45.9 at 450°C, and 1 mole of H_2 and 1 mole of I_2 are introduced into a 1-liter box at that temperature, what would be the concentration of H_2 at equilibrium?
    (A) 0.114 mole/liter
    (B) 0.228 mole/liter
    (C) 0.456 mole/liter
    (D) 0.516 mole/liter
    (E) 0.772 mole/liter

63. What is the gram-molecular weight of a nonionizing solid if 10 grams of this solid, dissolved in 100 grams of water, formed a solution which froze at −1.22°C?
    (A) 0.65 g
    (B) 6.5 g
    (C) 130 g
    (D) 154 g
    (E) 265 g

64. What is the pH of a solution with a hydroxide ion concentration of 0.00001 mole/liter?
    (A) −5
    (B) −1
    (C) 5
    (D) 9
    (E) 14

65. Electrolysis of a dilute solution of sodium chloride results in the cathode product
    (A) sodium
    (B) hydrogen
    (C) chlorine
    (D) oxygen
    (E) peroxide

66. Aqua regia is a common name for
    (A) nitric acid
    (B) hydrochloric acid
    (C) sulfuric acid
    (D) ammonia water
    (E) a mixture of (A) and (B)

67. Five liters of gas at STP weigh 12.5 grams. What is the molecular weight of the gas?
    (A) 12.5 g
    (B) 25.0 g
    (C) 47.5 g
    (D) 56.0 g
    (E) 125 g

68. A compound whose molecular weight is 90 grams contains 40.0% carbon, 6.67% hydrogen, and 53.33% oxygen. What is the true formula of the compound?
    (A) $C_2H_2O_4$
    (B) CH_2O_4
    (C) C_3H_6O
    (D) C_3HO_3
    (E) $C_3H_6O_3$

69. How many moles of CaO are needed to react with an excess of water to form 370 grams of calcium hydroxide?
    (A) 1
    (B) 2
    (C) 3
    (D) 4
    (E) 5

70. To what volume, in milliliters, must 50.0 milliliters of 3.50 M H_2SO_4 be diluted in order to make 2 M H_2SO_4?
    (A) 25
    (B) 60.1
    (C) 87.5
    (D) 93.2
    (E) 101

71. If K_e is small, it indicates that equilibrium occurs
    (A) at a low product concentration
    (B) at a high product concentration
    (C) after considerable time
    (D) with the help of a catalyst
    (E) with no forward reaction

72. A student measured 10.0 milliliters of an HCl solution into a beaker and titrated it with a standard NaOH solution that was 0.09 M. The initial NaOH buret reading was 34.7 milliliters while the final reading showed 49.2 milliliters.
    How many grams of HCl were present in 100 milliliters of the HCl solution?
    (A) 0.13
    (B) 0.47
    (C) 4.7
    (D) 14.5
    (E) 36.5

73. A student made the following observations in the laboratory:
    (a) Sodium metal reacted vigorously with water while a strip of magnesium did not seem to react at all.
    (b) The magnesium strip reacted with dilute hydrochloric acid faster than an iron strip.
    (c) A copper rivet suspended in silver nitrate solution was covered with silver-colored stalactites in several days, and the resulting solution had a blue color.
    (d) Iron filings dropped into the blue solution were coated with an orange color.

The order of *decreasing* strength as reducing agents is:
(A) Na, Mg, Fe, Cu, Ag
(B) Mg, Na, Fe, Cu, Ag
(C) Ag, Cu, Fe, Mg, Na
(D) Na, Fe, Mg, Cu, Ag
(E) Na, Mg, Fe, Cu, Ag

74. A student placed water, sodium chloride, potassium dichromate, sand, chalk, and hydrogen sulfide into a distilling flask and proceeded to distill. What ingredient besides water would be found in the distillate?
(A) sodium chloride
(B) chalk
(C) sand
(D) hydrogen sulfide
(E) chrome sulfate

75. Which test would you use to prove the presence of the other substance in the distillate flask of question 74?
(A) flame test for sodium
(B) barium chloride test for sulfate
(C) silver nitrate test for chloride
(D) acid test for carbonate in chalk
(E) lead acetate test for hydrogen sulfide

76. A student used a steam-jacketed eudiometer and filled it with 32 milliliters of oxygen and 4 milliliters of hydrogen over mercury. How much of which gas would be left uncombined after sparking?
(A) none of either
(B) 3 mL H_2
(C) 24 mL O_2
(D) 28 mL O_2
(E) 30 mL O_2

77. What would be the *total* volume, in milliliters, of gases in question 76 after sparking?
(A) 16
(B) 24
(C) 34
(D) 36
(E) 40

78. A student mixed a small amount of amyl alcohol, glacial acidic acid, and a few drops of concentrated H_2SO_4. She then gently heated the mixture until she observed an odor. What odor should she expect to be given off?
(A) banana oil
(B) oil of wintergreen
(C) pineapple
(D) apple
(E) formaldehyde

79. In which period is the most electronegative element found?
    (A) 1
    (B) 2
    (C) 3
    (D) 4
    (E) 5

80. What would be the equilibrium constant for this reaction: $a\mathrm{A} + b\mathrm{B} \rightleftharpoons c\mathrm{C} + d\mathrm{D}$, if A and D are solids?

    (A) $\dfrac{[C]^c[D]^d}{[A]^a[B]^b}$

    (B) $\dfrac{[A]^a[B]^b}{[C]^c[D]^d}$

    (C) $\dfrac{[C]^c}{[B]^b}$

    (D) $\dfrac{[C]^c[D]^d}{[A]^a}$

    (E) $[A]^a[B]^b[C]^c[D]^d$

81. Which of the following is NOT a commonly occurring sulfur compound?
    (A) H_2S
    (B) H_2SO_4
    (C) SO_2
    (D) AG_2S
    (E) SO_3

Questions 82–83 refer to the following setup:

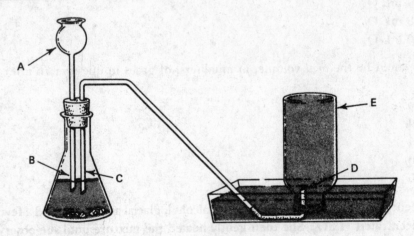

82. What letter designates an error in this laboratory setup?
    (A) A
    (B) B
    (C) C
    (D) D
    (E) E

83. If the reaction in question 82 created a gas, where would it be expelled under these conditions?
    (A) A
    (B) B
    (C) C
    (D) D
    (E) E

84. The most active nonmetal has
    (A) a high electronegativity
    (B) a low electronegativity
    (C) a medium electronegativity
    (D) large atomic radii
    (E) a deliquescent property

85. In the reaction $Fe + 2 S \rightarrow FeS_2$, which is true?
    (A) $Fe^0 + 2e^- \rightarrow Fe^{2+}$
    (B) $Fe^0 - 2e^- \rightarrow Fe^{2+}$
    (C) $Fe^{2+} - 2e^- \rightarrow Fe^0$
    (D) $S^0 - 2e^- \rightarrow S^{2-}$
    (E) $S^{-2} + 2e^- \rightarrow S^0$

STOP

IF YOU FINISH BEFORE ONE HOUR IS UP, YOU MAY GO BACK TO CHECK YOUR WORK OR COMPLETE UNANSWERED QUESTIONS.

*Answers and
Explanations to Test*

3

1. (B) The activation energy is the energy needed to begin the reaction.

2. (D) For the reverse reaction to occur, the sum of B + C is needed. This is shown by D.

3. (C) The heat of the reaction is the heat liberated between the level of potential energy of the reactants and that of the products. This is the quantity C on the diagram.

4. (A) The potential energy of the activated complex is the total of the original potential energy (A) and the activation energy (B).

5. (B) The spoon must be made the cathode to attract the Ag^+ ions.

6. (D) The silver plate is the anode.

7. (E) The solution of Ag^+ provides the silver for plating.

8. (A)
9. (B) } In a phase diagram, the zones are as shown below.
10. (C)

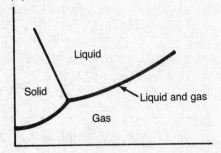

11. (E) The boundary of liquid and gas is shown on the diagram

12–14. The balanced equations with half-reactions are as follows:

12. (B)
$$3\ Cu^0 \rightarrow 3\ Cu^{2+} + 2e^-$$
$$\underline{2\ NO_3^- + 8\ H^+ + 3e^- \rightarrow 2\ NO + 4\ H_2O}$$
$$3\ Cu + 8\ HNO_3 \rightarrow 3\ Cu(NO_3)_2$$
$$+ 4\ H_2O + 2\ NO$$

13. (E)
$$2(MnO_4^{1-} + 8\ H^+ + 5e^- \rightarrow Mn^{2+} + 4\ H_2O)$$
$$\underline{5(SO_3^{2-} + H_2O \rightarrow SO_4^{2-} + 2\ H + 2e^-)}$$
$$2\ KMnO_4 + 5\ H_2SO_3 \rightarrow K_2SO_4$$
$$+ 2\ MnSO_4 + 2\ H_2SO_4 + 3\ H_2O$$

14. (B)
$$Br_2 + 2e^- \rightarrow 2\ Br^{1-}$$
$$\underline{SO_2 + 2\ H_2O \rightarrow SO_r^{2-} + 4\ H^+ + 2e^-}$$
$$Br_2 + SO_2 + 2\ H_2O \rightarrow H_2SO_4 + 2\ HBr$$

15. (A) KCl is ionic because it is the product of a very active metal combining with a very active nonmetal.

16. (B) The electronegativity difference between H and Cl is between .5 and 1.7. This indicates an unequal sharing of electrons, which result in a polar covalent bond.

17. (C) Because these polar bonds are symmetrically arranged in the methane molecule, the molecule is nonpolar covalent.

18. (E) Lithium is a metal.

19. (D) When hydrated copper sulfate is heated, the crystal crumples as the water is forced out of the structure, and a white powder is the result.

20. (A) Brownian movement is due to molecular collisions with colloidal particles, which knocks them about in a zigzag path noted by the reflected light from these particles.

21. (C) The indicator phenolphthalein turns pink in a basic solution.

22. (B) Litmus paper turns pink in an acid solution.

23. (E) A substance that is deliquescent draws water to its surface. At times it can draw enough water to form a water solution.

24. (B) The assertion is true because the nonmetals have a tendency to gain electrons readily. The reason is a true statement but does not explain the assertion.

25. (A) The assertion is explained by the reason. The graphic display of this is:

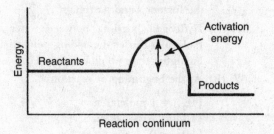

26. (C) The assertion is true but the reasoning is false. Transition elements have incomplete inner energy levels which are being filled with the additional electrons, thus leaving the outer shell the same in most cases. This gives these elements common oxidation numbers.

27. (D) The assertion is false and the reason is true. A supersaturated solution is holding more than its normal solubility, and the addition of a crystal causes crystallization to occur.

28. (D) Equilibrium is a dynamic condition because of the reason stated. The assertion is false; the reason, true.

29. (A) Both statements are true, and the reason explains why ionic bonding is the strongest.

30. (E) An equilibrium system must have a gaseous reactant or product for pressure to affect the equilibrium. Then increased pressure will cause the reaction to go in the direction that reduces the concentration of gaseous substances.

31. (A) The assertion is explained by the reason; both are true.

32. (B) While both assertion and reason are true, the reason does not apply because group VIIA ions have a nucleus with less positive charges than the negative charges of the electrons in their shells.

33. (A) The assertion is true; the reason is true and explains why nonmetals have the highest electronegativity.

34. (A) The assertion is explained by the reason; both are true.

35. (B) There are three moles of atoms in 18 grams of water because 18 grams is 1 mole of water molecules and each molecule has three atoms. The reason does not explain the assertion but is also true.

36. (D) Benzene is a nonionizing substance and therefore a nonelectrolyte. The reason is a true statement.

37. (C) Isomers have the same empirical formula but vary in their structural formulas.

38. (C) The reaction does go to completion, but a gaseous product is formed, not a precipitate. $2\ HCl + CaCO_3 \rightarrow CaCl_2 + H_2O + CO_2 \uparrow$

39. (D) In the Rutherford experiment relatively few alpha particles were deflected, indicating a great deal of empty space in the atom. The reason is a true statement.

40. (B) The correct coefficients are 2, 13, 8, and 10.

41. (D) $1\ K + 1\ Al + 2\ S + 8\ O = 12$ total

42. (D) Air is a mixture; all others are compounds. Washing soda (C) is sodium carbonate, and lime (E) is calcium oxide.

43. (C) One mole of a gas at STP occupies 22.4 L. Therefore, 1.5 moles $\times 22.4$ L $= 33.6$ L.

44. (D) The maximum number of electrons in each kind of orbitals is:
$s =$ 2 in one orbital
$p =$ 6 in three orbitals
$d =$ 10 in five orbitals
$f =$ 14 in seven orbitals

45. (E) An element with atomic number 11 will have an electron structure of 2, 8, 1. This gives it an oxidation number of +1, which would combine most readily with an oxidation number of −1. The atomic number 17 has an oxidation number −1.

46. (D) The only physical property named in the list is low melting point.

47. (A) If the gas is diatomic, then 6.02×10^{23} atoms will form $6.02 \times 10^{23}/2$ molecules. At STP, 6.02×10^{23} molecules occupy 22.4 liters. So half that amount would occupy 11.2 liters.

48. **(E)** Cascading excited electrons can fall only to the lowest energy level that is unfilled.

49. **(B)** The spectroscope uses a magnetic field to separate isotopes by bending their path. The lighter ones are bent further than the heavier ones.

50. **(A)** The pi bond is a bond between two *p* orbitals, like this:

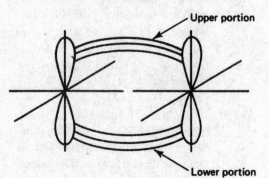

51. **(C)** At Pike's Peak (alt. approx. 14,000 ft.) the pressure is lower; therefore the vapor pressure at a lower temperature will equal the outside pressure and boiling will occur.

52. **(C)** The alkanes use the sp^3 hybrid orbitals.

53. **(A)** I, II, and III are correct.

54. **(D)** Only I is correct.

55. **(A)** I, II, and II are acid definitions.

56. **(C)** I is not necessary for the reactor, but often nuclear energy is used to operate an electrical generator. The others, II and III, are necessary for fuel and neutron-speed control, respectively.

57. **(E)** III is a salt from a strong base and a weak acid, which hydrolyzes to form a basic solution with water.

58. **(B)** Since the boiling point is increased by .51°C for each mole of particles, the 1 mole of NaCl → Na^+ + Cl^- gives 2 moles of particles. Therefore the boiling point will be 1.02° higher or 101.02°C.

59. **(B)** The sp^2 hybridized orbital is trigon, planar and so is related to the triangle.

60. **(D)** When hybridization forms the sp^3 orbitals, four entirely new orbitals are formed which are different from the former *s* and *p* orbitals.

61. **(A)** Hydrogen bonding between water molecules causes the boiling point to be higher than would be expected.

62. **(B)** At the beginning of reaction

$[H_2]$ = 1 mole/liter
$[I_2]$ = 1 mole/liter
$[HI]$ = 0

At equilibrium

$(H_2 + I_2 \rightleftarrows 2\ HI)$
(Let x = moles/liter o
H_2 and I_2 in HI form)

$[H_2]$ = $(1 - x)$ mole/liter
$[I_2]$ = $(1 - x)$ mole/liter
$[HI]$ = $2x$ mole/liter

$$K = \frac{[HI]^2}{[H_2][I_2]}$$

$$= \frac{(2x)^2}{(1-x)(1-x)} = 45.9$$

Taking the square root of both sides gives

$$\frac{2x}{1-x} = 6.8$$

$$x = .772 \text{ mole/liter}$$

Therefore $[H_2]$ = $1 - x = 1 - .772 = .228$ mole/liter

63. **(D)** 10 g/100 g water = 100 g/1000 g water. The freezing point depression, 1.22, is divided by 1.86, which is the depression caused by 1 mole in 1000 g of water, to find how many moles are dissolved.

1.22/1.86 = .65 mole present

So 100 g = .65 mole. Then 1 mole = $\frac{100 \text{ g}}{.65}$ = 153.8 or 154 g.

64. **(D)** The K_w of water = $[H^+][OH^-]$ = 10^{-14}. If $[OH^-]$ = 10^{-5} mole/liter, then
$[H^+]$ = $10^{-14}/10^{-5}$ = 10^{-9}
pH = $-\log [H^+]$ (by definition)
pH = $-[-9]$
pH = 9

65. (B) When dilute NaCl solution is electrolyzed, hydrogen is given off at the cathode, chlorine is given off at the anode, and sodium hydroxide is left in the container.

66. (E) Aqua regia is made by mixing HCl and HNO_3.

67. (D) One mole of a gas at STP occupies 22.4 liters.

If $\dfrac{5 \text{ liters}}{12.5 \text{ g}}$ then $\dfrac{22.4 \text{ liters}}{\text{gram-molecular wt.}}$

$$\frac{5}{12.5} = \frac{22.4}{x}$$

$x = 56$ grams

68. (E) To find the simple or empirical formula, divide each % by the element's atomic mass.

Carbon
$40 \div 12 = 3.333$
Hydrogen
$6.67 \div 1 = 6.67$
Oxygen
$53.33 \div 16 = 3.33$

Next, divide each quotient by the smallest quotient in an attempt to get small whole numbers.

$3.33 \div 3.33 = 1$ Carbon
$6.67 \div 3.33 = 2$ Hydrogen
$3.33 \div 3.33 = 1$ Oxygen

The simplest formula is CH_2O, which has a molecular weight of 30. The true molecular weight is given as 90, which is three times the simplest. Therefore the true formula is $C_3H_6O_3$.

69. (E) The reaction is

$$\overset{370\ g}{CaO} + H_2O \rightarrow Ca(OH)_2$$

$Ca(OH)_2$ molecular wt. = 74
So 370 g ÷ 74 = 5 moles of $Ca(OH)_2$ wanted. The reaction shows 1 mole of CaO produces 1 mole of $Ca(OH)_2$, so the answer is 5 moles.

70. (C) In dilution problems, this formula can be used:

$$M_{\text{before}} \times V_{\text{before}} = M_{\text{after}} \times V_{\text{after}}$$

Substituting gives:

$3.5 \times 50 = 2 \times (?x)$

$x = 87.5$ mL, new volume after dilution

71. (A) For the K_e to be small, the numerator, which is made up of the concentration(s) of the product(s) at equilibrium, must be smaller than the denominator. This generally means that equilibrium is reached rather rapidly.

72. (B) The amount of NaOH used is

$49.2 - 34.7 = 14.5$ mL

Using $M_1 = V_1 = M_2 \times V_2$

$.09\ M \times 14.5$ mL $= M_2 \times 10$ mL
$M_2 = .13\ M$

To find the grams of HCl present:
Multiply the $M \times$ gram-equivalent wt.

$$\frac{.13 \text{ gram-equiv. wt.}}{1000 \text{ mL}} \times \frac{36.5 \text{ g}}{\text{gram-equiv. wt.}}$$

$$= 4.7 \text{ g HCl/1000 mL}$$

If there are 4.7 g in 1000 mL, then there will be $4.7 \div 10 = .47$ g of HCl in 100 mL of .13 M.

73. (E) The reactions recorded indicated that the ease of losing electrons is greater in sodium than magnesium, greater in magnesium than iron, greater in iron than copper, and finally greater in copper than silver.

74. (D) Distillation removes only dissolved solids from the distillate. The volatile gases, like H_2S, will be carried into the distillate.

75. (E) The test for H_2S is to moisten lead acetate paper and look for brown-black precipitate to form, which is a positive result.

76. (E) H_2 to O_2 ratio by volume is $2:1$ in the formation of water. Therefore, 4 mL H_2 will react with 2 mL of O_2 to make 4 mL of steam.

$2\ H_2(g) + O_2(g) \rightarrow 2\ H_2O(g)$

This leaves 30 mL of O_2 uncombined.

77. (C) There will be 30 mL of O_2 + 4 mL of steam = 34 mL total.

78. (A) She would make a small amount of amyl acetate which is an ester with a banana-like odor.

79. (B) The most electronegative element is F, found in Period 2.

80. (C) Solids are incorporated into the K and therefore do not appear on the right side of the equation.

81. (E) SO_3 is not easily formed. The commercial process uses a catalyst.

82. (B) The thistle tube is below the fluid level in the flask and will cause liquid to be forced up the tube when gas is evolved in the reaction.

83. (A) At first the fluid will be expelled up the thistle tube by the gas generated and exerting pressure in the reaction flask. When the level of the fluid falls below the end of the thistle tube, the gas will then be released through the thistle tube.

84. (A) The most active nonmetal has a high attraction for another electron—thus high electronegativity.

85. (B) Fe loses electrons to form the Fe^{2+} ion.

DIAGNOSING YOUR NEEDS

After taking Practice Test 3, check your answers against the correct answers. Then fill in the chart below.

In the space under each question number, place a check if you answered that question correctly.

• EXAMPLE:

If your answer to question 5 was correct, place a check in the appropriate box:

Next, total the check marks for each section, and insert the number in the designated block. Now do the arithmetic indicated, and insert your percent for each area.

(✔) QUESTIONS ANSWERED CORRECTLY

SUBJECT AREA								
I. Atomic Theory and Structure, including periodic relationships	33	35	39	44	48	50	59	60
No. of checks ÷ 8 × 100 = _____%								

II. Nuclear Reactions	49	56
No. of checks ÷ 2 × 100 = _____%		

III. Chemical Bonding and Molecular Structure	15	16	17	18	29	41	45	54	61
No. of checks ÷ 9 × 100 = _____%									

IV. States of Matter and Kinetic Molecular Theory	8	9	10	11	43	47	51	67
No. of checks ÷ 8 × 100 = _____%								

V. Solutions, including concentration units, solubility, and colligative properties	19	20	27	58	70
No. of checks ÷ 5 × 100 = _____%					

VI. Acids and Bases	21	22	36	38	55	57	64	66
No. of checks ÷ 8 × 100 = _____%								

VII. Oxidation-Reduction and Electrochemistry	5	6	7	53	65	85
No. of checks ÷ 6 × 100 = _____%						

VIII. Stoichiometry	12	13	14	40	63	68	69	76	77
No. of checks ÷ 9 × 100 = _____%									

SUBJECT AREA | (✔) QUESTIONS ANSWERED CORRECTLY

IX. Reaction Rates							1	31

☐ No. of checks ÷ 2 × 100 = _____ %

X. Equilibrium				28	30	62	71	80

☐ No. of checks ÷ 5 × 100 = _____ %

XI. Thermodynamics: energy changes in chemical reactions, randomness, and criteria for spontaneity

						2	3	4	25

☐ No. of checks ÷ 4 × 100 = _____ %

XII. Descriptive Chemistry: physical and chemical properties of elements and their familiar compounds; organic chemistry; periodic properties

23	24	26	32	34	37	42	46	52
				78	79	81	84	

☐ No. of checks ÷ 13 × 100 = _____ %

XIII. Laboratory: equipment, procedures, observations, safety, calculations, and interpretation of results

72	73	74	75	82	83

☐ No. of checks ÷ 6 × 100 = _____ %

PLANNING YOUR STUDY

The percentages give you an idea of how you have done on the various major areas of the test. Because of the limited number of questions on some parts, these percentages may not be as reliable as the percentages for parts with larger numbers of questions. However, you should now have at least a rough idea of the areas in which you have done well and those in which you need more study.

Start your study with the areas in which you are the weakest. The corresponding chapters are indicated below:

Subject Area	Chapters to Review
I. Atomic Theory and Structure, including periodic relationships	2, 15
II. Nuclear Reactions	15
III. Chemical Bonding and Molecular Structure	3, 4

Subject Area	Chapters to Review
IV. States of Matter and Kinetic Molecular Theory	1
V. Solutions, including concentration units, solubility, and colligative properties	7
VI. Acids and Bases	11
VII. Oxidation-reduction and Electrochemistry	12
VIII. Stoichiometry	5, 6
IX. Reaction Rates	9
X. Equilibrium	10
XI. Thermodynamics, including energy changes in chemical reactions, randomness, and criteria for spontaneity	8
XII. Descriptive Chemistry: physical and chemical properties of elements and their familiar compounds; organic chemistry; periodic properties	1, 2, 13, 14
XIII. Laboratory: equipment, procedures, observations, safety, calculations, and interpretation of results	All lab diagrams, 16

ANSWER SHEET FOR PRACTICE TEST 4

Determine the correct answer for each question. Then, using a No. 2 pencil, blacken completely the oval containing the letter of your choice.

1. (A) (B) (C) (D) (E)   16. (A) (B) (C) (D) (E)   31. (A) (B) (C) (D) (E)
2. (A) (B) (C) (D) (E)   17. (A) (B) (C) (D) (E)   32. (A) (B) (C) (D) (E)
3. (A) (B) (C) (D) (E)   18. (A) (B) (C) (D) (E)   33. (A) (B) (C) (D) (E)
4. (A) (B) (C) (D) (E)   19. (A) (B) (C) (D) (E)   34. (A) (B) (C) (D) (E)
5. (A) (B) (C) (D) (E)   20. (A) (B) (C) (D) (E)   35. (A) (B) (C) (D) (E)
6. (A) (B) (C) (D) (E)   21. (A) (B) (C) (D) (E)   36. (A) (B) (C) (D) (E)
7. (A) (B) (C) (D) (E)   22. (A) (B) (C) (D) (E)   37. (A) (B) (C) (D) (E)
8. (A) (B) (C) (D) (E)   23. (A) (B) (C) (D) (E)   38. (A) (B) (C) (D) (E)
9. (A) (B) (C) (D) (E)   24. (A) (B) (C) (D) (E)   39. (A) (B) (C) (D) (E)
10. (A) (B) (C) (D) (E)   25. (A) (B) (C) (D) (E)   40. (A) (B) (C) (D) (E)
11. (A) (B) (C) (D) (E)   26. (A) (B) (C) (D) (E)   41. (A) (B) (C) (D) (E)
12. (A) (B) (C) (D) (E)   27. (A) (B) (C) (D) (E)   42. (A) (B) (C) (D) (E)
13. (A) (B) (C) (D) (E)   28. (A) (B) (C) (D) (E)   43. (A) (B) (C) (D) (E)
14. (A) (B) (C) (D) (E)   29. (A) (B) (C) (D) (E)   44. (A) (B) (C) (D) (E)
15. (A) (B) (C) (D) (E)   30. (A) (B) (C) (D) (E)   45. (A) (B) (C) (D) (E)

46. (A) (B) (C) (D) (E)   61. (A) (B) (C) (D) (E)   76. (A) (B) (C) (D) (E)
47. (A) (B) (C) (D) (E)   62. (A) (B) (C) (D) (E)   77. (A) (B) (C) (D) (E)
48. (A) (B) (C) (D) (E)   63. (A) (B) (C) (D) (E)   78. (A) (B) (C) (D) (E)
49. (A) (B) (C) (D) (E)   64. (A) (B) (C) (D) (E)   79. (A) (B) (C) (D) (E)
50. (A) (B) (C) (D) (E)   65. (A) (B) (C) (D) (E)   80. (A) (B) (C) (D) (E)
51. (A) (B) (C) (D) (E)   66. (A) (B) (C) (D) (E)   81. (A) (B) (C) (D) (E)
52. (A) (B) (C) (D) (E)   67. (A) (B) (C) (D) (E)   82. (A) (B) (C) (D) (E)
53. (A) (B) (C) (D) (E)   68. (A) (B) (C) (D) (E)   83. (A) (B) (C) (D) (E)
54. (A) (B) (C) (D) (E)   69. (A) (B) (C) (D) (E)   84. (A) (B) (C) (D) (E)
55. (A) (B) (C) (D) (E)   70. (A) (B) (C) (D) (E)   85. (A) (B) (C) (D) (E)
56. (A) (B) (C) (D) (E)   71. (A) (B) (C) (D) (E)
57. (A) (B) (C) (D) (E)   72. (A) (B) (C) (D) (E)
58. (A) (B) (C) (D) (E)   73. (A) (B) (C) (D) (E)
59. (A) (B) (C) (D) (E)   74. (A) (B) (C) (D) (E)
60. (A) (B) (C) (D) (E)   75. (A) (B) (C) (D) (E)

Practice Test

4

Note: For all questions involving solutions and/or chemical equations, assume that the system is in water unless otherwise stated.

Part A

Directions: Each set of lettered choices below refers to the numbered statements or formulas immediately following it. Select the one lettered choice that best fits each statement or formula and then fill in the corresponding oval on the answer sheet. A choice may be used once, more than once, or not at all in each set.

Questions 1–3 refer to the following graphs:

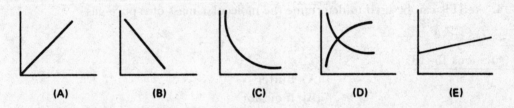

(A)   (B)   (C)   (D)   (E)

1. The graph that best shows the relationship of gas volume to temperature, with pressure held constant

2. The graph that best shows the relationship of gas volume to pressure, with temperature held constant

3. The graph that best shows the relationship of the number of grams of solute that is soluble in 100 grams of H_2O at varying temperatures if the solubility begins at a small quantity and increases in direct proportion to the increase in temperature

Questions 4–7

          (A) A molecule
          (B) A mixture of compounds
          (C) An isotope
          (D) An isomer
          (E) An acid salt

4. The simplest unit of water that retains its properties

5. A commercial cake mix

6. An atom with the same number of protons as another atom of the same element but a different number of neutrons

7. Classification of $NaHCO_3$

Questions 8–10

        (A) 1
        (B) 7
        (C) 9
        (D) 10
        (E) 14

8. The atomic number of an atom with an electron dot arrangement similar to $(: \overset{..}{\underset{..}{I}} :)$

9. The number of atoms represented in the formula $A(OH)_3$

10. The number that represents the most acid pH

Questions 11–14

        (A) Density
        (B) Equilibrium constant
        (C) Molecular weight
        (D) Freezing point
        (E) Molarity

11. Can be expressed in moles per liter of solution

12. Can be expressed in grams per liter of a gas

13. Will NOT be affected by changes in temperature and pressure

14. At STP, can be used to determine the molecular mass of a pure gas

Questions 15–18

        (A) Buffer
        (B) Indicator
        (C) Arrhenius acid
        (D) Arrhenius base
        (E) Neutral condition

15. Resists a rapid change of pH

16. Exhibits different colors in acidic and basic solutions

17. At 25°C, the aqueous solution has a pH < 7

18. At 25°C, the aqueous solution has a pH > 7

Questions 19–23

        (A) H_2
        (B) NH_3
        (C) CO_2
        (D) HCl
        (E) O_2

19. A gas produced by the reaction of zinc with hydrochloric acid

20. A gas that is heavier than air

21. A gas produced by the heating of potassium chlorate

22. A gas that is slightly soluble in water and gives a weakly acid solution

23. A gas that is very soluble in water and gives a weakly basic solution

Part B

Directions: Each question below consists of an <u>assertion</u> (statement) in the left-hand column and a <u>reason</u> in the right-hand column. On the appropriate line of the answer sheet fill in oval

A if both assertion and reason are true statements and the reason is <u>*a correct explanation*</u> of the assertion;

B if both assertion and reason are true statements but the reason is <u>*NOT a correct explanation*</u> of the assertion;

C if the assertion is true but the reason is a false statement;

D if the assertion is false but the reason is a true statement;

E if both assertion and reason are false statements.

Directions Summarized

	Assertion	Reason	
A	True	True	Reason is *a correct explanation*
B	True	True	Reason is *NOT a correct explanation*
C	True	False	
D	False	True	
E	False	False	

Assertion		*Reason*
24. According to the kinetic molecular theory, particles are in random motion	BECAUSE	the molecules undergo inelastic conditions.
25. The electron can have wave properties as well as corpuscular properties	BECAUSE	it depends on the design of the experiment as to which is verified.
26. The alkanes are considered a homologous series	BECAUSE	they are made up of only hydrogen and carbon atoms.
27. When an atom of an active metal becomes an ion, the radius of the ion is less than that of the atom	BECAUSE	the nucleus has less positive charge than the electron "cloud."
28. Petroleum products are a result of fractionally distilled oil	BECAUSE	the process makes available more gasoline than naturally occurs in the mixture.
29. Water is a polar substance	BECAUSE	the sharing of the bonding electrons is equal.
30. A catalyst accelerates a chemical reaction	BECAUSE	a catalyst lowers the activation energy of the reaction.

Directions Summarized

	Assertion	Reason	
A	True	True	Reason is *a correct explanation*
B	True	True	Reason is *NOT a correct explanation*
C	True	False	
D	False	True	
E	False	False	

	Assertion		Reason

31. Copper is an oxidizing agent in the re-action with silver nitrate solution — BECAUSE — copper loses electrons in a reaction with silver ions.

32. The rate of diffusion of hydrogen gas compared to that of helium gas is 1 : 4 — BECAUSE — the rate of diffusion of gases varies inversely as the square root of the molecular weights.

33. A gas heated from 10°C to 100°C at constant pressure will increase in volume — BECAUSE — Charles' Law states that, if the pressure remains constant, the volume varies directly as the absolute temperature.

34. When nonmetallic oxides are mixed with water, the resulting solution is basic — BECAUSE — red litmus turns blue in a basic solution.

35. The complete electrolysis of 45 grams of water will yield 40 grams of H_2 and 5 grams of O_2 — BECAUSE — water is composed of hydrogen and oxygen in a ratio of 8 : 1 by weight.

36. Ice is less dense than liquid water — BECAUSE — water molecules are non-polar.

37. Hydrochloric acid solutions have a high level of conductivity — BECAUSE — hydrogen chloride is completely ionized in solution.

38. Water is a good solvent — BECAUSE — water shows hydrogen bonding between oxygen atoms.

39. In the periodic chart, electronegativity decreases from left to right — BECAUSE — the number of protons in the nucleus of the atom increases from left to right.

Part C

Directions: Each of the questions or incomplete statements below is followed by five suggested answers or completions. Select the one that is best in each case and then fill in the corresponding oval on the answer sheet.

40. In this graphic representation of a chemical reaction, which arrow depicts the activation energy?
    (A) A
    (B) B
    (C) C
    (D) D
    (E) E

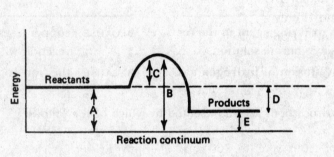

41. How many liters (STP) of O_2 can be produced by completely decomposing 1 mole of KCO_3?
    (A) 11.2
    (B) 22.4
    (C) 33.6
    (D) 44.8
    (E) 77.2

42. Which of the following is the correct structural representation of sodium?

    (A) 11 p
        11 n   )2)8)8)5

    (B) 11 p
        12 n   )2)8)11)2

    (C) 23 p
        23 n   )2)8)1

    (D) 23 p
        23 n   )2)8)11)2

    (E) 11 p
        12 n   )2)8)1

43. Which of the following statements is true?
    (A) A catalyst cannot lower the activation energy.
    (B) A catalyst can lower the activation energy.
    (C) A catalyst affects only the activation energy of the forward reaction.
    (D) A catalyst affects only the activation energy of the reverse reaction.
    (E) A catalyst is permanently changed after the activation energy is achieved.

44. If the molecular weight of NH_3 is 17, what is its density at STP?
    (A) 0.25 g/L
    (B) 0.76 g/L
    (C) 1.52 g/L
    (D) 3.04 g/L
    (E) 9.11 g/L

45. Which bond(s) is (are) ionic?
    I. H—Cl(g)
    II. S—Cl(g)
    III. Cs—F(s)
       (A) I, II, and III
       (B) I and II only
       (C) II and III only
       (D) I only
       (E) III only

46. Aromatic hydrocarbons are represented by which of the following?

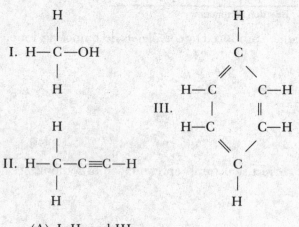

       (A) I, II, and III
       (B) I and II only
       (C) II and III only
       (D) I only
       (E) III only

47. According to placement in the periodic table, which statement(s) regarding the first ionization energies of certain elements should be true?
    I. Li has a higher value than Na.
    II. K has a higher value than Cs.
    III. Na has a higher value than Al.
       (A) I, II, and III
       (B) I and II only
       (C) II and III only
       (D) I only
       (E) III only

48. Correctly expressed half-reactions include which of the following?
    I. $CrO_4^{2-} + 8H^+ + 6e^- \rightarrow Cr^{3+} + 4\,H_2O$
    II. $I^- + 6\,OH^- \rightarrow IO_3^- + 3\,H_2O + 6e^-$
    III. $MnO_4^- + 2\,H_2O + 3e^- \rightarrow MnO_2 + 4\,OH^-$
    (A) I, II, and III
    (B) I and II only
    (C) II and III only
    (D) I only
    (E) III only

What is the apparent oxidation number of the underlined element in the compound

49. Na$\underline{N}$O$_3$?
    (A) +1
    (B) +2
    (C) +3
    (D) +4
    (E) +5

50. $\underline{Ca}SO_4$?
    (A) +1
    (B) −1
    (C) +2
    (D) −2
    (E) +3

51. $\underline{N}H_3$?
    (A) +2
    (B) −2
    (C) +3
    (D) −3
    (E) +5

52. An atom with an orbital notation of $1s^2\,2s^2\,2p^6\,3s^2\,3p^4$ will probably exhibit which oxidation state?
    (A) +2
    (B) −2
    (C) +3
    (D) −3
    (E) +5

53. In this Lewis dot structure, X:, what is the predictable oxidation number?
    (A) +1
    (B) −1
    (C) +2
    (D) −2
    (E) +3

Questions 54–57 refer to the following apparatus, assembled by a student:

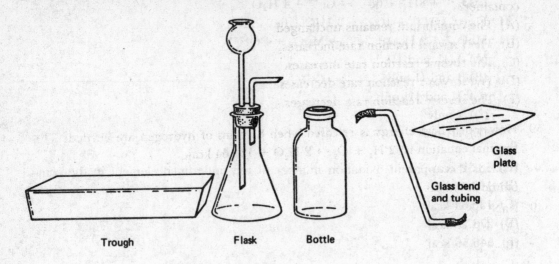

Trough   Flask   Bottle

Glass plate

Glass bend and tubing

54. The apparatus assembled is used to prepare a gas by a reaction that takes place when a solid
    (A) is heated
    (B) is exposed
    (C) reacts with a liquid
    (D) is heated in a vacuum
    (E) is decomposed by a catalyst

55. This apparatus suggests that the student plans to collect a gas that
    (A) supports combustion
    (B) is heavier than air
    (C) is not flammable
    (D) is insoluble in water
    (E) reacts with water

56. If the student planned to prepare hydrogen, what would also be needed?
    (A) mercuric oxide
    (B) acid
    (C) potassium chlorate
    (D) carbon disulfide
    (E) benzene

57. If this apparatus were to be modified to prepare oxygen by the decomposition of mercuric oxide, which piece would be replaced?
    (A) delivery tube
    (B) collecting bottle
    (C) trough of water
    (D) flask with thistle tube
    (E) glass plate

58. What occurs when a reaction is at equilibrium and more reactant is added to the container?
    (A) The equilibrium remains unchanged.
    (B) The forward reaction rate increases.
    (C) The reverse reaction rate increases.
    (D) The forward reaction rate decreases.
    (E) The reverse reaction rate decreases.

59. How much heat energy is released when 8 grams of hydrogen are burned? The thermal equation is: $2 H_2 + O_2 \rightarrow 2 H_2O + 136.64$ kcal.
    (A) 68.32 kcal
    (B) 102.48 kcal
    (C) 136.64 kcal
    (D) 273.28 kcal
    (E) 546.56 kcal

60. Would a spontaneous reaction occur between zinc *ions* and gold *atoms*?

    $Zn° \rightleftharpoons Zn^{2+} + 2e^-$ $\quad$ $E° = 0.76$ volt
    $Au° \rightleftharpoons Au^{3+} + 3e^-$ $\quad$ $E° = -1.42$ volt

    (A) yes – Reaction potential 2.18 volts
    (B) no – Reaction potential -2.18 volt
    (C) yes—Reaction potential 0.66 volt
    (D) no— Reaction potential -0.66 volt
    (E) yes—Reaction potential 0.56 volt

61. Four faradays of electrons $(4 \times 6.02 \times 10^{23}$ electrons) would electroplate how many grams of silver from a silver nitrate solution?
    (A) 108
    (B) 216
    (C) 324
    (D) 432
    (E) 540

62. A 5 M solution of HCl would have how many moles of H^+ ion in 1 liter?
    (A) 0.5
    (B) 1.0
    (C) 2.0
    (D) 2.5
    (E) 5.0

63. What is the K_{sp} for silver acetate if a saturated solution contains 2×10^{-3} mole of silver ion/liter of solution?
    (A) 2×10^{-3}
    (B) 2×10^{-6}
    (C) 4×10^{-3}
    (D) 4×10^{-6}
    (E) 4×10^6

64. The following data were obtained for H_2O and H_2S:

	Formula Wt.	Freezing Point (°C)	Boiling Point (°C)
H_2O	18	0	100
H_2S	34	−83	−60

What is the best explanation for the variation of physical properties between these two compounds?

(A) The H_2S has stronger bonds between molecules.

(B) The H_2O has a great deal of hydrogen bonding.

(C) The bond angles differ by about 15°.

(D) The formula weight is of prime importance.

(E) The oxygen atom has a smaller radius and thus cannot bump into other molecules as often as the sulfur.

65. What is the pH of a solution that has 0.00001 mole of H^+/liter of solution?

(A) 2

(B) 3

(C) 4

(D) 5

(E) 9

66. How many grams of sulfur are present in 1 mole of H_2SO_4?

(A) 2

(B) 32

(C) 49

(D) 64

(E) 98

67. What is the approximate weight, in grams, of 1 liter of nitrous oxide, N_2O, at STP?

(A) 1

(B) 2

(C) 11.2

(D) 22

(E) 44

68. If the simplest formula of a substance is CH_2 and its molecular weight is 56, what is its true formula?

(A) CH_2

(B) C_2H_4

(C) C_3H_4

(D) C_4H_8

(E) C_5H_{10}

Questions 69–70 refer to the following diagrams of two methods of collecting gases:

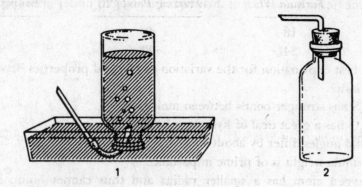

69. Method 1 is BEST suited to collect
    (A) a gas heavier than air
    (B) a gas lighter than air
    (C) a gas that is insoluble in water
    (D) a gas that is soluble in water
    (E) a gas that has a distinct color

70. Which of these gases should be collected by Method 2?
    (A) NH_3
    (B) H_2
    (C) HCl
    (D) CO_2
    (E) He

71. What is the gram-formula weight of $CaCO_3$?
    (A) 68 g
    (B) 75 g
    (C) 82 g
    (D) 100 g
    (E) 116 g

72. What volume, in liters, will be occupied at STP by 4 grams of H_2?
    (A) 11.2
    (B) 22.4
    (C) 33.6
    (D) 44.8
    (E) 56.0

73. What weight, in grams, of KOH is need to neutralize 20 grams of sulfuric acid?
    (A) 22.8
    (B) 45.6
    (C) 68.4
    (D) 110
    (E) 205

74. What volume, in liters, of ammonia is produced when 22.4 liters of nitrogen are made to combine with a sufficient quantity of hydrogen under appropriate conditions?
    (A) 11.2
    (B) 22.4
    (C) 44.8
    (D) 67.0
    (E) 78.2

75. What volume, in liters, of SO_2 will result from the complete burning of 64 grams of sulfur?
    (A) 2
    (B) 11.2
    (C) 44.8
    (D) 126
    (E) 158

76. The amount of energy to melt 5 grams of ice at 0°C would also heat 1 gram of water at 4°C to what condition?
    (A) water at 90°C
    (B) water at 100°C
    (C) steam at 100°C
    (D) 0.2 g water and 0.8 g steam at 100°C
    (E) 0.44 g water and 0.56 g steam at 100°C

77. How many faradays of electricity are needed to electroplate a deposit of 0.5 mole of silver from a silver nitrate solution?
    (A) 0.5
    (B) 1
    (C) 27
    (D) 54
    (E) 108

78. All of the following statements about carbon dioxide are true EXCEPT:
    (A) It can be prepared by the action of acid on limestone.
    (B) It is used in fire extinguishers.
    (C) It dissolves in water at room temperature.
    (D) It sublimes rather than melts at 20°C and 1 atm pressure.
    (E) It is a product of photosynthesis in plants.

79. Three moles of H_2 and 3 moles of I_2 are introduced into a liter box at a temperature of 490°C. What will the K_e expression be for this reaction? ($K_e = 45.9$)

    (A) $K_e = \dfrac{[H_2][I_2]}{[HI]}$

    (B) $K_e = \dfrac{[HI]}{[H_2][I_2]}$

    (C) $K_e = \dfrac{2x}{(x)(x)}$

    (D) $K_e = \dfrac{(2x)^2}{(3-x)^2}$

    (E) $K_e = \dfrac{(3-x)^2}{(2x)^2}$

80. What is the concentration, in moles/liter, of H_2 at equilibrium in question 79?
    (A) 2.3
    (B) 0.68
    (C) 1.6
    (D) 2.1
    (E) 4.64

81. Which of the following correctly completes this nuclear reaction: $^{14}_{7}N + ^{4}_{2}He \rightarrow \cdots$ $+ ^{1}_{1}H$?

    (A) $^{17}_{8}O$
    (B) $^{16}_{9}O$
    (C) $^{17}_{8}N$
    (D) $^{17}_{7}N$
    (E) $^{16}_{8}O$

82. How many grams of NaCl will be needed to make 100 cm³ of 2 M solution?
    (A) 5.85
    (B) 11.7
    (C) 29.2
    (D) 58.5
    (E) 117

83. How many grams of H_2SO_4 are in 1000 g of a 10% solution? (H_2SO_4 mol. wt. = 98 g)
    (A) 1.0
    (B) 9.8
    (C) 10
    (D) 98
    (E) 100

84. If 1 mole of ethyl alcohol in 1000 grams of water depresses the freezing point by 1.86° Celsius, what will be the freezing point of a solution of 1 mole of ethyl alcohol in 500 grams of water?
    (A) −0.93°C
    (B) −1.86°C
    (C) −2.79°C
    (D) −3.72°C
    (E) −5.58°C

85. Which substance may be used as fuel in a nuclear reactor?
    (A) ^{226}Ra
    (B) ^{235}U
    (C) ^{220}Fr
    (D) ^{204}Pb
    (E) ^{14}C

STOP

IF YOU FINISH BEFORE ONE HOUR IS UP, YOU MAY GO BACK TO CHECK YOUR WORK OR COMPLETE UNANSWERED QUESTIONS.

304 • PRACTICE TEST 4

*Answers and
Explanations to Test*

4

1. **(A)** The volume of a gas increases as temperature increases provided that pressure remains constant. This is a direct proportion. Heating a balloon is a good example of this.

2. **(C)** The volume of a gas decreases as the pressure is increased provided that the temperature is held constant. This is shown by the inversely proportional curve in (C). Pressure increase on a closed cylinder is a good example.

3. **(E)** The chart shows that there is a starting quantity in solution, and as you move to the right a slight positive slope indicates a directly proportional change.

4. **(A)** This is the definition of any molecule.

5. **(B)** A commercial cake mix is a mixture of ingredients.

6. **(C)** This is the definition of an isotope.

7. **(E)** An acid salt contains one or more H atoms in the salt formula separating a positive ion and the hydrogen-bearing negative ion. For example, Na_2SO_4 is a *normal* salt and $NaHSO_4$ is an *acid* salt because of the presence of H in the hydrogen sulfate ion.

8. **(C)** An atom with atomic number 9 would have a 2,7 electron configuration, which matches the outer shell of iodine.

9. **(B)** Three $(OH)^-$ each have 2 atoms = 6 atoms plus one Al = 7.

10. **(A)** pH from 1 to 6 is acid, 7 neutral, 8 to 14 basic. Most acid is 1.

11. **(E)** Molarity is defined as moles of solute/liter of solution.

12. **(A)** Gas densities can be expressed in grams/liter.

13. **(C)** Molecular weight is not affected by pressure and temperature.

14. **(A)** Knowing the density of a gas, the weight (mass) of 1 L can be multiplied by 22.4 to find the molecular

mass because 1 mole occupies 22.4 L at STP.

15. **(A)** Buffers resist changes in pH.

16. **(B)** Color change is the function of indicators.

17. **(C)** On the pH scale, from 1 to 6 is acid and 7 is neutral.

18. **(D)** On the pH scale, from 8 to 14 is basic.

19. **(A)** $Zn + 2\,HCl \rightarrow ZnCl_2 + H_2(g)\uparrow$ is the reaction that occurs.

20. **(C)** Only CO_2, with a molecular mass of 44, is heavier than air, for which the molecular mass is 29.

21. **(E)** $2\,KClO_3 \rightarrow 2\,KCl + 3O_2(g)\uparrow$ is the reaction that occurs.

22. **(C)** CO_2 is slightly soluble in water, forming carbonic acid, H_2CO_3, which is a weak acid.

23. **(B)** NH_3 is very soluble in water, and forms a solution of the weak ammonium hydroxide base.

24. **(C)** The assertion is true, but the particles undergo elastic collisions.

25. **(A)** Assertion and reason are true; the electron can be treated as either an electromagnetic wave or a bundle of negative charge.

26. **(B)** An homologous series increases each member by a constant number of carbons and hydrogens. An example is the alkane, alkene, alkyne series, which all increase the chain by a CH_2 group. The reason is true but does not explain the assertion.

27. **(C)** The nuclear charge is greater than the electron cloud. The reason is false.

28. **(C)** The reason should be "Because the various products can be separated by differences in boiling points."

29. **(C)** Water is a polar molecule because there is unequal, not equal, sharing of bonding electrons.

30. **(A)** This is a function of a catalyst—to

either speed up or slow down a reaction without permanent change to itself. Assertion and reason are true.

31. (D) The Cu is losing electrons and thus being oxidized; the assertion is false. It is furnishing electrons and thus is a reducing agent; the reason is true.

32. (D) $H_2 = 2$, He = 4 (molecular weight); then inversely $\sqrt{4} : \sqrt{2} = \sqrt{2} : \sqrt{1}$ is the rate of diffusion of hydrogen to oxygen. The assertion is false; the reason, true.

33. (A) Since the volume is being heated at constant pressure, it expands. The temperatures are converted to kelvins (K) by adding 273° to the Celsius readings. The fraction must be $\frac{373}{273}$ and this will increase the volume.

34. (D) Nonmetallic oxides are acid anhydrides in a water solution and thus test acid with litmus. Assertion is false.

35. (E) Water is $\frac{1}{9}$ hydrogen and $\frac{8}{9}$ oxygen by weight. Both assertion and reason are false.

36. (B) Both assertion and reason are true, but the reason does not explain the density difference.

37. (A) HCl in solution ionizes virtually completely, forming a very conductive solution.

38. (C) The reason why water is a good solvent is false.

39. (E) Both assertion and reason are false. The opposite is true for both statements.

40. (C) The energy necessary to get the reaction started, which is the activation energy, is shown at C.

41. (C) $2 \text{ KClO}_3 \rightarrow 2 \text{ KCl} + 3\text{O}_2 \uparrow$ shows that 2 moles of $KClO_3$ yield 3 moles of O_2. 1 mole will produce $\frac{1}{2} \times 3 = 1.5$ moles
$$1.5 \text{ moles} \times \frac{22.4 \text{ L}}{1 \text{ mole}} = 33.6 \text{ L}$$

42. (E) The atomic number gives the number of protons in the nucleus and the total number of electrons. The atomic mass indicates the total number of protons and neutrons in the nucleus—for Na, 23.

43. (B) A catalyst can speed up a reaction by lowering the activation energy needed to start the reaction and then keep it going.

44. (B) Density = $\frac{\text{Mass}}{\text{Volume}}$. For gases this is expressed as grams per liter. Since 1 gram-molecular weight of a gas occupies 22.4 L, 17 g/22.4 L = .76 g/L.

45. (E) Choice III is made up of elements from extreme sides of the periodic table and will therefore form ionic bonds.

46. (E) Only III is a ring hydrocarbon of the aromatic series.

47. (B) Since Li is higher in Group 1A than Na, and K is higher than Cs, they have smaller radii and hence higher ionization energies. A is to the right of Na and there has a higher ionization energy.

48. (C) Only II and III are correctly balanced. To be correct, I should only have $3e^-$.

49. (E) ⎫ These answers are based on the fact
50. (A) ⎪ that the total of the assigned oxida-
       ⎬ tion numbers for all the atoms in a
51. (D) ⎭ compound is zero.

52. (B) This orbital notation shows 6 electrons in the third shell. The atom would like to gain $2e^-$ to fill the $3p$ and thereby gain a -2 charge.

53. (A) With this structure, the atom would tend to lose these electrons and get a $+2$ charge.

54. (C) Assembled, the apparatus would look like this:

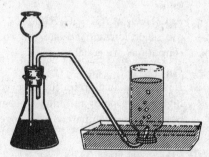

and could be used to prepare a gas by reacting a solid with a liquid.

55. (D) The setup depends on the property of insolubility of the gas collected over water.

56. (B) This apparatus would need an acid to cause the reaction.

57. (D) A test tube would replace the thistle tube and flask portion:

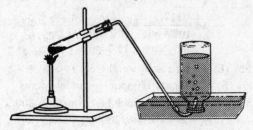

58. (B) The equilibrium shifts in the direction that tends to relieve the stress and thus regain equilibirum.

59. (D) The thermal reaction shows 2 moles of hydrogen reacting or 4 grams. Therefore 8 grams would release twice the amount of energy; 2×136.64 kcal = 273.28 kcal.

60. (B) The reaction potential calculation would be:

$$Zn^{2+} + 2e^- \rightarrow Zn^0 \qquad E^0 = -.76V$$
$$\underline{Au^0 \rightarrow Au^{3+} + 3e^- \qquad E^0 = -1.42V}$$
$$Zn^{2+} + Au^0 \rightarrow Zn^0 + Au^{3+} \quad E^0 = -2.18V$$

61. (D) Since $Ag^+ + 1e^- \rightarrow Ag^0$, 1 mole of electrons yields 1 mole of silver;

1 mole silver = 6.02×10^{23} atoms
4×108 g/mole = 432 g

62. (E) $5 M = \dfrac{5\ moles}{liter}$, and since HCl ionizes completely there would be 5 moles of H^+ and 5 moles of Cl^- ions.

63. (D) $K_{sp} = [Ag^+][C_2H_3O_2^-]$
$\qquad = [2 \times 10^{-3}][2 \times 10^{-3}]$

(Since
$AgC_2H_3O_2 \rightleftarrows Ag^+ + C_2H_3O_2^-$
the silver ion and acetate ion concentrations are equal.)
$K_{sp} = 4 \times 10^{-6}$

64. (B) It is the explanation for the observed high boiling point and high freezing point of water compared to hydrogen sulfide.

65. (D) $pH = -\log [H^+]$
$\qquad = -\log [10^{-5}] = -[-5] = 5$

66. (B) 1 mole H_2SO_4 contains 1 gram-atomic weight of sulfur, that is, 32 g.

67. (B) N_2O = 44g/mole
$\qquad (2 \times 14 + 16 = 44)$

1 mole of a gas occupies 22.4 liters. So 44 g/22.4 liters = 1.99 g/liter.

68. (D) $CH_2 = 14$
$\qquad (12 + 2 = 14$ molecular wt.$)$
$56 \div 14 = 4$
So $4 \times CH_2 = C_4H_8$

69. (C) Only insoluble gases can be collected this way.

70. (C) HCl is very soluble in water and heavier than air, so it is suited to the No. 2 collection method.

71. (D) Ca = 40
$\quad$ C = 12
$\quad\underline{30 = 48}$
$\qquad 100$ g

72. (D) The gram-molecular weight of H_2 is 2 g. 4 g is 2 moles, and each mole occupies 22.4 L. $2 \times 22.4 = 44.8$ L.

73. (A) $\qquad x$ g $\qquad\qquad$ 20 g
$2\ KOH + H_2SO_4 \rightarrow K_2SO_4 + 2\ H_2O$
$\quad$ 112 g $\qquad$ 98 g $\qquad$ So $\dfrac{x\ g}{112\ g} = \dfrac{20\ g}{98\ g}$.
Then $x = 22.8$ g.

74. (C) $\qquad$ 22.4 L $\qquad\quad x$ L
$N_2 + 2\ H_3 \rightarrow 2\ NH_3$
$\quad$ 1 L $\qquad$ 2 L $\qquad$ So $\dfrac{22.4\ L}{1\ L} = \dfrac{x\ L}{2\ L}$
Then $x = 44.8$ L.

75. (C) $\qquad$ 64 g $\qquad\quad x$ L
$S + O_2 \rightarrow SO_2$
$\quad$ 32 g $\qquad$ 22.4 L $\qquad$ So $\dfrac{64\ g}{32\ g} = \dfrac{x\ L}{22.4\ L}$.
Then $x = 44.8$ L.

76. (E) 5 g ice to water = 5×80 cal = 400 cal. 1 g at 4° can go to 100° C as water and absorb 1 cal per °C. So 400 cal − $(100° − 4°) = 400 − 96 = 304$ cal. 304 cal can change $\dfrac{304\ cal}{540\ cal/g}$ or .56 g of water to steam. So .56 g steam and .44 g water would remain.

77. (A) Since 1 F = 1 mole of e^- and Ag^+ gains 1 e^- to become $Ag°$, .5 mole would require .5°F.

78. (E) CO_2 is a reaction in photosynthesis, not a product. The reaction is
$6CO_2 + 6H_2O \rightarrow \underset{simple\ sugar}{C_6 H_{12} O_6} + 6\ O_2 \uparrow$
$\qquad\qquad$ or
$6CO_2 + 5H_2O \rightarrow \underset{cellulose}{C_6H_{12}O_5} + 6O_2 \uparrow$

79. (D) $K_e = \dfrac{[HI]^2}{[H_2][I_2]}$

Let x = moles of H_2 and also I_2 that combine to form HI.

Then at equilibrium $[H_2] = 3 - x$, $[I_2] = 3 - x$, $[HI] = 2x$.

So $K_e = \dfrac{(2x)^2}{(3-x)(3-x)}$.

80. (B) $45.9 = \dfrac{(2x)^2}{(3-x)^2}$

Taking the square root of both sides gives $6.77 = \dfrac{2x}{3-x}$.

Then $x = 2.3$ or $[H_2] = 3 - x$ or .68.

81. (A) This is Rutherford's famous artificial transmutation experiment done in 1919.

82. (B) $2\,M = \dfrac{2\ \text{moles}}{1000\ \text{mL}}$

2 moles of NaCl = 2 × 58.5 g
= 117.0 g

$2\,M$ = 117 g/1000 mL,

so $\dfrac{117\ \text{g}}{1000\ \text{mL}} = \dfrac{x\ \text{g}}{100\ \text{mL}}$,

$x = 11.7$ g

83. (E) Percent is by weight. So 10% is .1 × 1000 g or 100 g.

84. (D) First find the molality. 1 mole in 500 g = 2 moles in 1000 g. Then 2 × 1.86°C = 3.72°C drop from 0°C or −3.72° C.

85. (B) Uranium is the chief fuel of nuclear reactors. The most abundant isotope of uranium found in the earth is $^{238}_{92}U$. However ^{235}U is needed to run a reactor.

DIAGNOSING YOUR NEEDS

After taking Practice Test 4, check your answers against the correct ones. Then fill in the chart below.

In the space under each question number, place a check if you answered that question correctly.

• EXAMPLE:

If your answer to question 5 was correct, place a check in the appropriate box.

Next, total the check marks for each section, and insert the number in the designated block. Now do the arithmetic indicated, and insert your percent for each area.

SUBJECT AREA — (✔) QUESTIONS ANSWERED CORRECTLY

I. Atomic Theory and Structure, including periodic relationships	6	8	25	27	42	47	52	53

No. of checks ÷ 8 × 100 = _____%

II. Nuclear Reactions	81	85

No. of checks ÷ 2 × 100 = _____%

III. Chemical Bonding and Molecular Structure	4	9	29	45	49	50	51	62	64

No. of checks ÷ 9 × 100 = _____%

IV. States of Matter and Kinetic Molecular Theory	1	2	14	24	32	33	36	44

No. of checks ÷ 8 × 100 = _____%

V. Solutions, including concentration units, solubility, and colligative properties	3	11	82	83	84

No. of checks ÷ 5 × 100 = _____%

VI. Acids and Bases	7	10	15	16	17	18	34	37

No. of checks ÷ 8 × 100 = _____%

VII. Oxidation-Reduction and Electrochemistry	31	35	48	60	61	77

No. of checks ÷ 6 × 100 = _____%

VIII. Stoichiometry	41	66	67	68	71	72	73	74	75

No. of checks ÷ 9 × 100 = _____%

SUBJECT AREA

(✔) QUESTIONS ANSWERED CORRECTLY

IX. Reaction Rates		30	43

☐ No. of checks ÷ 2 × 100 = _____ %

X. Equilibrium	58	63	65	79	80

☐ No. of checks ÷ 5 × 100 = _____ %

XI. Thermodynamics: energy changes in chemical reactions, randomness, and criteria for spontaneity	40	59	76

☐ No. of checks ÷ 3 × 100 = _____ %

XII. Descriptive Chemistry: physical and chemical properties of elements and their familiar compounds; organic chemistry; periodic properties	5	12	13	19	20	21	22	23	26	28
							38	39	46	78

☐ No. of checks ÷ 14 × 100 = _____ %

XIII. Laboratory: equipment, procedures, observations, safety, calculations, and interpretation of results	54	55	56	57	69	70

☐ No. of checks ÷ 6 × 100 = _____ %

PLANNING YOUR STUDY

The percentages give you an idea of how you have done on the various major areas of the test. Because of the limited number of questions on some parts, these percentages may not be as reliable as the percentages for parts with larger numbers of questions. However, you should now have at least a rough idea of the areas in which you have done well and those in which you need more study.

Start your study with the areas in which you are the weakest. The corresponding chapters are indicated below:

Subject Area	Chapters to Review
I. Atomic Theory and Structure, including periodic relationships	2
II. Nuclear Reactions	15
III. Chemical Bonding and Molecular Structure	3, 4

*Subject Area	Chapters to Review
IV. States of Matter and Kinetic Molecular Theory	1
V. Solutions, including concentration units, solubility, and colligative properties	7
VI. Acids and Bases	11
VII. Oxidation-reduction and Electrochemistry	12
VIII. Stoichiometry	5, 6
IX. Reaction Rates	9
X. Equilibrium	10
XI. Thermodynamics, including energy changes in chemical reactions, randomness, and criteria for spontaneity	8
XII. Descriptive Chemistry: physical and chemical properties of elements and their familiar compounds; organic chemistry; periodic properties	1, 2, 13, 14
XIII. Laboratory: equipment, procedures, observations, safety, calculations, and interpretation of results	All lab diagrams, 16

Page 12   *No. 12*

To find the volume of 8 oz. sulfur, subtract the volume of water before from the volume after.

$$150\,mL - 50\,mL = 100\ mL$$

Then $8\ oz. \times \dfrac{454\ g}{8\ oz.} = 227\ g$

So, $\dfrac{227\ g}{100\ mL} = \underline{\underline{2.27\ g/mL}}$

Page 12   *No. 13*

Converting 5.08 in. to cm³:

$$5.08\ \cancel{in.} \times \frac{2.54\ cm}{1\ \cancel{in.}} = 12.9\ cm$$

(a) Then $20\ cm \times 20\ cm \times 12.9\ cm = \underline{\underline{5160\ cm^3}}$

Since 1 cm³ of water @ 4°C weighs 1 gram:

(b) $5160\ cm^3 = \underline{\underline{5160\ g}}$

Page 55   *No. 11*

H_2SO_4 is composed of:

$$
\begin{array}{ll}
2H = 4 & \text{Percentage of S:} \\
1S = 32 & \\
4O = \underline{64} & \dfrac{S = 32}{Total = 98} \times 100\% = \underline{\underline{32.6}}\ \text{or}\ \underline{\underline{33\%}} \\
Total = 98 &
\end{array}
$$

Page 55   *No. 12*

$$C = 12\overline{)85.7\%\ C} \qquad H = 1\overline{)14.3\%\ H}$$
$$7.14 \qquad\qquad 14.3$$

To find the lowest ratio of the whole numbers:

$$7.14\overline{)7.14} \qquad\qquad 7.14\overline{)14.3}$$
$$1.0 \qquad\qquad\qquad 2.0$$

So the empirical formula is $\underline{\underline{CH_2}}$.

Since the formula weight is given as 42, the empirical formula CH_2, which totals 14 amu, divides into 42 three times. Therefore, the true formula is $\underline{\underline{C_3H_6}}$.

Page 79   *No. 1*

$$
\begin{array}{cc}
x\ g & 20\ L
\end{array}
$$
Equation: $2\ H_2O \rightarrow 2\ H_2 + O_2$

Calculate the equation weight or volume so that the units match the ones above the equation.

$$
\begin{array}{cc}
x\ g & 20\ L \\
2\ H_2O \rightarrow 2\ H_2 & + O_2 \\
(2 \times mol.\ wt.\ of\ water) & (1 \times gram\text{-}molecular\ volume) \\
36\ g & 22.4\ liters
\end{array}
$$

Then

$$\frac{x\text{ g}}{36\text{ g}} = \frac{20\text{ L}}{22.4\text{ L}}$$

$$\underline{\underline{x\text{ g} = 32.2\text{ g}}}$$

Page 79 *No. 2*

$$x\text{ g} \qquad 44.8\text{ L}$$
$$4\text{ Al} + 3\text{ O}_2 \rightarrow 2\text{ Al}_2\text{O}_3$$

Calculate the equation weight or volume so that the units match the ones above the equation.

$$x\text{ g} \qquad\qquad 44.8\text{ L}$$
$$4\text{ Al} \quad + \quad 3\text{ O}_2 \quad \rightarrow 2\text{ Al}_2\text{O}_3$$
$$(4 \times 27\text{ g}) \quad (3 \times 22.4\text{ L})$$
$$108\text{ g} \qquad\quad 67.2\text{ L}$$

Then $\dfrac{x\text{ g}}{108\text{ g}} = \dfrac{44.8\text{ L}}{67.2\text{ L}}$

$$\underline{\underline{x\text{ g} = \underline{72\text{ g}}}}$$

Page 81 *No. 1*

The equation:

$$x\text{ g} \qquad\qquad\qquad 100\text{ g}$$
$$\text{MnO}_2 + 4\text{ HCl} \rightarrow \text{MnCl}_2 + \text{Cl}_2 + 2\text{ H}_2\text{O}$$

Calculate the equation weight for those substances involved.

$$x\text{ g} \qquad\qquad\qquad\qquad 100\text{ g}$$
$$\text{MnO}_2 \quad + 4\text{ HCl} \rightarrow \text{MnCl}_2 + \quad \text{Cl}_2 \quad + 2\text{ H}_2\text{O}$$
$$(1 \times 87\text{ g}) \qquad\qquad\qquad (1 \times 71\text{ g})$$
$$87\text{ g} \qquad\qquad\qquad\qquad 71\text{ g}$$

Then $\dfrac{x\text{ g}}{87\text{ g}} = \dfrac{100\text{ g}}{71\text{ g}}$

$$\underline{\underline{x\text{ g} = 122.5\text{ g}}}$$

Page 81 *No. 2*

The equation:

$$20\text{ g} \quad 20\text{ g} \qquad x\text{ g}$$
$$2\text{ Mg} + \text{O}_2 \rightarrow 2\text{ MgO}$$
$$48\text{ g} \quad 32\text{ g} \qquad 80\text{ g}$$

Since the equation weights indicate you must have more Mg than O_2, there is an excess of oxygen.

$$\frac{20\text{ g oxygen}}{48\text{ g oxygen}} = \frac{x\text{ g MgO}}{80\text{ g MgO}}$$

$$\underline{\underline{x = 33.3\text{ g MgO}}}$$

Page 83   *No. 1*

The equation:

$$\begin{array}{ccc} x \text{ liters} & & 12 \text{ liters} \\ H_2 & + Cl_2 \rightarrow & 2\,HCl \end{array}$$

Set up the proportion using the coefficients of the substances that have something indicated above them as denominators.

$$\frac{x \text{ liters}}{1 \text{ liter}} = \frac{12 \text{ liters}}{2 \text{ liters}}$$

$$\underline{x = 6 \text{ liters}}$$

Page 83   *No. 2*

The equation:

$$\begin{array}{ccc} 100 \text{ liters} & x \text{ liters} & \\ 2\,CO & + \quad O_2 & \rightarrow 2\,CO_2 \end{array}$$

Set up the proportion using the coefficients of the substances that have something indicated above them as denominators.

$$\frac{100 \text{ liters}}{2 \text{ liters}} = \frac{x \text{ liters}}{1 \text{ liter}}$$

$$\underline{x = 50 \text{ liters}}$$

Page 104   *No. 4*

Each gram of ice that dissolves at 0°C absorbs 80 calories.

So,   10 g $\times$ 80 calories/gram = $\underline{800 \text{ calories}}$

Page 104   *No. 5*

By definition, 1 calorie is the amount of heat needed to heat 1 gram of water 1° on the Celsius scale.

To heat 10 g from 4° to 14° would require

$$\overset{\text{Temperature difference}}{\underset{\downarrow}{}}$$
$$10 \text{ g} \times (14° - 4°) = \underline{100 \text{ calories}}$$

Page 104   *No. 12*

A 10% solution contains 10 g of solute/100 g of solution because

$$10\% \times 100 \text{ g} = \underline{10 \text{ g}}$$

Page 104   *No. 14*

98 grams of H_2SO_4 = 1 mole H_2SO_4
$$500\,mL = \tfrac{1}{2} \text{ liter}$$

So 1 mole in $\frac{1}{2}$ liter would be 2 moles in 1 liter or 2 molar solution.

Page 104   *No. 15*

$$684 \text{ grams of sucrose} = \frac{684}{342 \text{ g/mole}} = 2 \text{ moles of sucrose}$$

$$\frac{2 \text{ moles}}{2000 \text{ g } H_2O} = 1 \text{ molal solution}$$

So the freezing point is lowered only −1.86°C.

Page 137   *No. 10*

The formula:

$$M_{acid} \times V_{acid} = M_{base} \times V_{base}$$
$$1\,M \times 10\,mL = x\,M \times 20\,mL$$
$$\underline{\underline{x = .5M}}$$

Page 147 *Example*

The equation with only the oxidation numbers of the elements which show a change in oxidation number is:

$$\overset{-2}{}\quad\overset{+5}{}\qquad\qquad\qquad\overset{+6}{}\quad\overset{-1}{}$$
$$As_2S_5 + KClO_3 + H_2O \rightarrow H_3AsO_4 + H_2SO_4 + KCl$$

The electron change is: $4 \times (Cl^{5+} + 6e^- \rightarrow Cl^{1-})$
$$\underline{3 \times (S^{2-} \qquad\qquad \rightarrow S^{6+} + 8e^-)}$$

or

$$\begin{array}{l} 4\ Cl^{5+} + 24e^- \rightarrow 4\ Cl^- \\ \underline{3\ S^{2-} \qquad\qquad \rightarrow 3\ S^{6+} + 24e^-} \\ 4\ Cl^- + 3\ S^{2-} \rightarrow 4\ Cl^- + 3\ S^{6+} \end{array}$$

When you try to fit these numbers to the equation, you will find that, because the formula As_2S_5 has sulfur occurring 5 atoms per molecule, only three sulfurs can't be used. You must multiply the simplest equation by an appropriate number to get a multiple of 5.

$$5 \times (3\ S^{2-} + 4\ Cl^- \rightarrow 4\ Cl^- + 3\ S^{6+})$$

or

$$15\ S^{2-} + 20\ Cl^- \rightarrow 20\ Cl^- + 15\ S^{6+}$$

Transfer these to the equation:

$$\underline{3}\ As_2S_5 + \underline{20}\ KClO_3 + H_2O \rightarrow H_3AsO_4 + \underline{15}\ H_2SO_4 + \underline{20}\ KCl$$

Now set the coefficients of H_2O and H_3AsO_4 by inspection:

$$\underline{3}\ As_2S_5 + \underline{20}\ KClO_3 + \underline{24}\ H_2O \rightarrow \underline{6}\ H_3AsO_4 + \underline{15}\ H_2SO_4 + \underline{20}\ KCl$$

Page 149 *Example*

$$KMnO_4 + H_2SO_3 \rightarrow H_2SO_4 + MnSO_4 + K_2SO_4 + H_2O$$

The half-reactions are:

$$MnO_4^- \rightarrow Mn^{2+}$$
$$SO_3^{2-} \rightarrow SO_4^{2-}$$

Balancing these half-reactions chemically using H_2O and H^+, you get

$$MnO_4^- + 8\ H^+ \rightarrow Mn^{2+} + 4\ H_2O$$
$$SO_3^{2-} + H_2O \rightarrow SO_4^{2-} + 2\ H^+$$

Balancing them electrically and equalizing the electrons lost and gained:

$$2(MnO_4^- + 8\ H^+ + 5e^- \rightarrow Mn^{2+} + 4\ H_2O)$$
$$5(SO_3^{2-} + H_2O \rightarrow SO_4^{2-} + 2\ H^+ + 2e^-)$$

Adding the multiplied half-reactions:

$$2\ MnO_4^- + 5\ SO_3^{2-} + 6\ H^+ \rightarrow 2\ Mn^{2+} + 5\ SO_4^{2-} + 3\ H_2O$$

Carrying these coefficients into the equation, you get:

$$2\ KMnO_4 + 5\ H_2SO_3 \rightarrow H_2SO_4 + 2\ MnSO_4 + K_2SO_4 + 3\ H_2O$$

|_____ This sets this other coefficient as 1 _____↑

Now this far you have used 3 SO_4^{2-} in 2 $MnSO_4 + K_2SO_4$. Since you must have 5 SO_4^{2-}, you need 2 more, which sets the coefficient of H_2SO_4 at 2.

So

$$2\ KMnO_4 + 5\ H_2SO_3 \rightarrow 2\ H_2SO_4 + 2\ MnSO_4 + K_2SO_4 + 3\ H_2O$$

Page 150 *No. 1*

The half-reactions balanced chemically:

$$Zn \rightarrow Zn^{2+}$$
$$NO_3^- + 10\ H^+ \rightarrow NH_4^+ + 3\ H_2O$$

Balancing electrically and adding:

$$4(Zn \rightarrow Zn^{2+} + 2e^-)$$
$$\underline{NO_3^- + 10\ H^+ + 8e^- \rightarrow NH_4^+ + 3\ H_2O}$$
$$4\ Zn + NO_3^- + 10\ H^+ \rightarrow 4\ Zn^{2+} + NH_4^+ + 3\ H_2O$$

In the equation:

$$4\ Zn + 10\ HNO_3 \rightarrow 4\ Zn(NO_3)_2 + NH_4NO_3 + 3\ H_2O$$

Page 150 *No. 2*

The half-reactions balanced chemically:

$$Cu^0 \rightarrow Cu^{2+}$$
$$NO_3^{1-} + 4\ H^+ \rightarrow NO + 2\ H_2O$$

Balancing electrically and adding:

$$3(Cu^0 \rightarrow Cu^{2+} + 2e^-)$$
$$\underline{2(NO_3^{1-} + 4\ H^+ + 3e^- \rightarrow NO + 2\ H_2O)}$$
$$3\ Cu + 8\ H^+ + 2\ NO_3^{3-} \rightarrow 3\ Cu^{2+} + 2\ NO + 4\ H_2O$$

In the equation:

$$3\ Cu + 8\ HNO_3 \rightarrow 3\ Cu(NO_3)_2 + 4\ H_2O + 2\ NO$$

Page 150 *No. 3*

The half-reactions balanced chemically:

$$MnO_4^- + 8\ H^+ \rightarrow Mn^{2+} + 4\ H_2O$$
$$2\ Cl^- \rightarrow Cl_2$$

Balancing electrically and adding:

$$2(MnO_4^- + 8\ H^+ + 5e^- \rightarrow Mn^{2+} + 4\ H_2O)$$
$$\underline{5(2\ Cl^- \rightarrow Cl_2^0 + 2e^-)}$$
$$2\ MnO_4^- + 16\ H^+ + 10\ Cl^- \rightarrow 2\ Mn^{2+} + 8\ H_2O + 5\ Cl_2$$

In the equation:

$$2\ KMnO_4 + 16\ HCl \rightarrow 2\ KCl + 2\ MnCl_2 + 8\ H_2O + 5\ Cl_2$$

sets this as 2
by inspection

Often there are some equations on a chemistry test to either complete, or complete and balance. The following list of equations are some you should be capable of writing and balancing. On this page are the word equations. Try to write and balance them; then turn the page for the answers.

1. zinc + sulfuric acid → $Zn SO_4 + H_2 \uparrow$
   $Zn + H_2SO_4$

2. carbon + steam (H_2O) → carbon monoxide ↑ + hydrogen ↑ .
   $C \quad H_2O \quad CO \quad H_2$

3. calcium carbonate + hydrochloric acid → calcium chloride + water + carbon dioxide ↑ . $CaCO_3 \quad 2HCl \quad CaCl_2 \quad H_2O \quad CO_2$

4. iron (III) chloride + sodium hydroxide → iron (III) hydroxide ↓ + sodium chloride.
   $FeCl_3 \quad 3NaOH \quad Fe(OH)_3 \quad 3\,NaCl$

5. calcium phosphate + sulfuric acid → calcium sulfate + phosphoric acid (H_3PO_4).
   $Ca_3(PO_4)_2 \quad 3H_2SO_4 \quad 3\,CaSO_4 \quad 2H_3PO_4$

6. silver nitrate + hydrochloric acid → $AgCl \downarrow + HNO_3$
   $AgNO_3 \quad HCl$

7. aluminum hydroxide + sulfuric acid → aluminum sulfate + water.
   $2\,Al(OH)_3 \quad 3H_2SO_4 \quad Al_2(SO_4)_3 + 6H_2O$

✳ 8. potassium + water → $2KOH + H_2$
   $2K \quad +2H_2O$

9. copper + sulfuric acid → copper (II) sulfate + water + sulfur dioxide ↑ .
   $Cu \quad 2H_2SO_4 \quad CuSO_4 \quad 2H_2O \quad SO_2$

✳ 10. magnesium + oxygen → MgO
   $Mg \quad +O_2$

11. iron + sulfur → iron (II) sulfide.

12. silver nitrate + copper → copper (II) nitrate + silver ↓ .

13. sodium chloride + sulfuric acid (H_2SO_4) → sodium hydrogen sulfate + hydrogen chloride ↑ .

14. sodium peroxide + water →

15. barium chloride + sodium sulfate → sodium chloride + barium sulfate ↓ .

16. sodium hydroxide + carbon dioxide → sodium carbonate + water.

17. magnesium bromide + chlorine → magnesium chloride + bromine.

18. sodium chloride + silver nitrate → silver chloride ↓ + sodium nitrate.

19. lead (II) acetate + hydrogen sulfide → lead (II) sulfide ↓ + acetic acid ($HC_2H_3O_2$).

20. sodium sulfite + sulfuric acid → sodium sulfate + water + sulfur dioxide ↑ .

21. hydrogen + chlorine → hydrogen chloride ↑ .

22. iron (II) sulfide + hydrochloric acid → hydrogen sulfide ↑ + iron (II) chloride.

23. lead (II) nitrate + sulfuric acid → lead (II) sulfate + nitric acid.

24. zinc chloride + ammonium sulfide → zinc sulfide ↓ + ammonium chloride.

25. sodium iodide + bromine → sodium bromide + iodine.
   $2NaI \quad +Br_2 \quad 2NaBr + I_2$

1. zinc + sulfuric acid → zinc sulfate + hydrogen ↑
   $Zn + H_2SO_4 → ZnSO_4 + H_2 ↑$

2. $C + H_2O → CO ↑ + H_2 ↑$

3. $CaCO_3 + 2 HCl → CaCl_2 + H_2O + CO_2 ↑$

4. $FeCl_3 + 3 NaOH → Fe(OH)_3 ↓ + 3 NaCl$

5. $Ca_3(PO_4)_2 + 3 H_2SO_4 → 3 CaSO_4 + 2 H_3PO_4$

6. silver nitrate + hydrochloric acid → silver chloride + nitric acid
   $AgNO_3 + HCl → AgCl ↓ + HNO_3$

7. $2 Al(OH)_3 + 3 H_2SO_4 → Al_2(SO_4)_3 + 6 H_2O$

8. potassium + water → potassium hydroxide + hydrogen ↑
   $2 K + 2 H_2O → 2 KOH + H_2 ↑$

9. $Cu + 2 H_2SO_4 → CuSO_4 + 2 H_2O + SO_2 ↑$

10. magnesium + oxygen → magnesium oxide
    $2 Mg + O_2 → 2 MgO$

11. $Fe + S → FeS$

12. $2 AgNO_3 + Cu → Cu(NO_3)_2 + 2 Ag ↓$

13. $NaCl + H_2SO_4 → NaHSO_4 + HCl ↑$

14. sodium peroxide + water → sodium hydroxide + oxygen ↑
    $2 Na_2O_2 + 2 H_2O → 4 NaOH + O_2 ↑$

15. $BaCl_2 + Na_2SO_4 → 2 NaCl + BaSO_4 ↓$

16. $2 NaOH + CO_2 → Na_2CO_3 + H_2O$

17. $MgBr_2 + Cl_2 → MgCl_2 + Br_2$

18. $NaCl + AgNO_3 → AgCl ↓ + NaNO_3$

19. $Pb(C_2H_3O_2)_2 + H_2S → PbS ↓ + 2 HC_2H_3O_2$

20. $Na_2SO_3 + H_2SO_4 → Na_2SO_4 + H_2O + SO_2 ↑$

21. $H_2 + Cl_2 → 2 HCl ↑$

22. $FeS + 2 HCl → H_2S ↑ + FeCl_2$

23. $Pb(NO_3)_2 + H_2SO_4 → PbSO_4 ↓ + 2 HNO_3$

24. $ZnCl_2 + (NH_4)_2S → ZnS ↓ + 2 NH_4Cl$

25. $2 NaI + Br_2 → 2 NaBr + I_2$

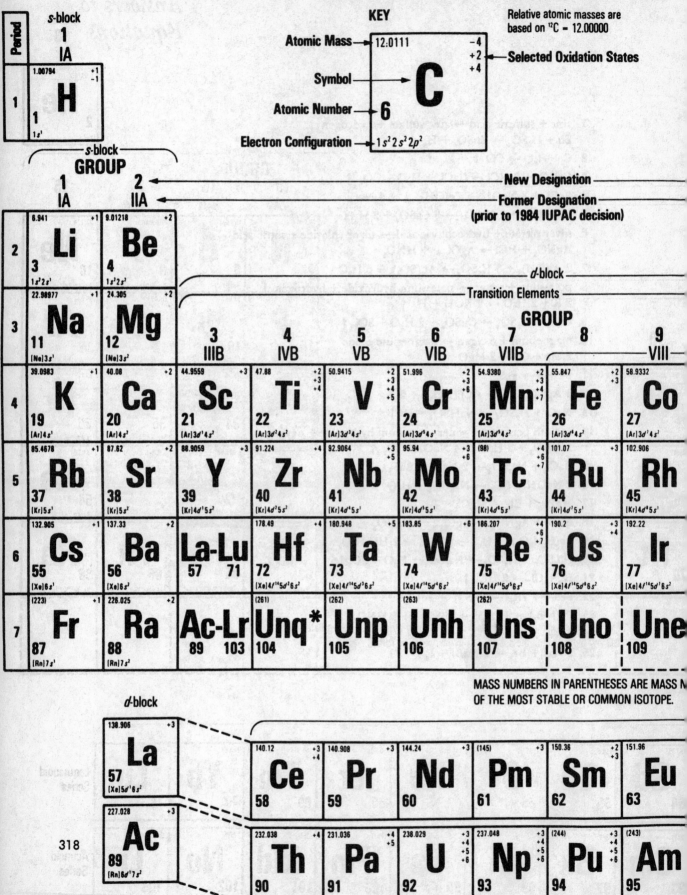

the Elements

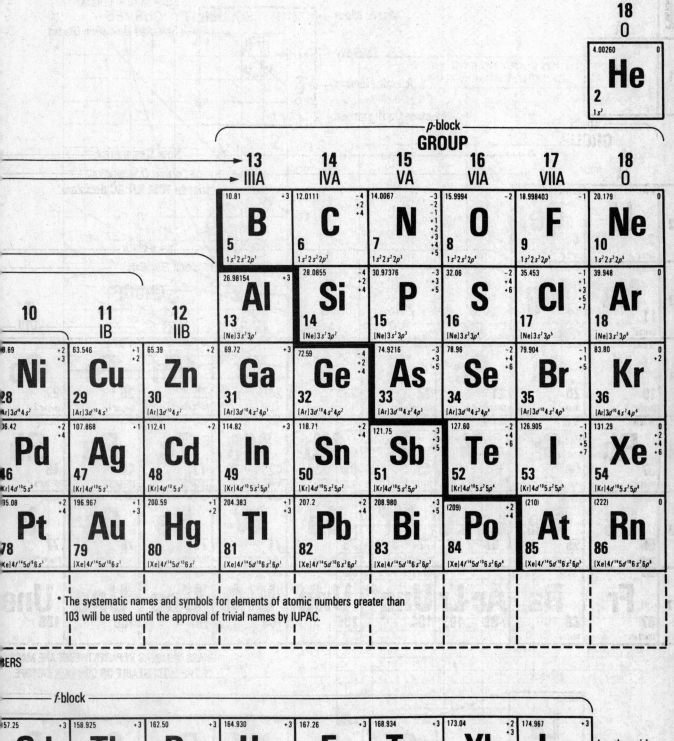

s-block

18	
0	

4.00260	0
He	
2	
$1s^2$	

p-block

GROUP

13	14	15	16	17	18
IIIA	IVA	VA	VIA	VIIA	0

10.81 +3	12.0111 −4 +2 +4	14.0067 −3 −2 −1 +1 +2 +3 +4 +5	15.9994 −2	18.998403 −1	20.179 0
B	**C**	**N**	**O**	**F**	**Ne**
5	6	7	8	9	10
$1s^2 2s^2 2p^1$	$1s^2 2s^2 2p^2$	$1s^2 2s^2 2p^3$	$1s^2 2s^2 2p^4$	$1s^2 2s^2 2p^5$	$1s^2 2s^2 2p^6$

26.98154 +3	28.0855 −4 +2 +4	30.97376 −3 +3 +5	32.06 −2 +4 +6	35.453 −1 +1 +5 +7	39.948 0
Al	**Si**	**P**	**S**	**Cl**	**Ar**
13	14	15	16	17	18
[Ne]$3s^2 3p^1$	[Ne]$3s^2 3p^2$	[Ne]$3s^2 3p^3$	[Ne]$3s^2 3p^4$	[Ne]$3s^2 3p^5$	[Ne]$3s^2 3p^6$

10	11	12						
	IB	IIB						

8.69 +2 +3	63.546 +1 +2	65.39 +2	69.72 +3	72.59 −4 +2 +4	74.9216 −3 +3 +5	78.96 −2 +4 +6	79.904 −1 +1 +5	83.80 0 +2
Ni	**Cu**	**Zn**	**Ga**	**Ge**	**As**	**Se**	**Br**	**Kr**
28	29	30	31	32	33	34	35	36
[Ar]$3d^8 4s^2$	[Ar]$3d^{10}4s^1$	[Ar]$3d^{10}4s^2$	[Ar]$3d^{10}4s^2 4p^1$	[Ar]$3d^{10}4s^2 4p^2$	[Ar]$3d^{10}4s^2 4p^3$	[Ar]$3d^{10}4s^2 4p^4$	[Ar]$3d^{10}4s^2 4p^5$	[Ar]$3d^{10}4s^2 4p^6$

06.42 +2 +4	107.868 +1	112.41 +2	114.82 +3	118.71 +2 +4	121.75 −3 +3 +5	127.60 −2 +4 +6	126.905 −1 +1 +5 +7	131.29 0 +2 +4 +6
Pd	**Ag**	**Cd**	**In**	**Sn**	**Sb**	**Te**	**I**	**Xe**
46	47	48	49	50	51	52	53	54
[Kr]$4d^{10}5s^0$	[Kr]$4d^{10}5s^1$	[Kr]$4d^{10}5s^2$	[Kr]$4d^{10}5s^2 5p^1$	[Kr]$4d^{10}5s^2 5p^2$	[Kr]$4d^{10}5s^2 5p^3$	[Kr]$4d^{10}5s^2 5p^4$	[Kr]$4d^{10}5s^2 5p^5$	[Kr]$4d^{10}5s^2 5p^6$

95.08 +2 +4	196.967 +1 +3	200.59 +1 +2	204.383 +1 +3	207.2 +2 +4	208.980 +3 +5	(209) +2 +4	(210)	(222) 0
Pt	**Au**	**Hg**	**Tl**	**Pb**	**Bi**	**Po**	**At**	**Rn**
78	79	80	81	82	83	84	85	86
[Xe]$4f^{14}5d^9 6s^1$	[Xe]$4f^{14}5d^{10}6s^1$	[Xe]$4f^{14}5d^{10}6s^2$	[Xe]$4f^{14}5d^{10}6s^2 6p^1$	[Xe]$4f^{14}5d^{10}6s^2 6p^2$	[Xe]$4f^{14}5d^{10}6s^2 6p^3$	[Xe]$4f^{14}5d^{10}6s^2 6p^4$	[Xe]$4f^{14}5d^{10}6s^2 6p^5$	[Xe]$4f^{14}5d^{10}6s^2 6p^6$

* The systematic names and symbols for elements of atomic numbers greater than 103 will be used until the approval of trivial names by IUPAC.

f-block

57.25 +3	158.925 +3	162.50 +3	164.930 +3	167.26 +3	168.934 +3	173.04 +2 +3	174.967 +3	
Gd	**Tb**	**Dy**	**Ho**	**Er**	**Tm**	**Yb**	**Lu**	Lanthanoid Series
64	65	66	67	68	69	70	71	

247) +3	(247) +3 +4	(251) +3	(252)	(257)	(258)	(259)	(260)	
Cm	**Bk**	**Cf**	**Es**	**Fm**	**Md**	**No**	**Lr**	Actinoid Series
96	97	98	99	100	101	102	103	

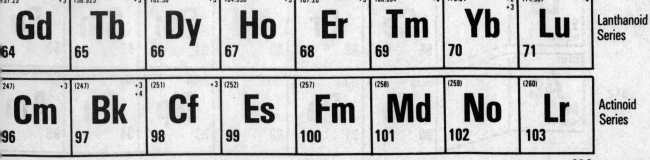

Tables for Reference

DENSITY AND BOILING POINTS OF SOME COMMON GASES

Name		Density grams/liter at STP*	Boiling Point (at 1 atm) K
Air	—	1.29	—
Ammonia	NH_3	0.771	240
Carbon dioxide	CO_2	1.98	195
Carbon monoxide	CO	1.25	82
Chlorine	Cl_2	3.21	238
Hydrogen	H_2	0.0899	20
Hydrogen chloride	HCl	1.64	188
Hydrogen sulfide	H_2S	1.54	212
Methane	CH_4	0.716	109
Nitrogen	N_2	1.25	77
Nitrogen (II) oxide	NO	1.34	121
Oxygen	O_2	1.43	90
Sulfur dioxide	SO_2	2.92	263

*STP is defined as 273 K and 1 atm

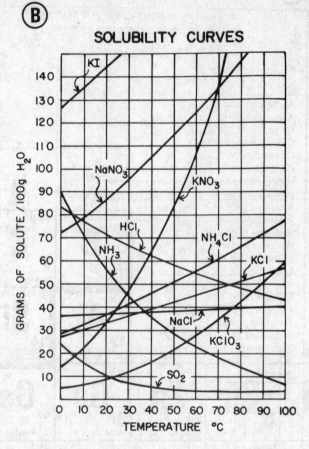

SOLUBILITY CURVES

TABLE OF SOLUBILITIES IN WATER

i — nearly insoluble
ss — slightly soluble
s — soluble
d — decomposes
n — not isolated

	acetate	bromide	carbonate	chloride	chromate	hydroxide	iodide	nitrate	phosphate	sulfate	sulfide
Aluminum	ss	s	n	s	n	i	s	s	i	s	d
Ammonium	s	s	s	s	s	s	s	s	s	s	s
Barium	s	s	i	s	i	s	s	s	i	i	d
Calcium	s	s	i	s	s	ss	s	s	i	ss	d
Copper II	s	s	i	s	i	i	n	s	i	s	i
Iron II	s	s	i	s	n	i	s	s	i	s	i
Iron III	s	s	n	s	i	i	n	s	i	ss	d
Lead	s	ss	i	ss	i	i	ss	s	i	i	i
Magnesium	s	s	i	s	s	i	s	s	i	s	d
Mercury I	ss	i	i	i	ss	n	i	s	i	ss	i
Mercury II	s	ss	i	s	ss	i	i	s	i	d	i
Potassium	s	s	s	s	s	s	s	s	s	s	s
Silver	ss	i	i	i	ss	n	i	s	i	ss	i
Sodium	s	s	s	s	s	s	s	s	s	s	s
Zinc	s	s	i	s	s	i	s	s	i	s	i

SELECTED POLYATOMIC IONS

Hg_2^{2+}	dimercury (I)	CrO_4^{2-}	chromate
NH_4^+	ammonium	$Cr_2O_7^{2-}$	dichromate
$C_2H_3O_2^-$	acetate	MnO_4^-	permanganate
CH_3COO^-		MnO_4^{2-}	manganate
CN^-	cyanide	NO_2^-	nitrite
CO_3^{2-}	carbonate	NO_3^-	nitrate
HCO_3^-	hydrogen carbonate	OH^-	hydroxide
		PO_4^{3-}	phosphate
$C_2O_4^{2-}$	oxalate	SCN^-	thiocyanate
ClO^-	hypochlorite	SO_3^{2-}	sulfite
ClO_2^-	chlorite	SO_4^{2-}	sulfate
ClO_3^-	chlorate	HSO_4^-	hydrogen sulfate
ClO_4^-	perchlorate	$S_2O_3^{2-}$	thiosulfate

E

STANDARD ENERGIES OF FORMATION OF COMPOUNDS AT 1 atm AND 298 K

Compound	Heat (Enthalpy) of Formation* kcal/mol ($\triangle H_f^\circ$)	Free Energy of Formation kcal/mol ($\triangle G_f^\circ$)
Aluminum oxide $Al_2O_3(s)$	−400.5	−378.2
Ammonia $NH_3(g)$	−11.0	−3.9
Barium sulfate $BaSO_4(s)$	−352.1	−325.6
Calcium hydroxide $Ca(OH)_2(s)$	−235.7	−214.8
Carbon dioxide $CO_2(g)$	−94.1	−94.3
Carbon monoxide $CO(g)$	−26.4	−32.8
Copper (II) sulfate $CuSO_4(s)$	−184.4	−158.2
Ethane $C_2H_6(g)$	−20.2	−7.9
Ethene (ethylene) $C_2H_4(g)$	12.5	16.3
Ethyne (acetylene) $C_2H_2(g)$	54.2	50.0
Hydrogen fluoride $HF(g)$	−64.8	−65.3
Hydrogen iodide $HI(g)$	6.3	0.4
Iodine chloride $ICl(g)$	4.3	−1.3
Lead (II) oxide $PbO(s)$	−51.5	−45.0
Magnesium oxide $MgO(s)$	−143.8	−136.1
Nitrogen (II) oxide $NO(g)$	21.6	20.7
Nitrogen (IV) oxide $NO_2(g)$	7.9	12.3
Potassium chloride $KCl(s)$	−104.4	−97.8
Sodium chloride $NaCl(s)$	−98.3	−91.8
Sulfur dioxide $SO_2(g)$	−70.9	−71.7
Water $H_2O(g)$	−57.8	−54.6
Water $H_2O(\ell)$	−68.3	−56.7

* Minus sign indicates an exothermic reaction.

Sample equations:

$$2Al(s) + \frac{3}{2}O_2(g) \rightarrow Al_2O_3(s) + 400.5 \text{ kcal}$$

$$2Al(s) + \frac{3}{2}O_2(g) \rightarrow Al_2O_3(s) \quad \triangle H = -400.5 \text{ kcal/mol}$$

G

HEATS OF REACTION AT 1 atm and 298 K

Reaction	$\triangle H$ (kcal)
$CH_4(g) + 2O_2(g) \rightarrow CO_2(g) + 2H_2O(\ell)$	−212.8
$C_3H_8(g) + 5O_2(g) \rightarrow 3CO_2(g) + 4H_2O(\ell)$	−530.6
$CH_3OH(\ell) + \frac{3}{2}O_2(g) \rightarrow CO_2(g) + 2H_2O(\ell)$	−173.6
$C_6H_{12}O_6(s) + 6O_2(g) \rightarrow 6CO_2(g) + 6H_2O(\ell)$	−669.9
$CO(g) + \frac{1}{2}O_2(g) \rightarrow CO_2(g)$	−67.7
$C_8H_{18}(\ell) + \frac{25}{2}O_2(g) \rightarrow 8CO_2(g) + 9H_2O(\ell)$	−1302.7
$KNO_3(s) \xrightarrow{H_2O} K^+(aq) + NO_3^-(aq)$	+8.3
$NaOH(s) \xrightarrow{H_2O} Na^+(aq) + OH^-(aq)$	−10.6
$NH_4Cl(s) \xrightarrow{H_2O} NH_4^+(aq) + Cl^-(aq)$	+3.5
$NH_4NO_3(s) \xrightarrow{H_2O} NH_4^+(aq) + NO_3^-(aq)$	+6.1
$NaCl(s) \xrightarrow{H_2O} Na^+(aq) + Cl^-(aq)$	+0.9
$KClO_3(s) \xrightarrow{H_2O} K^+(aq) + ClO_3^-(aq)$	+9.9
$LiBr(s) \xrightarrow{H_2O} Li^+(aq) + Br^-(aq)$	−11.7
$H^+(aq) + OH^-(aq) \rightarrow H_2O(\ell)$	−13.8

F

SELECTED RADIOISOTOPES

Nuclide	Half-Life	Decay Mode
^{198}Au	2.69 d	β^-
^{14}C	5730 y	β^-
^{60}Co	5.26 y	β^-
^{137}Cs	30.23 y	β^-
^{220}Fr	27.5 s	α
^{3}H	12.26 y	β^-
^{131}I	8.07 d	β^-
^{37}K	1.23 s	β^+
^{42}K	12.4 h	β^-
^{85}Kr	10.76 y	β^-
^{85m}Kr*	4.39 h	γ
^{16}N	7.2 s	β^-
^{32}P	14.3 d	β^-
^{239}Pu	2.44×10^4 y	α
^{226}Ra	1600 y	α
^{222}Rn	3.82 d	α
^{90}Sr	28.1 y	β^-
^{99}Tc	2.13×10^5 y	β^-
^{99m}Tc*	6.01 h	γ
^{232}Th	1.4×10^{10} y	α
^{233}U	1.62×10^5 y	α
^{235}U	7.1×10^8 y	α
^{238}U	4.51×10^9 y	α

y=years; d=days; h=hours; s=seconds
*m = meta stable or excited state of the same nucleus. Gamma decay from such a state is called an isomeric transition (IT).
Nuclear isomers are different energy states of the same nucleus, each having a different measurable lifetime.

H

SYMBOLS USED IN NUCLEAR CHEMISTRY

alpha particle	$^4_2$He	α
beta particle (electron)	$^0_{-1}$e	β^-
gamma radiation		γ
neutron	$^1_0$n	n
proton	$^1_1$H	p
deuteron	$^2_1$H	
triton	$^3_1$H	
positron	$^0_{+1}$e	β^+

I

IONIZATION ENERGIES AND ELECTRONEGATIVITIES

1								18
313 — First Ionization Energy (kcal/mol of atoms) **H** **2.2** — Electronegativity*								**567** **He**
	2	**13**	**14**	**15**	**16**	**17**		
125 **Li** **1.0**	**215** **Be** **1.5**	**191** **B** **2.0**	**260** **C** **2.6**	**336** **N** **3.1**	**314** **O** **3.5**	**402** **F** **4.0**		**497** **Ne**
119 **Na** **0.9**	**176** **Mg** **1.2**	**138** **Al** **1.5**	**188** **Si** **1.9**	**242** **P** **2.2**	**239** **S** **2.6**	**300** **Cl** **3.2**		**363** **Ar**
100 **K** **0.8**	**141** **Ca** **1.0**	**138** **Ga** **1.6**	**182** **Ge** **1.9**	**226** **As** **2.0**	**225** **Se** **2.5**	**273** **Br** **2.9**		**323** **Kr**
96 **Rb** **0.8**	**131** **Sr** **1.0**	**133** **In** **1.7**	**169** **Sn** **1.8**	**199** **Sb** **2.1**	**208** **Te** **2.3**	**241** **I** **2.7**		**280** **Xe**
90 **Cs** **0.7**	**120** **Ba** **0.9**	**141** **Tl** **1.8**	**171** **Pb** **1.8**	**168** **Bi** **1.9**	**194** **Po** **2.0**	**At** **2.2**		**248** **Rn**
Fr **0.7**	**122** **Ra** **0.9**							

* Arbitrary scale based on fluorine = 4.0

J

RELATIVE STRENGTHS OF ACIDS IN AQUEOUS SOLUTION AT 1 atm AND 298 K

Conjugate Pairs		K_a
ACID	*BASE*	
$HI = H^+ + I^-$		very large
$HBr = H^+ + Br^-$		very large
$HCl = H^+ + Cl^-$		very large
$HNO_3 = H^+ + NO_3^-$		very large
$H_2SO_4 = H^+ + HSO_4^-$		large
$H_2O + SO_2 = H^+ + HSO_3^-$		1.5×10^{-2}
$HSO_4^- = H^+ + SO_4^{2-}$		1.2×10^{-2}
$H_3PO_4 = H^+ + H_2PO_4^-$		7.5×10^{-3}
$Fe(H_2O)_6^{3+} = H^+ + Fe(H_2O)_5(OH)^{2+}$		8.9×10^{-4}
$HNO_2 = H^+ + NO_2^-$		4.6×10^{-4}
$HF = H^+ + F^-$		3.5×10^{-4}
$Cr(H_2O)_6^{3+} = H^+ + Cr(H_2O)_5(OH)^{2+}$		1.0×10^{-4}
$CH_3COOH = H^+ + CH_3COO^-$		1.8×10^{-5}
$Al(H_2O)_6^{3+} = H^+ + Al(H_2O)_5(OH)^{2+}$		1.1×10^{-5}
$H_2O + CO_2 = H^+ + HCO_3^-$		4.3×10^{-7}
$HSO_3^- = H^+ + SO_3^{2-}$		1.1×10^{-7}
$H_2S = H^+ + HS^-$		9.5×10^{-8}
$H_2PO_4^- = H^+ + HPO_4^{2-}$		6.2×10^{-8}
$NH_4^+ = H^+ + NH_3$		5.7×10^{-10}
$HCO_3^- = H^+ + CO_3^{2-}$		5.6×10^{-11}
$HPO_4^{2-} = H^+ + PO_4^{3-}$		2.2×10^{-13}
$HS^- = H^+ + S^{2-}$		1.3×10^{-14}
$H_2O = H^+ + OH^-$		1.0×10^{-14}
$OH^- = H^+ + O^{2-}$		$< 10^{-36}$
$NH_3 = H^+ + NH_2^-$		very small

Note: $H^+(aq) = H_3O^+$

Sample equation: $HI + H_2O = H_3O^+ + I^-$

K

CONSTANTS FOR VARIOUS EQUILIBRIA AT 1 atm AND 298 K

$H_2O(\ell) = H^+(aq) + OH^-(aq)$	$K_w = 1.0 \times 10^{-14}$
$H_2O(\ell) + H_2O(\ell) = H_3O^+(aq) + OH^-(aq)$	$K_w = 1.0 \times 10^{-14}$
$CH_3COO^-(aq) + H_2O(\ell) = CH_3COOH(aq) + OH^-(aq)$	$K_b = 5.6 \times 10^{-10}$
$Na^+F^-(aq) + H_2O(\ell) = Na^+(OH)^- + HF(aq)$	$K_b = 1.5 \times 10^{-11}$
$NH_3(aq) + H_2O(\ell) = NH_4^+(aq) + OH^-(aq)$	$K_b = 1.8 \times 10^{-5}$
$CO_3^{2-}(aq) + H_2O(\ell) = HCO_3^-(aq) + OH^-(aq)$	$K_b = 1.8 \times 10^{-4}$
$Ag(NH_3)_2^+(aq) = Ag^+(aq) + 2NH_3(aq)$	$K_{eq} = 8.9 \times 10^{-8}$
$N_2(g) + 3H_2(g) = 2NH_3(g)$	$K_{eq} = 6.7 \times 10^5$
$H_2(g) + I_2(g) = 2HI(g)$	$K_{eq} = 3.5 \times 10^{-1}$

Compound	K_{sp}	Compound	K_{sp}
AgBr	5.0×10^{-13}	Li_2CO_3	2.5×10^{-2}
AgCl	1.8×10^{-10}	$PbCl_2$	1.6×10^{-5}
Ag_2CrO_4	1.1×10^{-12}	$PbCO_3$	7.4×10^{-14}
AgI	8.3×10^{-17}	$PbCrO_4$	2.8×10^{-13}
$BaSO_4$	1.1×10^{-10}	PbI_2	7.1×10^{-9}
$CaSO_4$	9.1×10^{-6}	$ZnCO_3$	1.4×10^{-11}

(L)

STANDARD ELECTRODE POTENTIALS

Ionic Concentrations 1 M Water At 298 K, 1 atm

Half-Reaction	E^0 (volts)
$F_2(g) + 2e^- \rightarrow 2F^-$	+2.87
$8H^+ + MnO_4^- + 5e^- \rightarrow Mn^{2+} + 4H_2O$	+1.51
$Au^{3+} + 3e^- \rightarrow Au(s)$	+1.50
$Cl_2(g) + 2e^- \rightarrow 2Cl^-$	+1.36
$14H^+ + Cr_2O_7^{2-} + 6e^- \rightarrow 2Cr^{3+} + 7H_2O$	+1.23
$4H^+ + O_2(g) + 4e^- \rightarrow 2H_2O$	+1.23
$4H^+ + MnO_2(s) + 2e^- \rightarrow Mn^{2+} + 2H_2O$	+1.22
$Br_2(\ell) + 2e^- \rightarrow 2Br^-$	+1.09
$Hg^{2+} + 2e^- \rightarrow Hg(\ell)$	+0.85
$Ag^+ + e^- \rightarrow Ag(s)$	+0.80
$Hg_2^{2+} + 2e^- \rightarrow 2Hg(\ell)$	+0.80
$Fe^{3+} + e^- \rightarrow Fe^{2+}$	+0.77
$I_2(s) + 2e^- \rightarrow 2I^-$	+0.54
$Cu^+ + e^- \rightarrow Cu(s)$	+0.52
$Cu^{2+} + 2e^- \rightarrow Cu(s)$	+0.34
$4H^+ + SO_4^{2-} + 2e^- \rightarrow SO_2(aq) + 2H_2O$	+0.17
$Sn^{4+} + 2e^- \rightarrow Sn^{2+}$	+0.15
$2H^+ + 2e^- \rightarrow H_2(g)$	0.00
$Pb^{2+} + 2e^- \rightarrow Pb(s)$	−0.13
$Sn^{2+} + 2e^- \rightarrow Sn(s)$	−0.14
$Ni^{2+} + 2e^- \rightarrow Ni(s)$	−0.26
$Co^{2+} + 2e^- \rightarrow Co(s)$	−0.28
$Fe^{2+} + 2e^- \rightarrow Fe(s)$	−0.45
$Cr^{3+} + 3e^- \rightarrow Cr(s)$	−0.74
$Zn^{2+} + 2e^- \rightarrow Zn(s)$	−0.76
$2H_2O + 2e^- \rightarrow 2OH^- + H_2(g)$	−0.83
$Mn^{2+} + 2e^- \rightarrow Mn(s)$	−1.19
$Al^{3+} + 3e^- \rightarrow Al(s)$	−1.66
$Mg^{2+} + 2e^- \rightarrow Mg(s)$	−2.37
$Na^+ + e^- \rightarrow Na(s)$	−2.71
$Ca^{2+} + 2e^- \rightarrow Ca(s)$	−2.87
$Sr^{2+} + 2e^- \rightarrow Sr(s)$	−2.89
$Ba^{2+} + 2e^- \rightarrow Ba(s)$	−2.91
$Cs^+ + e^- \rightarrow Cs(s)$	−2.92
$K^+ + e^- \rightarrow K(s)$	−2.93
$Rb^+ + e^- \rightarrow Rb(s)$	−2.98
$Li^+ + e^- \rightarrow Li(s)$	−3.04

(N)

VAPOR PRESSURE OF WATER

°C	torr (mmHg)	°C	torr (mmHg)
0	4.6	26	25.2
5	6.5	27	26.7
10	9.2	28	28.3
15	12.8	29	30.0
16	13.6	30	31.8
17	14.5	40	55.3
18	15.5	50	92.5
19	16.5	60	149.4
20	17.5	70	233.7
21	18.7	80	355.1
22	19.8	90	525.8
23	21.1	100	760.0
24	22.4	105	906.1
25	23.8	110	1074.6

(M)

PHYSICAL CONSTANTS AND CONVERSION FACTORS

Name	Symbol	Value(s)	Units
Angstrom unit	Å	1×10^{-10} m	meter
Avogadro number	N_A	6.02×10^{23} per mol	
Charge of electron	e	1.60×10^{-19} C	coulomb
Electron volt	eV	1.60×10^{-19} J	joule
Speed of light	c	3.00×10^8 m/s	meters/second
Planck's constant	h	6.63×10^{-34} J·s	joule-second
		1.58×10^{-37} kcal·s	kilocalorie-second
Universal gas constant	R	0.0821 L·atm/mol·K	liter-atmosphere/mole-kelvin
		1.98 cal/mol·K	calories/mole-kelvin
		8.31 J/mol·K	joules/mole-kelvin
Atomic mass unit	μ(amu)	1.66×10^{-24} g	gram
Volume standard, liter	L	1×10^3 cm^3 = 1 dm^3	cubic centimeters, cubic decimeter
Standard pressure, atmosphere	atm	101.3 kPa	kilopascals
		760 mmHg	millimeters of mercury
		760 torr	torr
Heat equivalent, kilocalorie	kcal	4.18×10^3 J	joules

Physical Constants for H$_2$O

Molal freezing point depression	1.86°C
Molal boiling point elevation	0.52°C
Heat of fusion	79.72 cal/g
Heat of vaporization	539.4 cal/g

STANDARD UNITS

Symbol	Name	Quantity
m	meter	length
kg	kilogram	mass
Pa	pascal	pressure
K	kelvin	thermodynamic temperature
mol	mole	amount of substance
J	joule	energy, work, quantity of heat
s	second	time
C	coulomb	quantity of electricity
V	volt	electric potential, potential difference
L	liter	volume

Selected Prefixes

Factor	Prefix	Symbol
10^6	mega	M
10^3	kilo	k
10^{-1}	deci	d
10^{-2}	centi	c
10^{-3}	milli	m
10^{-6}	micro	μ
10^{-9}	nano	n

O

RADII OF ATOMS

KEY

Symbol →	F
Covalent Radius, Å →	0.64
Atomic Radius in Metals, Å →	(−)
Van der Waals Radius, Å →	1.35

A dash (−) indicates data are not available.

H																	He
H 0.37 (−) 1.2																	**He** (−) (−) 1.22
Li 1.23 1.52 (−)	**Be** 0.89 1.13 (−)											**B** 0.88 (−) 2.08	**C** 0.77 (−) 1.85	**N** 0.70 (−) 1.54	**O** 0.66 (−) 1.40	**F** 0.64 (−) 1.35	**Ne** (−) (−) 1.60
Na 1.57 1.54 2.31	**Mg** 1.36 1.60 (−)											**Al** 1.25 1.43 (−)	**Si** 1.17 (−) 2.0	**P** 1.10 (−) 1.90	**S** 1.04 (−) 1.85	**Cl** 0.99 (−) 1.81	**Ar** (−) (−) 1.91
K 2.03 2.27 2.31	**Ca** 1.74 1.97 (−)	**Sc** 1.44 1.61 (−)	**Ti** 1.32 1.45 (−)	**V** 1.22 1.32 (−)	**Cr** 1.17 1.25 (−)	**Mn** 1.17 1.24 (−)	**Fe** 1.17 1.24 (−)	**Co** 1.16 1.25 (−)	**Ni** 1.15 1.25 (−)	**Cu** 1.17 1.28 (−)	**Zn** 1.25 1.33 (−)	**Ga** 1.25 1.22 (−)	**Ge** 1.22 1.23 (−)	**As** 1.21 (−) 2.0	**Se** 1.17 (−) 2.0	**Br** 1.14 (−) 1.95	**Kr** 1.89 (−) 1.98
Rb 2.16 2.48 2.44	**Sr** 1.92 2.15 (−)	**Y** 1.62 1.81 (−)	**Zr** 1.45 1.60 (−)	**Nb** 1.34 1.43 (−)	**Mo** 1.29 1.36 (−)	**Tc** (−) 1.36 (−)	**Ru** 1.24 1.33 (−)	**Rh** 1.25 1.35 (−)	**Pd** 1.28 1.38 (−)	**Ag** 1.34 1.44 (−)	**Cd** 1.41 1.49 (−)	**In** 1.50 1.63 (−)	**Sn** 1.40 1.41 (−)	**Sb** 1.41 (−) 2.2	**Te** 1.37 (−) 2.20	**I** 1.33 (−) 2.15	**Xe** 2.09 (−) (−)
Cs 2.35 2.65 2.62	**Ba** 1.98 2.17 (−)	**La-Lu**	**Hf** 1.44 1.56 (−)	**Ta** 1.34 1.43 (−)	**W** 1.30 1.37 (−)	**Re** 1.28 1.37 (−)	**Os** 1.26 1.34 (−)	**Ir** 1.26 1.36 (−)	**Pt** 1.29 1.38 (−)	**Au** 1.34 1.44 (−)	**Hg** 1.44 1.60 (−)	**Tl** 1.55 1.70 (−)	**Pb** 1.54 1.75 (−)	**Bi** 1.52 1.55 (−)	**Po** 1.53 1.67 (−)	**At** (−) (−) (−)	**Rn** 2.14 (−) (−)
Fr (−) 2.7 (−)	**Ra** (−) 2.20 (−)	**Ac-Lr**															

La	Ce	Pr	Nd	Pm	Sm	Eu	Gd	Tb	Dy	Ho	Er	Tm	Yb	Lu
La 1.69 1.88 (−)	**Ce** 1.65 1.83 (−)	**Pr** 1.65 1.83 (−)	**Nd** 1.64 1.82 (−)	**Pm** (−) 1.81 (−)	**Sm** 1.66 1.80 (−)	**Eu** 1.85 2.04 (−)	**Gd** 1.61 1.80 (−)	**Tb** 1.59 1.78 (−)	**Dy** 1.59 1.77 (−)	**Ho** 1.58 1.77 (−)	**Er** 1.57 1.76 (−)	**Tm** 1.56 1.75 (−)	**Yb** 1.70 1.94 (−)	**Lu** 1.56 1.73 (−)
Ac (−) 1.88 (−)	**Th** (−) 1.80 (−)	**Pa** (−) 1.61 (−)	**U** (−) 1.39 (−)	**Np** (−) 1.31 (−)	**Pu** (−) 1.51 (−)	**Am** (−) 1.84 (−)	**Cm** (−) (−) (−)	**Bk** (−) (−) (−)	**Cf** (−) (−) (−)	**Es** (−) (−) (−)	**Fm** (−) (−) (−)	**Md** (−) (−) (−)	**No** (−) (−) (−)	**Lr** (−) (−) (−)

GLOSSARY OF COMMON TERMS
(See Index for additional references.)

absolute temperature   Temperature measured on the absolute scale, which has its origin at absolute zero. *See also* **Kelvin scale.**

absorption   The process of taking up by capillary, osmotic, chemical, or solvent action, as a sponge absorbs water.

acid   A water solution that has an excess of hydrogen ions; an acid turns litmus paper pink or red, has a sour taste, and neutralizes bases to form salts.

acid anhydride   A nonmetallic oxide that, when placed in water, reacts to form an acid solution.

acid salt   A salt formed by replacing part of the hydrogen ions of a dibasic or tribasic acid with metallic ions.
Examples: $NaHSO_4$, NaH_2PO_4.

actinide series   The series of radioactive elements starting with actinium, No. 89, and ending with lawrencium, No. 103.

activated charcoal   A specially treated and finely divided form of carbon, which possesses a high degree of adsorption.

activation energy   The minimum energy necessary to start a reaction.

adsorption   The adhesion (in an extremely thin layer) of the molecules of gases, of dissolved substances, or of liquids to the surfaces of solid or liquid bodies with which they come into contact.

alcohol   An organic hydroxyl compound formed by replacing one or more hydrogen atoms of a hydrocarbon with an equal number of hydroxyl (OH) groups.

aldehyde   An organic compound formed by dehydrating oxidized alcohol; contains the characteristic –CHO group.

alkali   Usually, a strong base, such as sodium hydroxide or potassium hydroxide.

alkaline   Referring to any substance that has basic properties.

alkyl   A radical obtained from a saturated hydrocarbon by removing one hydrogen atom.
Examples: methyl (CH_3-), ethyl (C_2H_5-).

alkylation   The combining of a saturated hydrocarbon with an unsaturated one.

allotropic forms   Forms of the same element that differ in their crystalline structures.

alloy   A substance composed of two or more metals, which are intimately mixed; usually made by melting the metals together.

alpha particles   Positively charged helium nuclei.

alum   A double sulfate of a monovalent metal and a trivalent metal, such as $K_2SO_4 \cdot Al_2(SO_4)_3 \cdot 24H_2O$; also, a common name for commercial aluminum sulfate.

amalgam   An alloy of mercury and another metal.

amine   A compound such as CH_3NH_2, derived from ammonia by substituting one or more hydrocarbon radicals for hydrogen atoms.

amino acid   One of the "building blocks" of proteins; contains one or more NH_2- groups that have replaced the same number of hydrogen atoms in an organic acid.

amorphous   Having no definite crystalline structure.

amphoteric   Referring to a hydroxide that may have either acidic or basic properties, depending on the substance with which it reacts.

analysis   The breaking down of a compound into two or more simpler substances.

angstrom   A unit of length used in atomic measurements; equal to 10^{-8}cm (abbreviation: Å).

anhydride   A compound derived from another compound by the removal of water; it will combine with water to form an acid (acid anhydride) or a base (basic anhydride).

anhydrous   Containing no water.

anion   An ion or particle that has a negative charge and thus is attracted to a positively charged anode.

anode   The electrode in an electrolytic cell that has a positive charge and attracts negative ions.

antichlor   A substance used to remove the excess of chlorine in the bleaching process.

aqua regia   A mixture of three volumes of concentrated HCl and one volume of concentrated HNO_3 that can dissolve gold or platinum.

aromatic compounds   A compound whose basic structure contains the benzene ring; it usually has an odor.

aryl   A radical obtained from a ring hydrocarbon by removing one hydrogen atom from the ring.

atmosphere   The layer of gases surrounding the earth; also, a unit of pressure (1 atm = approx. 760 mm or torr).

atom   The smallest particle of an element that retains the properties of that element and can enter into a chemical reaction.

atomic energy   *See* **Nuclear energy**, a more accurate term.

atomic mass   The average mean value of the isotopic masses of the atoms of an element. It indicates the relative mass of the element as compared with that of carbon-12, which is assigned a mass of exactly 12 atomic mass units.

atomic mass unit   One twelfth of the mass of a carbon-12 atom; equivalent to 1.660531×10^{-27} kilogram (abbreviation: amu).

atomic number   The number that indicates the order of an element in the periodic system; numerically equal to the number of protons in the nucleus of the atom, or the number of negative electrons located outside the nucleus of the atom.

atomic radius   One-half the distance between adjacent nuclei in the crystalline or solid phase of an element; the distance from the atomic nucleus to the valence electrons.

atomic weight.   *See* **Atomic mass.**

Avogadro's hypothesis   *See under* **Laws.**

Avogadro's number   The number of molecules in 1 gram-molecular volume of a substance, or the number of atoms in 1 gram-atomic mass of an element; equal to 6.022169 x 10^{23}. *See* **Mole.**

barometer   An instrument, invented by Torricelli in 1643, used for measuring atmospheric pressure.

base   A water solution that contains an excess of hydroxide ions; a proton acceptor; a base turns litmus paper blue and neutralizes acids to form salts.

basic anhydride   A metallic oxide that forms a base when placed in water.

beta particles   High-speed, negatively charged electrons emitted in radiation.

binary   Referring to a compound composed of two elements, such as H_2O.

boiling point   The temperature at which the vapor pressure of a liquid equals the atmospheric pressure.

bond energy   The energy needed to break a chemical bond and form a neutral atom.

bonding   The union of atoms to form compounds or molecules by filling their outer shells of electrons. This can be done through giving and taking electrons (ionic) or by sharing electrons (covalent).

Boyle's law   *See under* **Laws.**

brass   An alloy of copper and zinc.

breeder reactor   A nuclear reactor in which more fissionable material is produced than is used up during operation.

British thermal unit   A unit of heat; the amount of heat necessary to raise the temperature of 1 pound of water 1°F (abbreviation: Btu).

Brownian movement   Continuous zigzagging movement of colloidal particles in a dispersing medium, as viewed through an ultramicroscope.

buffer   A substance that, when added to a solution, makes changing the pH of the solution more difficult.

calorie   A unit of heat; the amount of heat needed to raise the temperature of 1 gram of water 1 degree on the Celsius scale.

calorimeter   An instrument used to measure the amount of heat liberated or absorbed during a change.

carat   A unit of weight (200 mg) for diamonds and precious stones; also the fineness of gold, based on a standard of 24 carats for pure gold.

carbohydrate   A compound of carbon, hydrogen, and oxygen, with hydrogen and oxygen usually present in the ratio of 2 to 1, as in H_2O.

carbonated water   Water containing dissolved carbon dioxide.

carbon dating   The use of radioactive carbon-14 to estimate the age of ancient materials, such as archeological or paleontological specimens.

catalyst   A substance that speeds up or slows down a reaction without being permanently changed itself.

cathode   The electrode in an electrolytic cell that is negatively charged and attracts positive ions.

cathode rays   Streams of electrons given off by the cathode of a vacuum tube.

cation   An ion that has a positive charge.

Celsius scale   A temperature scale divided into 100 equal divisions and based on water freezing at 0° and boiling at 100°. Synonymous with centigrade.

chain reaction   A reaction produced during nuclear fission when at least one neutron from each fission produces another fission, so that the process becomes self-sustaining without additional external energy.

Charles' law   *See under* **Laws.**

chemical change   A change that alters the atomic structures of the substances involved and results in different properties.

chemical property   A property that determines how a substance will behave in a chemical reaction.

chemistry   The science concerned with the compositions of substances and the changes that they undergo.

colligative properties   Properties of a solution that depend primarily on the concentration, not the type, of particles present.

colloids   Particles larger than those found in a solution but smaller than those in a suspension.

Combining volumes   *See under* **Laws.**

combustion   A chemical action in which both heat and light are given off.

compound   A substance composed of elements chemically united in definite proportions by weight.

condensation   (a) A change from gaseous to liquid state; (b) the union of like or unlike molecules with the elimination of water, hydrogen chloride, or alcohol.

Conservation of energy   *See under* **Laws.**

Conservation of matter   *See under* **Laws.**

control rods   In a nuclear reactor, a rod of a certain metal such as cadmium, which controls the speed of the chain reaction by absorbing neutrons.

coordinate covalence   Covalence in which both electrons in a pair come from the same atom.

covalent bonding   Bonding accomplished through the sharing of electrons so that atoms can fill their outer shells.

critical mass   The smallest amount of fissionable material that will sustain a chain reaction.

critical temperature   The temperature above which no gas can be liquefied, regardless of the pressure applied.

crystalline   Having a definite molecular or ionic structure.

crystallization   The process of forming definitely shaped crystals when water is evaporated from a solution of the substance.

cyclotron   A device used to accelerate charged particles to high energies for bombarding the nuclei of atoms.

Dalton's law of partial pressures   *See under* **Laws.**

decomposition   The breaking down of a compound into simpler substances or into its constituent elements.

definite composition   *See under* **Laws.**

dehydrate   To take water from a substance.

dehydrating agent   A substance able to withdraw water from another substance, thereby drying it.

deliquescence   The absorption by a substance of water from the air, so that the substance becomes wet.

denatured alcohol   Ethyl alcohol that has been "poisoned" in order to produce (by avoiding federal tax) a cheaper alcohol for industrial purposes.

density   The mass per unit volume of a substance; the mathematical formula is $D = m/V$, where D = density, m = mass, and V = volume.

destructive distillation   The process of heating an organic substance, such as coal, in the absence of air to break it down into solid and volatile products.

deuterium   An isotope of hydrogen, sometimes called heavy hydrogen, with an atomic weight of 2.

dew point   The highest temperature at which water vapor condenses out of the air.

dialysis   The process of separation of a solution by diffusion through a semipermeable membrane.

diffusion   The process whereby gases or liquids intermingle freely of their own accord.

dipole-dipole attraction   relatively weak force of attraction between polar molecules; a component of van der Waals forces.

disaccharide   A sugar, as $C_{12}H_{22}O_{11}$, that hydrolyzes to form two molecules of simpler sugars.

displacement   A change by which an element takes the place of another element in a compound.

dissociation (ionic)   The separation of the ions of an electrovalent compound due to the action of a solvent.

distillation   The process of first vaporizing a liquid and then condensing the vapor back into a liquid, leaving behind the nonvolatile impurities.

double bond   A bond between atoms involving two electron pairs. In organic chemistry: unsaturated.

double displacement   A reaction in which two chemical substances exchange ions with the formation of two new compounds.

"dry ice"   Solid carbon dioxide.

ductile   Capable of being drawn into a thin wire.

Dulong and Petit's law   *See under* **Laws.**

effervescence   The rapid escape of excess gas that has been dissolved in a liquid.

efflorescence   The loss by a substance of its water of hydration on exposure to air at ordinary temperatures.

Einstein equation   The equation $E = mc^2$, which relates mass to energy; E is energy in ergs, m is mass in grams, and c is velocity of light, 3×10^{10} cm/sec.

electrode   A terminal of an electrolytic cell.

electrode potential   The difference in potential between an electrode and the solution in which it is immersed.

electrolysis   The process of separating the ions in a compound by means of electrically charged poles.

electrolysis (electrolytic) cell   A cell in which electrolysis is carried out.

electrolyte   A liquid that will conduct an electric current.

electron   A negatively charged particle found outside the nucleus of the atom; it has a mass of 9.109×10^{-28}g.

electron dot symbol   *See* **Lewis dot symbol.**

electronegativity   The numerical expression of the relative strength with which the atoms of an element attract valence electrons to themselves; the higher the number, the greater the attraction.

electron volt   A unit for expressing the kinetic energy of subatomic particles; the energy acquired by an electron when it is accelerated by a potential difference of 1 volt: equals 1.6×10^{-12} erg or 23.1 kcal per mole (abbreviation: eV).

electroplating   Depositing a thin layer of (usually) a metallic element on the surface of another metal by electrolysis.

electrovalence   The bond of attraction between negatively and positively charged ions.

element   One of the more than 100 "building blocks" of which all matter is composed. An element consists of atoms of only one kind and cannot be decomposed further by ordinary chemical means.

empirical (molecular) formula   A simple formula that shows only the simplest ratio of the numbers and kinds of atoms such as CH_4.

emulsifying agent   A colloidal substance that forms a film about the particles of two immiscible liquids, so that one liquid remains suspended in the other.

emulsion   A suspension of fine particles or droplets of one liquid in another, the two liquids being immiscible in each other; the droplets are surrounded by a colloidal (emulsifying) agent.

endothermic   Referring to a chemical reaction that results in an overall absorption of heat from its surroundings.

energy   The capacity to do work. In every chemical change energy is either given off or taken in; forms of energy are heat, light, motion, sound, and electrical, chemical, and nuclear energy.

enthalpy   The heat content of a chemical system.

entropy   The measure of the randomness that exists in a system.

equation   A shorthand method of showing the changes that take place in a chemical reaction.

equilibrium   The point in a reversible reaction at which the forward reaction is occurring at the same rate as the opposing reaction.

erg   A unit of energy or work done by a force of 1 dyne (1/980 gram of force) acting through a distance of 1 centimeter; equals 2.4×10^{-11} kcal.

ester   An organic salt formed by the reaction of an alcohol with an organic (or inorganic) acid.

esterification   A chemical reaction between an alcohol and an acid, in which an ester is formed.

ether   An organic compound containing the –O– group.

eudiometer   A graduated glass tube into which gases are placed and subjected to an electric spark; used to measure the individual volumes of combining gases.

evaporation   The process in which molecules of a liquid (or a solid) leave the surface in the form of vapor.

exothermic   Referring to a chemical reaction that results in the giving off of heat to its surroundings.

Fahrenheit scale   The temperature scale that has 32° as the freezing point of water and 212° as the boiling point.

fallout   The residual radioactivity from an atmospheric nuclear test, which eventually settles on the surface of the earth.

faraday   A unit of electric charge that deposits by electrolysis one equivalent weight of an element; equals 96,500 coulombs.

Faraday's law   *See under* **Laws.**

filtration   The process by which suspended matter is removed from a liquid by passing the liquid through a porous material.

First Law of Thermodynamics   *See under* **Laws.**

fission   A nuclear reaction that releases energy because of the splitting of large nuclei into smaller ones.

fixation of nitrogen   Any process for converting atmospheric nitrogen into compounds, such as ammonia and nitric acid.

flame   The glowing mass of gas and luminous particles produced by the burning of a gaseous substance.

flammable   Capable of being easily set on fire; combustible (same as inflammable).

fluorescence   Emission by a substance of electromagnetic radiation, usually visible, as the immediate result of (and only during) absorption of energy from another source.

fluoridation   Addition of small amounts of fluoride (usually NaF) to drinking water to help prevent tooth decay.

flux   In metallurgy: a substance that helps to melt and remove the solid impurities as slag. In soldering: a substance that cleans the surface of the metal to be soldered. In nucleonics: the concentration of nuclear particles or rays.

formal solution   A solution that contains 1 gram formula-weight of a substance in 1 liter of solution.

formula   An expression that uses the symbols for elements and subscripts to show the basic makeup of a substance.

formula mass   The sum of the atomic masses of all the atoms (or ions) contained in a formula.

fractional crystallization   The separation of the components in a mixture of dissolved solids by evaporation according to individual solubilities.

fractional distillation   The separation of the components in a mixture of liquids having different boiling points by vaporization.

freezing point   The specific temperature at which a given liquid and its solid form are in equilibrium.

fuel   Any substance used to furnish heat by combustion. *See also* **Nuclear fuel.**

fuel cell   A device for converting an ordinary fuel such as hydrogen or methane directly into electricity.

functional group   A group of atoms that characterizes certain types of organic compounds, such as –OH for alcohols, and that reacts more or less independently.

fusion   A nuclear reaction that releases energy because of the union of smaller nuclei to form larger ones.

fusion melting   Changing a solid to the liquid state by heating.

galvanizing   Applying a coating of zinc to iron or steel to protect the latter from rusting.

gamma rays   A type of radiation consisting of high-energy waves that can pass through most materials.

gas   A phase of matter that has neither definite shape nor definite volume.

Gay-Lussac's law   *See under* **Laws.**

glass   An amorphous, usually translucent substance consisting of a mixture of silicates. Ordinary glass is made by fusing together silica sand and sodium carbonate and lime; the various forms of glass contain many other silicates.

glyceride   An ester of glycerol.

Graham's law   *See under* **Laws.**

gram   A unit of weight in the metric system; the weight of 1 ml of water at 4°C (abbreviation: g).

gram-atomic mass   The atomic mass, in grams, of an element.

gram-formula weight   The formula weight, in grams, of a substance.

gram-molecular mass   The molecular mass, in grams, of a substance.

gram-molecular volume   The volume occupied by 1 gram-molecular mass of a gaseous substance at standard conditions; equals 22.4 l.

group   A vertical column of elements in the periodic table that generally have similar properties.

Haber process   A catalytic method for the union of atmospheric nitrogen with hydrogen to form ammonia.

half-life   The time required for half of the mass of a radioactive substance to disintegrate.

half-reaction   One of the two parts, either the reduction part or the oxidation part, of a redox reaction.

halogen   Any of the five nonmetallic elements (flourine, chlorine, bromine, iodine, astatine) that form part of group VII of the periodic table.

heat   A form of molecular energy; it passes from a warmer body to a cooler one.

heat capacity (specific heat)   The quantity of heat, in calories, needed to raise the temperature of 1 gram of a substance 1 degree centigrade.

heat of formation   The quantity of heat either given off or absorbed in the formation of 1 mole of a substance from its elements.

heat of fusion   The amount of heat, in calories, required to melt 1 gram of a solid; for water, 80 cal.

heat of vaporization   The quantity of heat needed to vaporize 1 gram of a liquid at constant temperature and pressure; for water at 100 °C, 540 cal.

heavy water (deuterium oxide, D_2O)   Water in which the hydrogen atoms are replaced by atoms of the isotope of hydrogen, deuterium.

Henry's law   *See under* **Laws.**

homogeneous   Uniform; having every portion exactly like every other portion.

homologous   Alike in structure; referring to series of organic compounds, such as hydrocarbons, in which each member differs from the next by the addition of the same group.

humidity   The amount of moisture in the air.

hybridization   The combination of two or more orbitals to form new orbitals.

hydrate   A compound that has water molecules included in its crystalline makeup.

hydride   Any binary compound containing hydrogen, such as HCl.

hydrogenation   A process in which hydrogen is made to combine with another substance, usually organic, in the presence of a catalyst.

hydrogen bond   A weak chemical linkage between the hydrogen of one polar molecule and the oppositely charged portion of a closely adjacent molecule.

hydrolysis   Of carbohydrates: the action of water in the presence of a catalyst upon one carbohydrate to form simpler carbohydrates. Of salts: a reaction involving the splitting of water into its ions by the formation of a weak acid, a weak base, or both.

hydronium ion   A hydrated hydrogen ion, $H_2O \bullet H^+$ or H_3O^+.

hydroponics   Growing plants without the use of soil, as in nutrient solution or in sand irrigated with nutrient solution.

hydroxyl   Referring to the –OH radical.

hygroscopic   Referring to the ability of a substance to draw water vapor from the atmosphere to itself and become wet.

hypothesis   A possible explanation of the nature of an action or phenomenon; a hypothesis is not as completely developed as a theory.

Ideal Gas Law   *See under* **Laws.**

immiscible   Referring to the inability of two liquids to mix.

indicator   A dye that shows one color in the presence of the hydrogen ion (acid) and a different color in the presence of the hydroxyl ion (base).

inert   Referring to an atom having a complete outer shell of electrons and consequently not reactive.

inert gas structure   The outer-shell electron configuration characteristic of the inert gases—two electrons for helium; eight electrons for all others.

inertia   The property of matter whereby it remains at rest or, if in motion, remains in motion in a straight line unless acted upon by an outside force.

ion   An atom or a group of combined atoms that carries one or more electric charges.
Examples: NH_4^+, OH^-.

ionic bonding   The bonding of ions due to their opposite charges.

ionic equation   An equation showing a reaction among ions.

ionization   The process in which ions are formed from neutral atoms.

ionization equation   An equation showing the ions set free from an electrolyte.

isomerization   The rearrangement of atoms in a molecule to form isomers.

isomers   Two or more compounds having the same percentage composition but different arrangements of atoms in their molecules and hence different properties.

isotope   Two or more forms of an element that differ only in the number of neutrons in the nucleus and hence in their mass numbers.

Kelvin scale   A temperature scale based on water freezing at 273 and boiling at 373 Kelvin units; its origin is absolute zero. Synonymous with absolute scale.

kernel (atomic)   The nucleus and all the electron shells of an atom except the outer one; usually designated by the symbol for the atom.

ketone   An organic compound containing the –CO– group.

kilocalorie   A unit of heat; the amount of heat needed to raise the temperature of 1 kilogram of water 1 degree on the Celsius scale.

kindling temperature   The temperature to which a given substance must be raised before it ignites.

kinetic-molecular theory   The theory that all molecules are in motion; this motion is most rapid in gases, less rapid in liquids, and very slow in solids.

lanthanide series   The "rare earth" series of elements starting with lanthanum, No. 57, and ending with lutetium, No. 71.

law (in science)   A generalized statement about the uniform behavior in natural processes.

laws
Avogadro's   Equal volumes of gases under identical conditions of temperature and pressure contain equal numbers of particles (atoms, molecules, ions, or electrons).

Boyle's   The volume of a confined gas is inversely proportional to the pressure to which it is subjected, provided that the temperature remains the same.

Charles's   The volume of a confined gas is directly proportional to the absolute temperature, provided that the pressure remains the same.

Combining volumes   See Gay-Lussac's law.

Conservation of energy   Energy can be neither created nor destroyed, so that the energy of the universe is constant.

Conservation of matter   Matter can be neither created nor destroyed (or weight remains constant in an ordinary chemical change).

Dalton's   When a gas is made up of a mixture of different gases, the pressure of the mixture is equal to the sum of the partial pressures of the components.

Definite composition   A compound is composed of two or more elements chemically combined in a definite ratio by weight.

Dulong and Petit's   Most metals require 6.2 calories of heat to raise the temperature of 1 gram-atomic mass of the element 1 °C.

Faraday's   During electroylsis, the weight of any element liberated is proportional (1) to the quantity of electricity passing through the cell, and (2) to the equivalent weight of the element.

First Law of Thermodynamics   The total energy of the universe is constant and cannot be created or destroyed.

Gay-Lussac's   The ratio between the combining volumes of gases and the product, if gaseous, can be expressed in small whole numbers.

Graham's   The rate of diffusion of a gas is inversely proportional to the square root of its molecular weight.

Henry's   The solubility of a gas (unless the gas is very soluble) is directly proportioned to the pressure applied to the gas.

Ideal Gas   Any gas that obeys the gas laws perfectly. No such gas actually exists.

Multiple Proportions   When any two elements, A and B, combine to form more than one compound, the different weights of B that unite with a fixed weight of A bear a small whole-number ratio to each other.

Periodic   The chemical properties of elements vary periodically with their atomic numbers.

Second Law of Thermodynamics   Heat cannot, of itself, pass from a cold body to a hot body.

Le Châtelier's principle   If a stress is placed on a system in equilibrium, the system will react in the direction that relieves the stress.

leptons   Elementary particles that are believed to make up the electron and neutrino.

Lewis dot symbol   The chemical symbol (kernel) for an atom, surrounded by dots to represent its outer-shell electrons.
Examples: K, Sr.

liquid   A phase of matter that has a definite volume but takes the shape of the container.

liquid air   Air that has been cooled and compressed until it liquefies.

litmus   An organic substance, obtained from the lichen plant and used as an indicator; it turns red in acidic solution and blue in basic solution.

London force   The weakest of the van der Waal forces between molecules. These weak attractive forces become apparent only when the molecules approach one another closely (usually at low temperatures and high pressure). They are due to the way the positive charges of one molecule attract the negative charges of another molecule because of the charge distribution at any one instant.

luminous   Emitting a steady, suffused light.

malleable   Capable of being hammered or pounded into thin sheets.

manometer   A U-tube (containing mercury or some other liquid) used to measure the pressure of a confined gas.

mass   The quantity of matter that a substance possesses; it can be measured by its resistance to a change in position or motion, and is not related to the force of gravity.

mass number   The nearest whole number to the combined atomic mass (weight) of the individual atoms of an isotope when that mass is expressed in atomic mass units.

mass spectograph   A device for determining the masses (weights) of electrically charged particles by separating them into distinct streams by means of magnetic deflection.

matter   A substance that occupies space, has weight, and cannot be created or destroyed easily.

melting   The change in phase of a substance from solid to liquid.

melting point   The specific temperature at which a given solid changes to a liquid.

meson   Any unstable, elementary nuclear particle having a mass between that of an electron and that of a proton.

metal   (a) An element whose oxide combines with water to form a base; (b) an element that readily loses electrons and acquires a positive valence.

metallurgy   The process involved in obtaining a metal from its ores.

meter   The basic unit of length in the metric system; defined as 1,650,763.73 times the wavelength of krypton-86 when excited to give off an orange-red spectral line.

MeV   A unit for expressing the kinetic energy of subatomic particles; equals 10^6 electron volts.

micron   One thousandth of a millimeter (abbreviation: μ).

millibar   A unit of pressure used in meteorology; equals 1000 dynes per square centimeter.

millimicron   One millionth of a millimeter (abbreviation: mμ).

mineral   An inorganic substance of definite composition found in nature.

miscible Referring to the ability of two liquids to mix with one another.

mixture A substance composed of two or more components, each of which retains its own properties.

moderator A substance such as graphite, paraffin, or heavy water used in a nuclear reactor to slow down neutrons.

molal solution A solution containing 1 mole of solute in 1,000 grams of solvent (indicated by m).

molar mass The mass arrived at by the addition of the atomic masses of the units that make up a molecule of an element or compound. Expressed in grams, the molar mass of a gaseous substance at S.T.P. occupies a molar volume equal to 22.4 liters.

molar solution A solution containing 1 mole of solute in 1,000 ml of solution (indicated by M).

mole A unit of quantity that consists of 6.02×10^{23} particles.

molecular mass The sum of the atomic masses of all the atoms in a molecule of a substance.

molecular theory *See* **Kinetic-molecular theory.**

molecule The smallest particle of a substance that retains the physical and chemical properties of that substance.

monobasic acid An acid having only one hydrogen atom which can be replaced by a metal or a positive radical.

monosaccharide A simple sugar, such as $C_6H_{12}O_6$.

mordant A chemical, such as aluminum sulfate, used for fixing colors on textiles.

nascent (atomic) Referring to an element in the atomic form as it has just been liberated in a chemical reaction.

neutralization The union of the hydrogen ion of an acid and the hydroxyl ion of a base to form water.

neutron A subatomic particle found in the nucleus of the atom; it has no charge and has the same mass as the proton.

neutron capture A nuclear reaction in which a neutron attaches itself to a nucleus; a gamma ray is usually emitted simultaneously.

nitriding A process in which ammonia or a cyanide is used to produce case-hardened steel; a nitride is formed instead of a carbide.

nitrogen fixation Any process by which atmospheric nitrogen is converted into a compound such as ammonia or nitric acid.

noble gas A gaseous element that has a complete outer shell of electrons; any of a group of rare gases (helium, neon, argon, krypton, etc.) that exhibit great stability and very low reaction rates.

nonelectrolyte A substance whose solution does not conduct a current of electricity.

nonmetal (a) An element whose oxide reacts with water to form an acid; (b) an element that takes on electrons and acquires a negative valence.

nonpolar compound A compound in whose molecules the atoms are arranged symmetrically so that the electric charges are uniformly distributed.

normal salt A salt in which all the hydrogen of the acid has been displaced by a metal.

normal solution A solution that contains 1 gram of H^+ (or its equivalent: 17 grams of OH^-, 23 grams of Na^+, 20 grams of Ca^{2+}, etc.) in 1 liter of solution (indicated by N).

nuclear energy The energy released by spontaneously or artificially produced fission, fusion, or disintegration of the nuclei of atoms.

nuclear fuel A substance that is consumed during nuclear fission or fusion.

nuclear reaction Any reaction involving a change in nuclear structure.

nuclear reactor A device in which a controlled chain reaction of fissionable material can be produced.

nucleonics The science that deals with the constituents and all the changes in the atomic nucleus.

nucleus The center of the atom, which contains protons and neutrons.

nuclide A species of atom characterized by the constitution of its nucleus.

octane number A conventional rating for gasoline based on its behavior as compared with isooctane as 100.

orbital A subdivision of a nuclear shell; it may contain none, one or two electrons.

ore A natural mineral substance from which an element, usually a metal, may be obtained with profit.

organic acid An organic compound that contains the –COOH group.

organic chemistry The branch of chemistry dealing with carbon compounds, usually those found in nature.

oxidation The chemical process by which oxygen is attached to a substance; the process of losing electrons.

oxidation number (state) A positive or negative number representing the charge that an ion has or an atom appears to have when its electrons are counted according to arbitrarily accepted rules: (1) electrons shared by two unlike atoms are counted with the more electronegative atom; (2) electrons shared by two like atoms are divided equally between the atoms.

oxidation potential An electrode potential associated with the oxidation half-reaction.

oxidizing agent A substance that (a) gives up its oxygen readily, (b) removes hydrogen from a compound, (c) takes electrons from an element.

ozone An allotropic and very active form of oxygen, having the formula O_3.

paraffin series The methane series of hydrocarbons.

pasteurization Partial sterilization of a substance, such as milk, by heating to approximately 65 °C for ½ hour.

Pauli exclusion principle   Each electron orbital of an atom can be filled by only two electrons, each with an opposite spin.

period   A horizontal row of elements in the periodic table.

Periodic law   *See under* **Laws.**

petroleum (meaning "oil from stone")   A complex mixture of gaseous, liquid, and solid hydrocarbons obtained from the earth.

pH   A numerical expression of the hydrogen or hydronium ion concentration in a solution; defined as $-\log [H^+]$, where $[H^+]$ is the concentration of hydrogen ions, in moles per liter.

phenolphthalein   An organic indicator; it is colorless in acid solution and red in the presence of OH ions.

photosynthesis   The reaction taking place in all green plants that produces glucose from carbon dioxide and water under the catalytic action of chlorophyll in the presence of light.

physical change   A change that does not involve any alteration in chemical composition.

physical property   A property of a substance arrived at through observation of its smell, taste, color, density, and so on, which does not relate to chemical activity.

pi bond   A bond between p orbitals.

pile   A general term for a nuclear reactor; specifically, a graphite-moderated reactor in which uranium fuel is distributed throughout a "pile" of graphite blocks.

pitchblende   A massive variety of uraninite that contains a small amount of radium.

plasma   Very hot ionized gases.

polar covalent bond   A bond in which electrons are closer to one atom than to another. *See* **Polar molecule.**

polar dot structure   Representation of the arrangements of electrons around the atoms of a molecule in which the polar characteristics are shown by placing the electrons closer to the more electronegative atom.

polar molecule   A molecule that has differently charged areas because of unequal sharing of electrons.

polymerization   The process of combining several molecules to form one large molecule (polymer). (a) Additional polymerization: The addition of unsaturated molecules to each other. (b) Condensation polymerization: The reaction of two molecules by loss of a molecule of water.

polysaccharide   A large, complex molecule with the general formula $(C_6H_{10}O_5)_n$.

positron   A positively charged particle of electricity with about the same weight as the electron.

potential energy   Energy due to the position of a body or to the configuration of its particles.

precipitate   An insoluble compound formed in the chemical reaction between two or more substances in solution.

protein   Large, complex organic molecules, with nitrogen an essential part, found in plants and animals.

proton   A subatomic particle found in the nucleus that has a positive charge.

qualitative analysis   A term applied to the methods and procedures used to determine any or all of the constituent parts of a substance.

quantitative analysis   A term applied to the methods and procedures used to determine the definite quantity or percentage of any or all of the constituent parts of a substance.

quark   A subatomic particle with a fractional charge of $\frac{1}{3}$ or $\frac{2}{3}$ the charge of an electron. Six quarks have been described.

quenching   Cooling a hot piece of metal rapidly, as in water or oil.

radiation   The emission of particles and rays from a radioactive source; usually alpha and beta particles and gamma rays.

radical   A group of chemically united atoms that react as a unit and have an electric charge.

radioactive   Referring to substances that have the ability to emit radiations (alpha or beta particles or gamma rays).

radioisotopes   Isotopes that are radioactive, such as uranium-235.

reactant   A substance involved in a reaction.

reaction   A chemical transformation or change. The four basic types are combination (synthesis), decomposition (analysis), single replacement or single displacement, and double replacement or double displacement.

reaction potential   The sum of the oxidation potential and reduction potential for a particular reaction.

reagent   Any chemical taking part in a reaction.

recrystallization   A series of crystallizations, repeated for the purpose of greater purification.

redox   A shortened name for a reaction that involves reduction and oxidation.

reducing agent   From an electron standpoint, a substance that loses its valence electrons to another element; a substance that is readily oxidized.

reducing sugar   A sugar that reduces copper or silver salts in an alkaline solution.

reduction   A chemical reaction that removes oxygen from a substance; a gain of electrons.

reduction potential   An electrode potential associated with a reduction half-reaction.

refraction (of light)   The bending of light rays as they pass from one material into another.

relative humidity   The ratio, expressed in percent, between the amount of water vapor in a given volume of air and the amount the same volume can hold when saturated at the same temperature.

resonance   The phenomenon in a molecular structure that exhibits properties between those of a single bond and those of a double bond and thus possesses two or more alternate structures.

reversible reaction   Any reaction that reaches an equilibrium, or that can be made to proceed from right to left as well as from left to right.

roasting   Heating an ore (usually a sulfide) in an excess of air to convert the ore to an oxide, which can then be reduced.

saccharin $C_6H_4COSO_2NH$, a white, crystalline, organic compound derived from coal tar, and 400 times sweeter than cane sugar.

salt   A compound, such as NaCl, made up of a positive metallic ion and a negative nonmetallic ion or radical.

saponification   The reaction taking place when an alkali reacts with a fat or vegetable oil; used to make soap.

saturated solution   A solution that contains the maximum amount of solute under the existing temperature and pressure.

secondary cell   A cell (battery) that can be regenerated or "charged."

Second Law of Thermodynamics   *See under* **Laws.**

shell   A level outside the nucleus of an atom that electrons can inhabit without expending energy.

sigma bond   A bond between *s* orbitals or between an *s* orbital and another kind of orbital.

significant figures   All the certain digits recorded in a measurement plus one uncertain digit.

slag   The product formed when the flux reacts with the impurities of an ore in a metallurgical process.

solid   A phase of matter that has a definite size and shape.

solubility   A measure of the amount of solute that will dissolve in a given quantity of solvent at a given temperature.

solute   The material that is dissolved to make a solution.

solution   A uniform mixture of a solute in a solvent.

solvent   The dispersing substance that allows the solute to go into solution.

specific gravity (weight)   The ratio between the weight of a certain volume of a substance and the weight of an equal volume of water (or, in the case of gases, an equal volume of air); expressed as a single number.

specific heat   The ratio between the number of calories needed to raise the temperature of a certain weight of a substance 1 °C and the number of calories needed to raise the temperature of the same weight of water 1 °C.

spectroscope   An instrument used to analyze light by separating it into its component wave lengths.

spectrum   The image formed when radiant energy is dispersed by a prism or grating into its various wave lengths.

spinthariscope   A device for viewing through a microscope the flashes of light made by particles from radioactive materials against a sensitized screen.

spontaneous combustion (ignition)   The process in which slow oxidation produces enough heat to raise the temperature of a substance to its kindling temperature.

stable   Referring to a substance not easily decomposed or dissociated.

standard conditions   An atmospheric pressure of 760 mm or torr or 1 atm (mercury pressure) and a temperature of 0 °C (273 K) (abbreviation: STP).

stratosphere   The upper portion of the atmosphere, in which the temperature changes but little with altitude, and clouds of water never form.

strong acid (or base)   An acid (or a base) capable of a high degree of ionization in water solution. Examples: sulfuric acid, sodium hydroxide.

structural (graphic) formula   A pictorial representation of the atomic arrangement of a molecule.

sublime   To vaporize directly from the solid to the gaseous state, and then condense back to the solid.

substance   A single kind of matter, element or compound.

substitution products   Products formed by the substitution of other elements or radicals for hydrogen atoms in hydrocarbons.

sulfation   An accumulation of lead sulfate on the plates and at the bottom of a (lead) storage cell.

supersaturated solution   A solution that contains a greater quantity of solute than is normally possible at a given temperature.

suspension   A mixture of finely divided solid material in a liquid, from which the solid settles on standing.

symbol   A letter or letters representing an element of the periodic table. Examples: O, Mn.

synthesis   The chemical process of forming a substance from its individual parts.

Systems International d'Unites   Modernized metric system of measurements universally used by scientists. There are seven base units: kilogram, meter, second, ampere, kelvin, mole, and candela.

temperature   The intensity or the degree of heat of a body, measured by a thermometer.

tempering   The heating and then rapid cooling of a metal to increase its hardness.

ternary   Referring to a compound composed of three different elements, such as H_2SO_4.

theory   An explanation used to interpret the "mechanics" of nature's actions; a theory is more fully developed than a hypothesis.

thermochemical equation   An equation that includes values for the calories absorbed or evolved.

thermoplastic   Capable of being softened by heat; may be remolded.

thermosetting   Capable of being permanently hardened by heat and pressure; resistant to the further effects of heat.

tincture   An alcoholic solution of a substance, such as a tincture of iodine.

tracer   A minute quantity of radioactive isotope used in medicine and biology to study chemical changes within living tissues.

transmutation   Conversion of one element into another, either by bombardment or by radioactive disintegration.

tribasic acid   An acid that contains three replaceable hydrogen atoms in its molecule, such as H_3PO_4.

tritium   A very rare, unstable, "triple-weight" hydrogen isotope (H^3) that can be made synthetically.

Tyndall effect   The scattering of a beam of light as it passes through a colloidal material.

ultraviolet light   The portion of the spectrum that lies just beyond the violet; therefore of short wave length.

U.S.P. (United States Pharmacopeia) chemicals   Chemicals certified as having a standard of purity that demonstrates their fitness for use in medicine.

valence   The combining power of an element; the number of electrons gained, lost, or borrowed in a chemical reaction.

valence electrons   The electrons in the outermost shell or shells of an atom that determine its chemical properties.

Van der Waals forces   Weak attractive forces existing between molecules.

vapor   The gaseous phase of a substance that normally exists as a solid or liquid at ordinary temperatures.

vapor density   The density of a gas in relation to another gas taken as a standard under identical conditions of temperature and pressure.

vapor pressure   The pressure exerted by a vapor given off by a confined liquid or solid when the vapor is in equilibrium with its liquid or solid form.

volatile   Easily changed to a gas or a vapor at relatively low temperature.

volt   A unit of electrical potential or voltage, equal to the difference of potential between two points in a conducting wire carrying a constant current of 1 ampere when the power dissipated between these two points is equal to 1 watt (abbreviation: V).

volume   The amount of three-dimensional space occupied by a substance.

vulcanization   The process of combining sulfur or other additives, using heat, with rubber to improve its properties, particularly to give it added hardness.

water gas   A poisonous fuel gas consisting chiefly of hydrogen and carbon monoxide, produced by blowing steam over glowing coal or coke.

water of hydration   Water that is held in chemical combination in a hydrate and can be removed without essentially altering the composition of the substance. *See* **Hydrate**.

weak acid (or base)   An acid (or base) capable of being only slightly ionized in an aqueous solution. Examples: Acetic acid, ammonium hydroxide.

weak electrolyte   A substance that, when dissolved in water, ionizes only slightly and hence is a poor conductor of electricity.

weight   The measure of the force with which a body is attracted toward the earth by gravity.

work   The product of the force exerted on a body and the distance through which the force acts; expressed mathematically by the equation $W = Fs$, where W = work, F = force, and s = distance.

X-rays   Penetrating radiations, of extremely short wave length, emitted when a stream of electrons strikes a solid target in a vacuum tube.

zeolite   A natural or synthesized silicate used to soften water.

INDEX